AF389439

Greenhouse Management for Better Vegetable Quality, Higher Nutrient Use Efficiency and Healthier Soil

Greenhouse Management for Better Vegetable Quality, Higher Nutrient Use Efficiency and Healthier Soil

Editors

Xiaohui Hu
Shiwei Song
Xun Li

MDPI • Basel • Beijing • Wuhan • Barcelona • Belgrade • Manchester • Tokyo • Cluj • Tianjin

Editors

Xiaohui Hu
Northwest Agriculture &
Forestry University
China

Shiwei Song
South China Agricultural
University
China

Xun Li
Chinese Academy of Sciences
China

Editorial Office
MDPI
St. Alban-Anlage 66
4052 Basel, Switzerland

This is a reprint of articles from the Special Issue published online in the open access journal *Horticulturae* (ISSN 2311-7524) (available at: https://www.mdpi.com/journal/horticulturae/special_issues/Greenhouse_Management).

For citation purposes, cite each article independently as indicated on the article page online and as indicated below:

LastName, A.A.; LastName, B.B.; LastName, C.C. Article Title. *Journal Name* **Year**, *Volume Number*, Page Range.

ISBN 978-3-0365-6301-5 (Hbk)
ISBN 978-3-0365-6302-2 (PDF)

Cover image courtesy of Xun Li

Contents

About the Editors

Xiaohui Hu

Dr. Xiaohui Hu has worked as a Professor at the College of Horticulture, Northwest A&F University since 2006. She obtained her bachelor's and master's degrees at Northeast Agricultural University, China, in 2000 and 2003, respectively, and obtained her PhD degree at Nanjing Agricultural University, China, in 2006. Her research areas cover the fields of effective greenhouse production, the anti-stress mechanism of protected crops and its technical application, soilless culture and automatic control of greenhouse environments, as well as water and fertilizers.

Shiwei Song

Dr. Shiwei Song has worked as a Professor in vegetable biology and protected horticulture science at the College of Horticulture, South China Agricultural University since 2008. He obtained his bachelor's degree at Shandong Agriculture University, China, in 2001, his master's degree at South China Agriculture University, China, in 2004, and his PhD degree in Vegetable Science at Shanghai Jiao Tong University, China, in 2008, respectively. His research areas cover the fields of vegetable physiology and molecular biology, environmental horticulture, vegetable nursery science, and plant factories.

Xun Li

Dr. Xun Li is currently working as an Associate Researcher at the Institute of Soil Science, Chinese Academy of Sciences. He received his bachelor's and PhD degrees in Chemistry from Nanjing University, China, in 2005 and 2010, respectively. He then worked at the Institute of Soil Science as an Assistant Researcher and Associate Researcher. His research interests involve the regulation of greenhouse environmental factors and their effect on greenhouse soil properties, nitrogen transformation, and vegetable growth and qualities.

 *horticulturae*

Editorial

Greenhouse Management for Better Vegetable Quality, Higher Nutrient Use Efficiency, and Healthier Soil

Xun Li [1], Xiaohui Hu [2], Shiwei Song [3] and Da Sun [4,*]

[1] State Key Laboratory of Soil and Sustainable Agriculture, Institute of Soil Science, Chinese Academy of Sciences, Nanjing 210008, China
[2] College of Horticulture, Northwest A&F University, Yangling 712100, China
[3] College of Horticulture, South China Agricultural University, Guangzhou 510642, China
[4] Technology Extension Station of Agriculture and Fisheries of Nanhu District of Jiaxing, Jiaxing 314051, China
* Correspondence: sddfj_725@163.com; Tel.: +86-0573-8283-8432

Citation: Li, X.; Hu, X.; Song, S.; Sun, D. Greenhouse Management for Better Vegetable Quality, Higher Nutrient Use Efficiency, and Healthier Soil. *Horticulturae* **2022**, *8*, 1192. https://doi.org/10.3390/horticulturae8121192

Received: 16 November 2022
Accepted: 21 November 2022
Published: 14 December 2022

Publisher's Note: MDPI stays neutral with regard to jurisdictional claims in published maps and institutional affiliations.

1. Introduction

Greenhouse cultivation provides an artificially controlled environment for the off-season production of vegetables, and has played an increasingly important role in agriculture production systems in recent decades. With the exception of soil-less cultivation, vegetables are directly cultured in the soil in most Asian, European, and North American greenhouses. Compared with open-field cultivation, more fertilizer is invested in intensive vegetable production in greenhouses to preserve and increase yields. Thus, after a few years of growth, deteriorated vegetable quality, decreased nutrient use efficiency, and degenerated soil property usually emerge. A well-known reason for these drawbacks is that the vegetables are grown under sub-/supra-optimal greenhouse conditions and suffer diverse abiotic stresses, including extreme temperature, irradiance, poor water condition, a lack of nutrient availability, inappropriate CO_2 concentration and salinity, etc. Recent works have shown that improving greenhouse conditions can promote the growth of vegetables and enhance the uptake of nutrients, leading to better vegetable quality. Meanwhile, greenhouse conditions not only directly influence soil nutrient cycling processes and properties, but also indirectly affect them by regulating vegetable root growth and plant–soil interactions.

This Special Issue (SI), entitled "Greenhouse Management for Better Vegetable Quality, Higher Nutrient Use Efficiency, and Healthier Soil", aims to highlight state-of-the-art greenhouse management that can contribute to increasing vegetable yield and quality, improving nutrient and water use efficiency, and achieving environmentally sustainable utilization of greenhouse soil.

2. Special Issue Overview

This SI features twelve original research articles dealing with the effects of novel greenhouse practices and strategies on the yield and quality of horticulture crops, as well as greenhouse soil properties. Among these publications, three studied the effects of fertilizers, including organic and macro- and micro-nutrient fertilizers, on the growth and nutrient uptake of vegetables [1–3]. Two articles described the effects of water and nutrient supply using irrigation or hydroponic supplying systems on the yield and quality of vegetables [4,5]. Four articles investigated the effects of environmental conditions (mainly light and temperature) on the growth and quality of vegetables [6–9]. In terms of degenerated greenhouse soil, three articles showed how reductive soil disinfestation (RSD) decreased soil salinity, improved soil quality, and inactivated soil-borne pathogens [10–12].

2.1. Fertilizers

Fertilizer is one of the most important factors affecting the yield and quality of greenhouse vegetables, and manure fertilizer coupled with mineral fertilizer is commonly used in greenhouse vegetable cultivation. Sallam et al. [1] compared the effects of various combination ratios of manure and mineral fertilizer on the productivity parameters of greenhouse cucumber. The authors found that the maximum cucumber yield was achieved when 30 kg poultry manure coupled with 3 kg mineral fertilizer was added to coco-peat per cubic meter. The vegetative growth, yield productivity, and fruit quality of cucumber were significantly enhanced when 3 kg mineral fertilizer was applied. Less mineral fertilizer inhibited yield formation, whereas more mineral fertilizer delayed the fruiting stage.

Concerning nitrogen fertilizer, the NH_4^+/NO_3^- ratio has a great impact on vegetable growth and nitrogen utilization. Wang et al. [2] determined the effects of different NH_4^+/NO_3^- ratios in nutrient solution on the growth and nitrogen uptake of Chinese kale. The optimum NH_4^+/NO_3^- ratio was found to be 25/75 to ensure the best growth, the highest fresh and dry weight, and the highest indices of root growth in Chinese kale. Meanwhile, the total N accumulation and N use efficiency were also highest at an NH_4^+/NO_3^- ratio of 25/75. A higher NH_4^+/NO_3^- ratio (50/50) promoted the growth of Chinese kale seedlings at the early growth stage but inhibited it at the late stage, mainly due to the excessive addition of NH_4^+-induced rhizosphere acidification and ammonia toxicity.

In addition to macro-nutrient fertilizers, micro-nutrient fertilizers also play an important role in greenhouse vegetable growth and quality. Xu et al. [3] investigated the effects of four boron levels and two application methods (leaf and root application) on the growth, fruit quality, and flavor of tomato under greenhouse conditions. The results revealed that both application methods significantly increased the net photosynthetic rate and chlorophyll content and stabilized the leaf structure of tomato, mainly due to the improved antioxidant capacity. Leaf spray of 1.9 mg L^{-1} H_3BO_3 was the most effective done for improving the plant growth and photosynthetic indices of tomato, and root application of 3.8 mg L^{-1} H_3BO_3 resulted in better fruit quality and flavor. This work also showed that the application of boron can directly or indirectly regulate the synthesis of volatile substances in tomato fruit.

2.2. Water and Nutrient Supply

Vegetables are highly water-dependent horticultural crops, so water supply could significantly affect the yield and quality of greenhouse vegetables. Ju et al. [4] regulated the greenhouse soil moisture by adding zeolite and alternate drip irrigation, and evaluated the coupling effect of water–zeolite on tomato growth, physiology, yield, quality, and water use efficiency. Using the principal component analysis method, the optimum water for tomato planting was 100% water surface evaporation and the addition of 6 t ha^{-1} zeolite, under alternate drip irrigation conditions with mulch.

The ratio of water and nutrients determines the strength of the nutrient supply, which is also a key factor in the yield forming and quality optimization of horticultural crops. Mouroutoglou et al. [5] compared the effects of four soil-less culture systems (i.e., aeroponic, floating, nutrient film technique, and aggregate systems) on growth, yield, and nutrient uptake in the Greek sweet onion landrace. The results showed that the highest plant biomass, onion yield, and water use efficiency were obtained in floating and aggregate systems, mainly due to the sufficient water and nutrient supply and decreased limit on root growth, when compared with the nutrient film technique. The highest tissue macronutrient concentrations were found in aeroponic and nutrient film technique systems, probably due to root prevalence and condensation effects, respectively. Moreover, the macronutrient uptake concentrations in this study provided a sound basis for the establishment of nutrient solution recommendations for sweet onion cultivation in different closed soil-less culture systems.

2.3. Environmental Conditions

One of the topics addressed in this SI is how to manipulate environmental conditions in a greenhouse to improve vegetable growth and quality. Light is one of the most important environmental factors controlling plant growth and development, so it is essential to determine the light intensity and photoperiod in greenhouse vegetable production. Cui et al. [6] investigated the effect of daily light integral, including the light intensity and photoperiod, on cucumber plug seedlings in an artificial-light plant factory. The optimal daily light integral was 6.35 mol m^{-2} d^{-1}, the optimal intensity was 110–125 μmol m^{-2} s^{-1}, and the optimal photoperiod was 14–16 h to achieve higher plant biomass, shoot dry matter rate, seedling index, and photochemical efficiency. Jiang et al. [7] studied different supplemental lighting modes on the fruit quality of cherry tomatoes in greenhouse production. The results showed that the appearance, flavor quality, nutrient indicators, and aroma of cherry tomato fruits under continuous supplemental lighting and dynamic altered supplemental lighting were generally higher. Considering the electrical cost, dynamic altered supplemental lighting was suggested as a cost-effective supplemental lighting mode for high-value greenhouse cherry tomato production.

The cover materials of greenhouses, such as films or screens, also have a great impact on the microclimate conditions in the greenhouse, and the consequent effects on the vegetable growth and quality are worth exploring. Yamaura et al. [8] examined the physiological and morphological changes in tomato growth and fruit quality in a high tunnel covered with near-infrared reflective film. They found a decrease in total dry matter, resulting from a lower transmitted photosynthetic photon flux density and leaf area index; moreover, they found lower photosynthetic capacity in single leaves because of a decrease in both total nitrogen and chlorophyll content under the near-infrared reflective film. However, the fruit cracking rate was significantly decreased under near-infrared reflective film due to the lower fruit temperature and decrease in fruit dry matter. Therefore, marketable tomato yields can still be guaranteed under near-infrared reflective film. Wen et al. [9] studied the effects of insect-proof screens on the microclimate, reference evapotranspiration, and growth of Chinese flowering cabbages. The results showed that insect-proof screens significantly decreased wind speed and slightly decreased total solar radiation; however, they increased the daily average air humidity, as well as air and soil temperature in the screenhouse. The microclimate improvement resulting from insect-proof screens caused the yield of Chinese flowering cabbages and the irrigation water use efficiency to significantly increase.

2.4. Restoration of Degenerated Soil

Keeping sustainable utilization in mind, the restoration of degenerated greenhouse soil due to RSD practice was investigated. Liu et al. [10] found that the electrical conductivity and available nutrient (NO$_3$$^-$-N, NH$_4$$^+$-N, available K, and available P) content, the abundance of fungi, potential fungal soil-borne pathogens (*F. oxysporum* and *F. solani*), and fungi/bacteria were significantly increased in the plastic-shed soil compared with those in the nearby open-air soil. Meanwhile, the organic fertilizer treatment could not effectively improve the plastic shed soil properties, in which the electrical conductivity and the abundance of potential fungal soil-borne pathogens were even higher. In contrast, soil EC, NO$_3$$^-$-N content, the abundance of the fungi *F. oxysporum* and *F. solani*, and the ratio of fungi to bacteria were remarkably decreased in the RSD-treated soil, while soil pH, the abundance of bacteria, total microbial activity, metabolic activity, and carbon source utilization were significantly increased. Further study revealed that the RSD treatment effectively enhanced microbial interactions and functions, and the microbial network was more complex and connected [11]. Specifically, hydrocarbon, nitrogen, and sulfur cycling functions were significantly increased in RSD-treated soil, whereas bacterial and fungal plant pathogen functions were decreased. On the other hand, Zhu et al. [12] studied how to improve the efficiency of RSD by optimizing the soil water content and organic amendment rate. The results showed that increasing the soil water content and maize straw application rate elevated the soil pH and accelerated the removal of excess sulfate and nitrate in greenhouse soil; moreover, it considerably increased the levels of organic acids that could strongly inhibit soil-borne pathogens. They found that holding 100% WHC and applying maize straw to 10 g kg^{-1} soil created the optimum conditions for RSD field operations for the effective restoration of degraded greenhouse soil.

Author Contributions: Writing—original draft preparation, X.L.; writing—review and editing, X.H., S.S. and D.S.; funding acquisition, D.S. All authors have read and agreed to the published version of the manuscript.

Funding: This research was funded by the Key-Area Research and Development Program of Guangdong Province (2020B0202010006).

Data Availability Statement: Not applicable.

Acknowledgments: We are grateful to the authors who contributed to this Special Issue and to the reviewers for their insightful and unbiased suggestions.

Conflicts of Interest: The authors declare no conflict of interest.

References

1. Sallam, B.N.; Lu, T.; Yu, H.; Li, Q.; Sarfraz, Z.; Iqbal, M.S.; Khan, S.; Wang, H.; Liu, P.; Jiang, W. Productivity Enhancement of Cucumber (*Cucumis sativus* L.) through Optimized Use of Poultry Manure and Mineral Fertilizers under Greenhouse Cultivation. *Horticulturae* **2021**, *7*, 256. [CrossRef]
2. Wang, Y.; Zhang, X.; Liu, H.; Sun, G.; Song, S.; Chen, R. High NH_4^+/NO_3^- Ratio Inhibits the Growth and Nitrogen Uptake of Chinese Kale at the Late Growth Stage by Ammonia Toxicity. *Horticulturae* **2021**, *8*, 8. [CrossRef]
3. Xu, W.; Wang, P.; Yuan, L.; Chen, X.; Hu, X. Effects of Application Methods of Boron on Tomato Growth, Fruit Quality and Flavor. *Horticulturae* **2021**, *7*, 223. [CrossRef]
4. Ju, X.; Lei, T.; Guo, X.; Sun, X.; Ma, J.; Liu, R.; Zhang, M. Evaluation of Suitable Water–Zeolite Coupling Regulation Strategy of Tomatoes with Alternate Drip Irrigation under Mulch. *Horticulturae* **2022**, *8*, 536. [CrossRef]
5. Mouroutoglou, C.; Kotsiras, A.; Ntatsi, G.; Savvas, D. Impact of the Hydroponic Cropping System on Growth, Yield, and Nutrition of a Greek Sweet Onion (*Allium cepa* L.) Landrace. *Horticulturae* **2021**, *7*, 432. [CrossRef]
6. Cui, J.; Song, S.; Yu, J.; Liu, H. Effect of Daily Light Integral on Cucumber Plug Seedlings in Artificial Light Plant Factory. *Horticulturae* **2021**, *7*, 139. [CrossRef]
7. Jiang, C.; Rao, J.; Rong, S.; Ding, G.; Liu, J.; Li, Y.; Song, Y. Fruit Quality Response to Different Abaxial Leafy Supplemental Lighting of Greenhouse-Produced Cherry Tomato (*Solanum lycopersicum* var. *Cerasiforme*). *Horticulturae* **2022**, *8*, 423. [CrossRef]
8. Yamaura, H.; Furuyama, S.; Takano, N.; Nakano, Y.; Kanno, K.; Ando, T.; Amasaki, I.; Watanabe, Y.; Iwasaki, Y.; Isozaki, M. Effects of NIR Reflective Film as a High Tunnel-Covering Material on Fruit Cracking and Biomass Production of Tomatoes. *Horticulturae* **2022**, *8*, 51. [CrossRef]
9. Wen, J.; Ning, S.; Wei, X.; Guo, W.; Sun, W.; Zhang, T.; Wang, L. The Impact of Insect-Proof Screen on Microclimate, Reference Evapotranspiration and Growth of Chinese Flowering Cabbage in Arid and Semi-Arid Region. *Horticulturae* **2022**, *8*, 704. [CrossRef]
10. Liu, L.; Long, S.; Deng, B.; Kuang, J.; Wen, K.; Li, T.; Bai, Z.; Shao, Q. Effects of Plastic Shed Cultivation System on the Properties of Red Paddy Soil and Its Management by Reductive Soil Disinfestation. *Horticulturae* **2022**, *8*, 279. [CrossRef]
11. Yan, Y.; Wu, R.; Li, S.; Su, Z.; Shao, Q.; Cai, Z.; Huang, X.; Liu, L. Reductive Soil Disinfestation Enhances Microbial Network Complexity and Function in Intensively Cropped Greenhouse Soil. *Horticulturae* **2022**, *8*, 476. [CrossRef]
12. Zhu, R.; Huang, X.; Zhang, J.; Cai, Z.; Li, X.; Wen, T. Efficiency of Reductive Soil Disinfestation Affected by Soil Water Content and Organic Amendment Rate. *Horticulturae* **2021**, *7*, 559. [CrossRef]

 *horticulturae*

Article

Effect of Daily Light Integral on Cucumber Plug Seedlings in Artificial Light Plant Factory

Jiawei Cui [1,2], Shiwei Song [1], Jizhu Yu [2,*] and Houcheng Liu [1,*]

[1] College of Horticulture, South China Agricultural University, Guangzhou 510642, China; cuijiawei@saas.sh.cn (J.C.); swsong@scau.edu.cn (S.S.)
[2] Shanghai Key Lab of Protected Horticultural Technology, Horticultural Research Institute, Shanghai Academy of Agricultural Sciences, Shanghai 201403, China
* Correspondence: yy2@saas.sh.cn (J.Y.); liuhch@scau.edu.cn (H.L.)

Abstract: In a controlled environment, in an artificial light plant factory during early spring or midsummer, vegetable seedlings can be uniform, compact, and high quality. Appropriate light parameters can speed up the growth of seedlings and save on production costs. Two experiments were carried out in this study: (1) cucumber seedling growth under different daily light integrals (DLIs) (5.41–11.26 mol·m^{-2}·d^{-1}) and optimum DLI for seedling production were explored (experiment 1: Exp. 1); (2) under the same DLI selected by Exp. 1, the effects of different light intensities and photoperiods on cucumber seedlings were investigated (experiment 2: Exp. 2). The root biomass, root-to-shoot ratio, seedling index, and shoot dry matter rate increased as the DLI increased from 5.41 to 11.26 mol·m^{-2}·d^{-1}, while the shoot biomass and leaf area decreased in Exp. 1. The cucumber seedlings became more compact as DLI increased, but more flowers developed after transplanting when the DLI was 6.35 mol·m^{-2}·d^{-1}. Under the optimal DLI (6.35 mol·m^{-2}·d^{-1}), the optimal intensity was 110–125 µmol·m^{-2}·s^{-1}, and the optimal photoperiod was 14–16 h, in which plant biomass, shoot dry matter rate, seedling index, and photochemical efficiency were higher.

Keywords: daily light integral; cucumber seedling; seedling quality; flower development

Citation: Cui, J.; Song, S.; Yu, J.; Liu, H. Effect of Daily Light Integral on Cucumber Plug Seedlings in Artificial Light Plant Factory. *Horticulturae* **2021**, *7*, 139. https://doi.org/10.3390/horticulturae7060139

Academic Editor: Othmane Merah

Received: 16 April 2021
Accepted: 20 May 2021
Published: 7 June 2021

Publisher's Note: MDPI stays neutral with regard to jurisdictional claims in published maps and institutional affiliations.

1. Introduction

The daily light integral (DLI) is the total amount of photosynthetic light delivered to plants each day [1,2]. The DLI is an important environmental parameter for plants and displays stronger correlation with vegetative growth parameters than with photoperiod and light intensity [3,4]. Many studies have shown that a suitable DLI could promote the growth and development of plants, such as the germination rate of photoblastic seeds [5], shoot and root biomass [6–9], stem diameter [10,11], leaf area and leaf number [12], and the differentiation of flower buds [4,13,14].

The production of high-quality seedlings with characteristics of higher biomass, compactness and adequate leaf area in a short cultivation period is the target of producers. However, this constitutes a significant challenge under insufficient scrutiny [15,16]. Many operations use supplemental lighting to increase DLI under low irradiance to accelerate growth and improve the morphology of seedlings in typical greenhouses [17–19]. Cucumber seedlings became more compact and vigorous, with characteristics of a short hypocotyl length, large leaf area, and high biomass under a high DLI [19,20]. Increases in DLI during the seedling stage have been also reported to accelerate subsequent floral initiation and decrease the time to the first flowering after transplant [13,21,22]. The number of tomato floral buds and days to flowering increased corresponding to the decrease in the shading before anthesis [23].

However, higher DLI is correlated with more illuminating apparatus and electric energy consumption. Additionally, superfluous DLI with a high light intensity or long photoperiod may cause damage to the photosynthetic system and inhibits the production of photosynthates [24,25]. For example, high DLI results in a decrease in photosystem II (PSII) activity, thus leading to the low production of lettuce [26]. The total fresh mass and total dry mass of spinach were higher when DLI was 17.3 $mol \cdot m^{-2} \cdot d^{-1}$ compared with 20.2 $mol \cdot m^{-2} \cdot d^{-1}$ [27]. These reductions in biomass may result in a reduced source of flower bud formation, since it has been previously demonstrated that flower initiation requires a certain amount of photoassimilates [8]. Extra DLI did not further increase the flowering rate or floral bud number in some bedding plants [28–30]. Therefore, an optimal DLI which can only just meet the demand of seedling growth and facilitate late flower bud differentiation should be selected by the producer for energy-saving purposes.

Under the same DLI, a longer photoperiod is more beneficial to the growth of plants. The plant height, shoot dry mass, root dry mass and chlorophyll content index of rudbeckia seedlings were larger with longer photoperiods [18]. A longer photoperiod with the same DLI increased the light interception, chlorophyll content index, quantum yield of photosystem II, and aboveground biomass of lettuce and mizuna [31,32].

Under artificial light conditions, DLI is completely supplied by lamps and varies via changing the light intensity or photoperiod. Due to the controllable illumination parameters, DLI parameters for plants are more reliable and can be used for plant cultivation and research with higher accuracy. Exploring the optimum DLI and illumination parameters is conducive to the scheduling and improvement of profitability.

Cucumber is cultivated worldwide, and its seedling stage is crucial for subsequent growth and development. Scarce research has been conducted on the effects of DLI from sole-source (SS) lighting on cucumber plug seedlings. Therefore, the objectives of our study were to determine the optimal DLI by investigating the morphology, biomass, chlorophyll fluorescence, and subsequent flower development of cucumber plug seedlings (Exp. 1); then the effects of different illumination intensities and times on the quality of cucumber seedlings were further investigated under the optimal DLI selected in Exp. 1, and the appropriate light parameters for cucumber plug seedlings were finally selected (Exp. 2).

2. Materials and Methods

2.1. Plant Material

In both Exp. 1 and 2, seeds of cucumber 'Yuexiu No.3' were place onto humid cotton in an incubator for 2 days. The temperature and relative humidity were set to 25 °C and 80%, respectively. The germinated seeds were sown into plug trays (50-cell size (28 mL)) in a substrate with peat, coconut coir and perlite (6: 3: 1, v: v: v) and placed into a walk-in growth chamber at the College of Horticulture of South China Agricultural University, Guangzhou. Each tray was a repeat. Each experiment conducted four treatments, in which three repeats were concluded. Before cotyledons were fully expanded after sowing, 800 ± 50 mL tap water was added to each tray every second day. Once cotyledons were fully expanded, 800 ± 50 mL 1/4 strength modified Hoagland and Arnon's nutrient solution was added to each tray every second day to provide (in $mg \cdot L^{-1}$) 53 nitrogen (N), 40 calcium (Ca), 59 potassium (K), 8 phosphorus (P), 25 magnesium (Mg), 33 sulphur (S), 0.56 iron (Fe), 0.5 boron (B), 0.5 manganese (Mn), 0.05 zinc (Zn), 0.01 molybdenum (Mo) and 0.005 copper (Cu) until the emergence of the second true leaf. Then, 800 ± 50 mL 1/2 strength modified Hoagland and Arnon's nutrient solution was added to each tray every second day until sampling.

2.2. Growth Chamber Environment

All the plug trays were placed on steel shelves in two vertical layers in a walk-in growth chamber, with an average daily temperature of 25 °C, relative humidity (RH) of 75%, and average carbon dioxide (CO_2) concentration of 400 $\mu mol \cdot mol^{-1}$.

2.3. Sole-Source Lighting Treatments

Adjustable LED lamps of red (655–660 nm):blue (455–460 nm) = 1:1 (R_1:B_1) (Unihero technology Co. Ltd., Huizhou, China) were placed above the steel shelves providing sole-source lighting with different photosynthetic photon flux densities (PPFDs) and photoperiods. The spectra for PPFD at 200 $\mu mol \cdot m^{-2} \cdot s^{-1}$ are shown in Figure 1. Two experiment lighting treatments were conducted as follows. Exp. 1: the average total photon flux was delivered from sole-source light-emitting diodes (LEDs) with light ratios (%) of red/blue 50:50 to achieve target light intensities of 125 to 200 $\mu mol \cdot m^{-2} \cdot s^{-1}$ with 12–16 h photoperiods. Four gradients of DLIs in each run set from 7.47 to 11.26 $mol \cdot m^{-2} \cdot s^{-1}$ (run 1) and from 5.41 to 8.43 $mol \cdot m^{-2} \cdot s^{-1}$ (run 2), respectively (Table 1). Exp. 2: target light intensities were of the same light ratios as Exp. 1, of 110–175 $\mu mol \cdot m^{-2} \cdot s^{-1}$, with 10–16 h photoperiods, and four lighting treatments were conducted with the same DLI (6.35 $mol \cdot m^{-2} \cdot d^{-1}$) selected by Exp. 1 (Table 2).

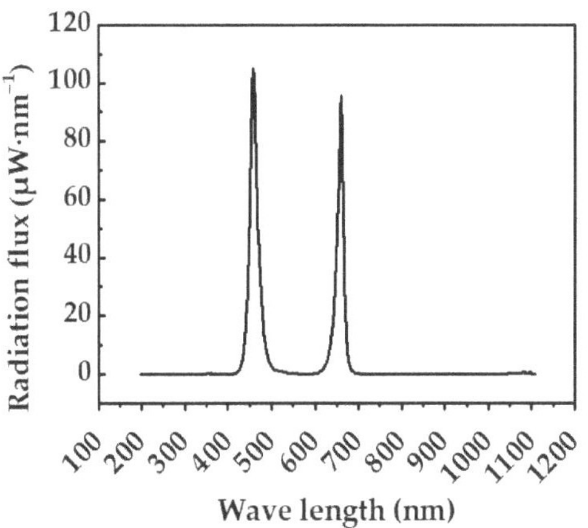

Figure 1. Spectral quality of 200 $\mu mol \cdot m^{-2} \cdot s^{-1}$ from SS LEDs with light ratios (%) of red/blue 50:50.

Table 1. Illumination parameters of different treatments in Exp. 1.

Treatments	Light Intensity * ($\mu mol \cdot m^{-2} \cdot s^{-1}$)	Photoperiod (h)	Average DLI * ($mol \cdot m^{-2} \cdot d^{-1}$)
	148.17 ± 2.75	14	7.47 ± 0.14
Run 1	148.00 ± 2.22	16	8.52 ± 0.13
	199.33 ± 2.50	14	10.05 ± 0.13
	195.43 ± 1.72	16	11.26 ± 0.10
	125.14 ± 1.06	12	5.41 ± 0.05
Run 2	126.00 ± 1.69	14	6.35 ± 0.09
	151.57 ± 1.13	14	7.64 ± 0.06
	146.43 ± 0.87	16	8.43 ± 0.05

* The data represent the mean ± SE of 10 spots under lights in each treatment.

Table 2. Four combinations with different light intensities and photoperiods that achieve the same DLI in Exp. 2.

Treatments	Light Intensity * ($\mu mol \cdot m^{-2} \cdot s^{-1}$)	Photoperiod (h)	Average DLI * ($mol \cdot m^{-2} \cdot d^{-1}$)
T1	111.17 ± 0.48	16	6.40 ± 0.03
T2	125.17 ± 1.90	14	6.31 ± 0.10
T3	146.00 ± 1.32	12	6.31 ± 0.06
T4	176.67 ± 1.33	10	6.36 ± 0.05

* The data represent the mean ± SE of 10 spots under lights in each treatment.

2.4. Morphology and Growth Measurement

The morphology and growth of cucumber plug seedlings in both Exp. 1 and 2 were measured on the 26th day after sowing. Six seedlings of each repeat (i.e., 18 seedlings per DLI treatment) were randomly selected and measured to determine the hypocotyl length, stem diameter, total leaf area, shoot fresh mass, shoot dry mass, root fresh mass, and root dry mass. The fresh mass of seedlings was recorded by an electronic balance, and the dry mass was recorded after seedlings were heated to de-enzyme at 105 °C, then dried at 80 °C until steady.

Seedling index = (stem diameter/hypocotyl length + root dry mass/shoot dry mass) × total dry mass × 10 [33], root-to-shoot ratio = root dry mass/shoot dry mass, leaf area ratio = total leaf area/shoot dry mass, and shoot dry matter rate = shoot dry mass/shoot fresh mass were calculated [11,34].

2.5. Physiological Characteristics and Photosynthesis Pigment Content

The soluble sugar content was extracted and analyzed by anthrone colorimetry. A 0.5 g amount of leaf fresh sample was added to a test tube with 10 mL of distilled water, and then it was extracted in boiling water for 30 min. The extract was filtered into a 25 mL volumetric flask, the test tube and residue were rinsed repeatedly, and the volume was maintained to scale. An amount of 0.1 mL of the sample extract was absorbed into a 20 mL graduated test tube, followed by adding 1.9 mL of distilled water, 0.5 mL of ethyl anthrone acetate reagent and 5 mL of concentrated sulfuric acid. The solution was fully oscillated and cooled down to room temperature. The absorbance was measured at 630 nm by using a UV-spectrophotometer [35].

The soluble protein content was extracted and analyzed by Coomassie Brilliant Blue methods. Amount of 0.5 g of fresh leaf sample and 5 mL of distilled water were ground into a homogenate and centrifuged at $10,000 \times g$ for 10 min at 4 °C. After mixing 0.3 mL of supernatant and 0.7 mL of distilled water, 5 mL of Coomassie Brilliant Blue G-250 solution (0.1 g/L) was added to the mixture. After 2 min, the absorbance at 595 nm was determined by using a UV-spectrophotometer [35].

The content of photosynthesis pigment was determined by an acetone–ethanol blend method. An amount of 0.2 g of leaf with removed veins was placed into a test tube containing 20 mL of acetone and ethanol (1:1, *v:v*), and extracted under dark conditions. The absorbance at 663, 645 and 440 nm was determined by using a UV-spectrophotometer after the leaves turned white. Chlorophyll a (Chl a) content $(mg \cdot g^{-1}) = (12.70 \times A_{663} - 2.69 \times A_{645}) \times 20\ mL/(1000 \times 0.2\ g)$; chlorophyll b (Chl b) content $(mg \cdot g^{-1}) = (22.90 \times A_{645} - 4.86 \times A_{663}) \times 20\ mL/(1000 \times 0.2\ g)$; total chlorophyll (Total Chl) content $(mg \cdot g^{-1}) = (8.02 \times A_{663} + 20.20 \times A_{645}) \times 20\ mL/(1000 \times 0.2\ g)$; carotenoid content $(mg \cdot g^{-1}) = (4.70 \times A_{440} - 2.17 \times A_{663} - 5.45 \times A_{645}) \times 20\ mL/(1000 \times 0.2\ g)$ [36].

2.6. Chlorophyll Fluorescence

Chlorophyll fluorescence was measured by MINI-PAM-II (WALZ company, Germany), and the actinic illumination was set at 286 $\mu mol \cdot m^{-2} \cdot s^{-1}$. Leaves were dark-acclimated for 20 min with the manufacturer's plastic and foam clips before measurements were recorded. Six seedlings of each repeat were randomly selected and measured, and the locus was the second true leaf of each seedling. The photochemical quenching indexes (qP and qL), the non-photochemical quenching indexes (qN and NPQ), photic injury non-photochemical quenching quantum yield (Y(NO)), photoprotection non-photochemical quenching quantum yield (Y(NPQ)) and PSII conversion efficiency of light energy (Φ_{PSII}), and maximal photochemical efficiency of PSII in the dark (Fv/Fm) were measured.

2.7. Flowering in Cucumber Plant after Transplant

As a supplementary experiment, only the run 2 in Exp. 1 was carried out for flowering statistics. After DLI treatments, nine cucumber seedlings from each DLI treatment were transplanted into a growth bag (15 cm diameter, 15 cm height) filled with substrate of peat, coconut coir and perlite (6:3:1, v:v:v). Plants were watered with 1/2 strength Hoagland and Arnon's nutrient solution before the first flower bloom, and then full strength nutrient solution. The days to first flower bloom from transplant, node of the first male flower, male flower number, node of first female flower, female flower number on main stem, number of lateral branches, and total female flower number under the 15th node were recorded at 60 days after the transplant.

2.8. Statistical Analysis

Data were analyzed by one-way ANOVA using PASW Statistics 18.0. Duncan's multiple range test at $p < 0.05$ was used to determine significant treatment differences. The plotting and fitting of monomial and binomial response functions were conducted using Origin 8.6.

3. Results

3.1. Experiment 1

3.1.1. Morphology and Growth of Cucumber Plug Seedlings

The morphology and growth of cucumber plug seedlings were affected by DLI. The hypocotyl length and leaf area of cucumber plug seedlings showed the same trend as DLI increased. Hypocotyl length decreased from 8.28 to 3.55 cm, while the leaf area decreased from 123.15 to 56.07 cm^2, as DLI increased from 5.41 to 11.26 mol·m^{-2}·d^{-1} (Figure 2a,c). The stem diameter of cucumber plug seedlings was not affected by DLI (Figure 2b). Shoot fresh mass and dry mass decreased as DLI increased from 5.41 to 11.26 mol·m^{-2}·d^{-1}. The root fresh mass and dry mass first increased and then decreased as the DLI increased, peaking at 8.43 mol·m^{-2}·d^{-1} (Figure 3a). The root-to-ratio and seedling index first increased and then remained steady with the increase in DLI. At a lower DLI, the leaf area ratio decreased significantly with an increased DLI, but it remained at a low value under higher DLIs. The shoot dry mass rate increased from 0.073 to 0.095 as the DLI increased from 5.41 to 11.26 mol·m^{-2}·d^{-1} (Figure 3b).

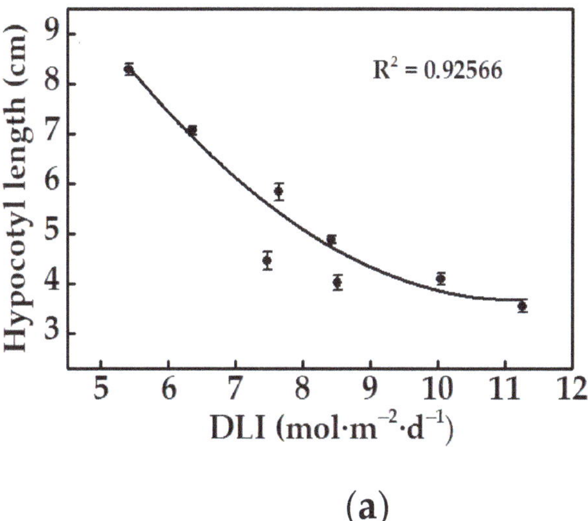

Figure 2. *Cont.*

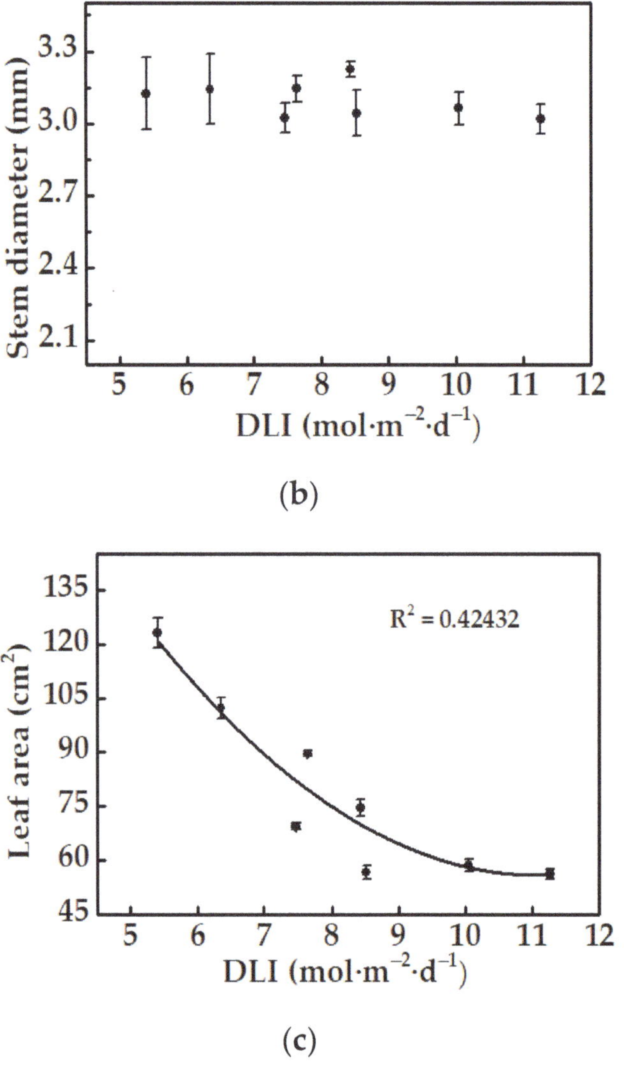

(b)

(c)

Figure 2. Relationships between DLI and (**a**) hypocotyl length, (**b**) stem diameter and (**c**) leaf area of cucumber seedlings. Each symbol represents the mean of 18 plants in each treatment, and error bars represent SEs of means. Fitting formula shown in supplementary section.

3.1.2. Physiological Characteristics and Photosynthesis Pigment Content of Cucumber Plug Seedlings

The soluble sugar content of cucumber plug seedlings increased by 8.72 mg·g^{-1} linearly as the DLI increased from 5.41 to 11.26 mol·m^{-2}·d^{-1} (Figure 4a). When the DLI was in the range of 6.35–8.52 mol·m^{-2}·d^{-1}, the soluble protein remained high but then fell dramatically beyond this range (Figure 4b). All the photosynthesis pigment contents were negatively correlated with DLI (Figure 5).

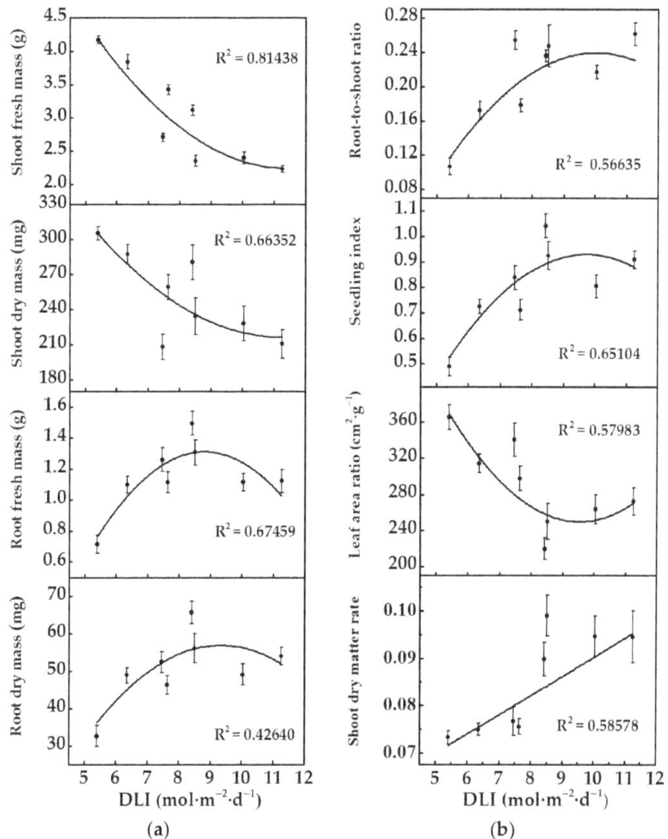

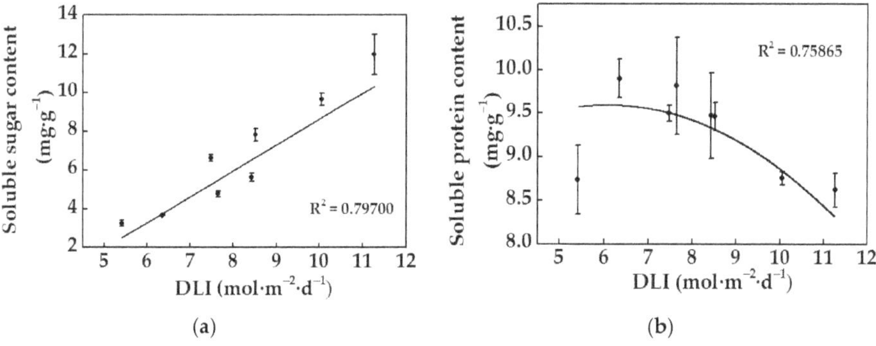

Figure 3. Relationships between DLI and (**a**) shoot fresh and dry mass, root fresh and dry mass, (**b**) root-to-shoot ratio, seedling index, leaf area ratio, and shoot dry matter rate of cucumber seedlings. Each symbol represents the mean of 18 plants in each treatment, and error bars represent SEs of means. Fitting formula shown in supplementary section.

Figure 4. Relationships between DLI and (**a**) soluble sugar content and (**b**) soluble protein content of cucumber seedlings. Each symbol represents the mean of three repeats in each treatment, and error bars represent SEs of means. Fitting formula shown in supplementary section.

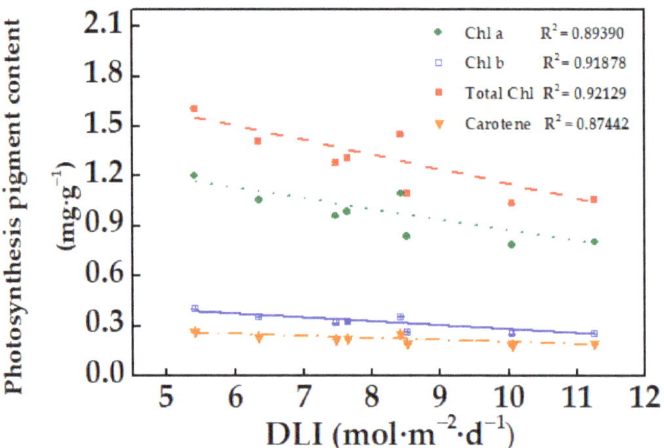

Figure 5. Relationships between DLI and photosynthesis pigment content of cucumber seedlings. Each symbol represents the mean of three repeats in each treatment, and error bars represent SEs of means. Fitting formula shown in supplementary section.

3.1.3. Flowering in Cucumber Plant after Transplant

The days to the first flower bloom from transplant, and the node of the first male and female flowers (data not shown) were unaffected by DLI. When DLI was about $6.35\ \text{mol·m}^{-2}\text{·d}^{-1}$, more male flowers, female flowers and lateral branches appeared (Table 3).

Table 3. The effect of DLI on cucumber plant flowering after transplant of cucumber plug seedlings in run 2 *.

DLI/mol·m^{-2}·d^{-1}	Days to First Flower Bloom from Transplant	Male Flower Number	Female Flower Number of Main Stem	Number of Lateral Branches	Total Female Flower Number
5.41	39.83 ± 0.75 [a]	46.00 ± 4.47 [b]	1.17 ± 0.17 [b]	1.00 ± 0.26 [b]	2.17 ± 0.31 [b]
6.35	39.00 ± 0.26 [a]	57.33 ± 2.60 [a]	1.17 ± 0.17 [b]	2.50 ± 0.43 [a]	3.67 ± 0.56 [a]
7.64	38.83 ± 0.31 [a]	50.33 ± 3.87 [a,b]	1.33 ± 0.21 [a]	2.17 ± 0.54 [a,b]	3.50 ± 0.43 [a,b]
8.43	40.00 ± 0.45 [a]	39.83 ± 2.24 [b]	1.33 ± 0.21 [a]	1.33 ± 0.42 [a,b]	2.67 ± 0.42 [a,b]

* Presented values are means $\pm$ SEs ($n = 9$). [a,b] Different lowercase letters indicate statistically differences among treatments ($p < 0.05$).

3.2. Experiment 2

3.2.1. Morphology and Growth of Cucumber Plug Seedlings

Under the same DLI, the hypocotyl length, stem diameter and shoot fresh mass were independent of changes in illumination strength and time (supplementary section and Figure 6a). The shoot dry mass, root fresh mass and root dry mass decreased by 50.24, 310.00 and 15.13 mg, respectively, as the light intensity increased from 111 to 177 $\mu\text{mol·m}^{-2}\text{·s}^{-1}$ and the photoperiod decreased from 16 to 10 h (Figure 6a). Two sets of low light intensities and long photoperiods (T1 and T2) had a higher root-to-shoot ratio, seedling index, shoot dry matter rate and lower leaf area ratio (Figure 6b).

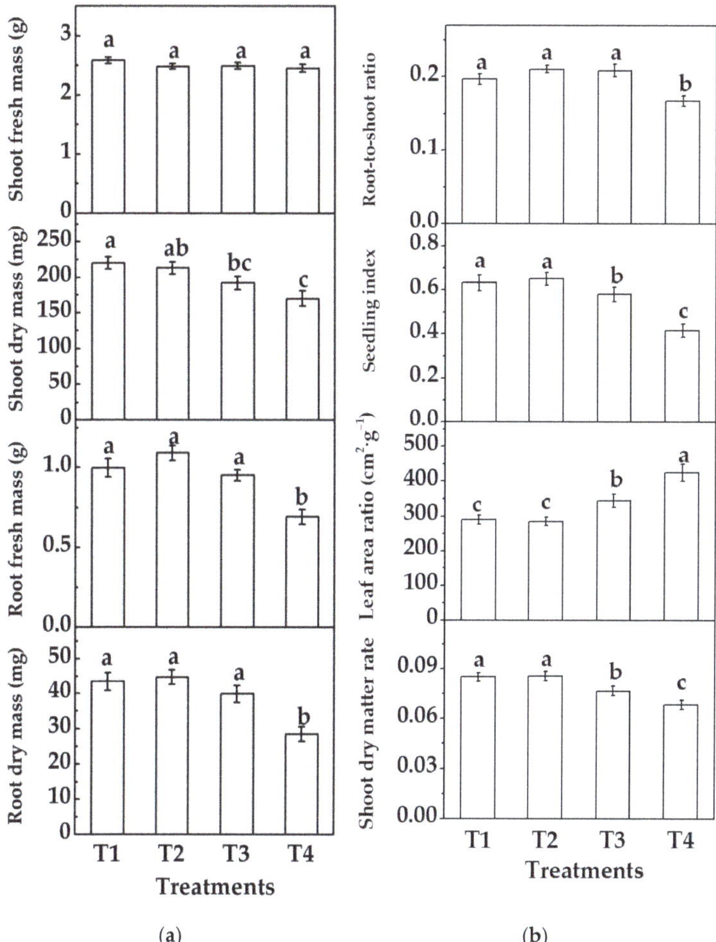

Figure 6. The effects of different light intensities and photoperiods on (**a**) shoot fresh and dry mass, root fresh and dry mass (**b**) root-to-shoot ratio, seedling index, leaf area ratio, and shoot dry matter rate of cucumber seedlings. Different letters indicate statistically significant differences among treatments ($p \leq 0.05$). Error bars show SEs ($n = 18$).

3.2.2. Physiological Characteristics of Cucumber Plug Seedlings

The soluble sugar content was the highest in T1, and there were no significant differences in the other three treatments. The soluble sugar content in T1 was 25.66%, 27.95%, and 35.42% higher than in T2, T3, and T4, respectively (Figure 7a). There was no obvious pattern between the content of soluble protein with the variation of illumination parameters under the same DLI (Figure 7b).

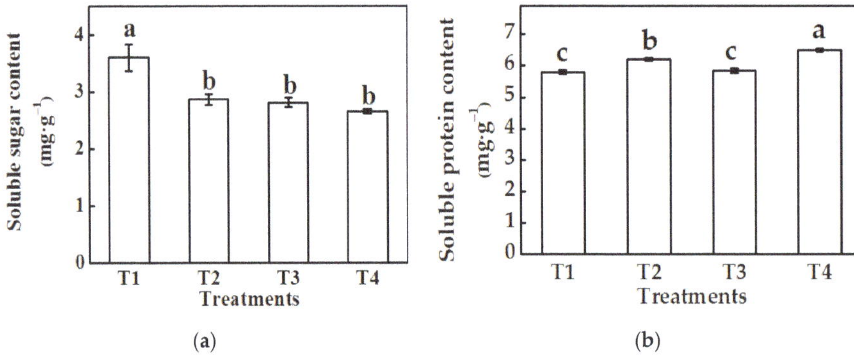

Figure 7. The effects of different light intensities and photoperiods on (**a**) soluble sugar and (**b**) soluble protein content of cucumber seedlings. Different letters indicate statistically significant differences among treatments ($p \leq 0.05$). Error bars show SEs ($n = 9$).

3.2.3. Photosynthesis Pigment Content and Chlorophyll Fluorescence of Cucumber Plug Seedlings

When the light intensity increased and the photoperiod decreased, the contents of chlorophyll and carotenoid tended to increase (data shown in supplementary section). As the light intensity increased and the photoperiod decreased, the photochemical quenching indexes (qP and qL) and Φ_{PSII} all showed a downwards trend, and the decreased range from T3 to T4 was sharp, and there were significant differences among the four treatments. The non-photochemical quenching indexes (qN and NPQ) and Y(NPQ) increased first then decreased later, which was the largest in T3. Y(NO) showed an increasing trend with the increase in light intensity and shortening of the photoperiod, and T4 was 146.84%, 144.89%, and 117.27% higher than T1, T2, and T3, respectively. The Fv/Fm of T4 was the smallest, and there was no difference among T1, T2, and T3 (Table 4).

Table 4. The effects of different light intensities and photoperiods on chlorophyll fluorescence of cucumber seedlings *.

Treatment	Photochemical Quenching		Non-Photochemical Quenching		Y(NO)	Y(NPQ)	Φ_{PSII}	Fv/Fm
	qP	qL	qN	NPQ				
T1	0.63 ± 0.01 [a]	0.39 ± 0.01 [a]	0.62 ± 0.01 [b]	1.11 ± 0.04 [b]	0.29 ± 0.00 [c]	0.32 ± 0.01 [c]	0.39 ± 0.01 [a]	0.78 ± 0.00 [a]
T2	0.58 ± 0.01 [b]	0.35 ± 0.00 [b]	0.65 ± 0.01 [a,b]	1.23 ± 0.05 [a,b]	0.29 ± 0.00 [c]	0.36 ± 0.01 [b]	0.35 ± 0.01 [b]	0.77 ± 0.00 [a]
T3	0.40 ± 0.02 [c]	0.21 ± 0.01 [c]	0.66 ± 0.01 [a]	1.31 ± 0.07 [a]	0.33 ± 0.01 [b]	0.42 ± 0.01 [a]	0.25 ± 0.01 [c]	0.78 ± 0.01 [a]
T4	0.12 ± 0.01 [d]	0.05 ± 0.00 [d]	0.31 ± 0.01 [c]	0.31 ± 0.02 [c]	0.71 ± 0.01 [a]	0.22 ± 0.01 [d]	0.07 ± 0.00 [d]	0.64 ± 0.02 [b]

* Presented values are means $\pm$ SEs ($n = 18$). [a–d] Different lowercase letters indicate statistically differences among treatments ($p < 0.05$).

4. Discussion

Under low irradiance conditions, plants extend stems and leaves to capture more light energy, and are prone to thin branches and leaves [34,37]. Increasing DLI along with low irradiance can effectively inhibit plants spindling growth (elongated hypocotyl, thin stem diameter) of cucumber seedling [17]. As the DLI increased, the hypocotyl length and leaf area decreased in this study, but the stem diameter remained at 3–3.3 cm.

With the increased DLI, cucumber seedling quality improved in some respects (Figure 3b). High-quality vegetable seedlings are usually compact and fully rooted transplants with a higher seedling index ratio and root-to-shoot ratio [15]. Hurt et al. [38] found that when DLI was increased to about 10 mol·m^{-2}·d^{-1}, the seedling quality index of five annual bedding plants was significantly improved. The root-to-shoot ratio of new guinea impatiens, geranium, and petunia cuttings were increased as DLI increased [11]. In this study, the root-to-shoot ratio and seedling index increased as DLI increasing from 5.4 to 8.5 mol·m^{-2}·d^{-1}, then, became stable due to the decrease of root dry mass with DLI continuing to increase. A lower leaf area ratio and a higher shoot dry matter rate are also regarded as high quality, which represents a higher proportion of structural materials and carbohydrates per of unit fresh weight [34]. As the DLI increased, the leaf area ratio of eight kinds of bedding plants (i.e., ageratum, begonia, impatiens, marigold, petunia, salvia, vinca, and zinnia) decreased, while the shoot dry matter rate increased [34]. This phenomenon also occurred with the *Tecoma stans* [16], impatiens cuttings [37], *Heuchera americana* [39], lettuce [40], and *brassica* microgreens [41].

High-quality seedlings exhibit luxuriant growth both before and post-transplant growth, which is mainly due to a higher shoot fresh and dry mass. Generally, increasing DLI can promote the accumulation of shoot biomass [41,42]. However, there was a negative correlation between DLI and biomass in this study. It is possible that higher DLIs exceeded the amount required for cucumber seedlings. As in Ji's [20] experiment, when DLI exceeded 14.4 mol·m^{-2}·d^{-1}, the correlation between the fresh mass and dry mass of the shoot in three cucumber seedlings and DLI changed from positive to negative. The promotion of DLI to flower development is usually accompanied by an increased shoot biomass [23,28,34,43]. Leaf area reached 500 cm^2, and dry mass reached 3 g, which were prerequisite for the flowering of tomato seedlings, revealing that some level of photoassimilates is necessary for floral initiation [23]. In this study, the decreased shoot biomass and leaf area with a higher DLI might have slowed down the development of cucumber seedlings, which stunted flower development after transplant. Therefore, the optimal DLI for cucumber seedlings at the nursery stage was 6.35 mol·m^{-2}·d^{-1}, in which there is a higher shoot biomass and better floral initiation.

Under the optimal DLI (6.35 mol·m^{-2}·d^{-1}) screened by Exp. 1, a further exploration of appropriate lighting parameters combining different light intensities and photoperiods of cucumber seedlings was conducted. Under the same DLI, the variation of the light intensity and photoperiod mainly influences dry matter accumulation [44]. In this study, the shoot dry mass decreased with the increased light intensity and the reduction in the photoperiod, while there was no significant change in shoot fresh mass. Additionally, the combined treatment of a high light intensity and short photoperiod was not conducive to thicker leaves and stronger seedlings (Figure 6b).

In Exp. 2, the chlorophyll fluorescence parameters were consistent with the growth conditions, and also demonstrated that the combination of a low light intensity and long photoperiod was more favorable for the growth of cucumber seedlings. Under the same photoperiod, as PPFD increased, Φ_{PSII} decreased in three species (*Ipomea batatas*, *Lactuca sativa*, and *Epipremnum aureum*) [45]. This phenomenon was also found in lettuce and mizuna under the same DLI, which meant that most PSII reaction centers were closed at high PPFDs [31,46]. In this study, Φ_{PSII} decreased as PPFD increased, resulting in a decline in photosynthate. NPQ and Y(NPQ) increased first and then decreased, indicating that with the increase in light intensity and the shortening of the photoperiod, the photoprotection mechanism of cucumber seedlings was activated, but the excessive light intensity destroyed the PSII and reduced the photoprotection ability of cucumber seedlings [47]. Meanwhile, Y(NO) showed that the degree of light damage was severe in T4. Due to the low light intensity requirements, these combinations (low light intensities and long light durations) can also reduce capital expenses by reducing the cost of light fixtures [18,31].

With the increased DLI, the photosynthetic pigment showed a linear declining trend, which confirmed that leaves gained more light capture ability by increasing the chlorophyll content under lower DLIs [48,49]. The relative chlorophyll content of cucumber leaves of different ages decreased upon increasing the DLI, and the apparent quantum yield decreased according to the light response curve [50]. The chlorophyll content was positively correlated with shoot biomass in Exp. 1 and negatively correlated with shoot biomass in Exp. 2, which reveal that the chlorophyll content only represents the adaptability of plants to environmental changes and had no direct relationship with the amount of photosynthate produced by plants [49].

Plant leaves adapt to light conditions not only anatomically and morphologically, but also biochemically [48]. The increase of soluble sugar content increased the plant resistance to some extent [51]. Previous research has shown that increasing DLI promotes the accumulation of soluble sugar in lettuce [40]. Higher DLIs and longer photoperiods may be more beneficial to improve the resistance of cucumber seedlings. These might be future research perspectives.

5. Conclusions

In conclusion, within the range of 5.41–11.26 mol·m^{-2}·d^{-1}, the increment in DLI was conducive to a compact morphology but inhibited the shoot biomass of cucumber plug seedlings. The optimal DLI in an artificial light condition was 6.35 mol·m^{-2}·d^{-1}, for which the leaf area and shoot biomass were relatively high, and floral initiation was accelerated in the subsequent stage. Under the condition of optimal DLI (6.35 mol·m^{-2}·d^{-1}), the combinations of low light intensities and long photoperiods are more conducive to the accumulation of root growth and whole plant dry matter, with a relatively high seedling index and thicker leaves. Therefore, the optimal light intensity was 110–125 μmol·m^{-2}·s^{-1}, and the optimal photoperiod was 14–16 h for cucumber plug seedlings.

Supplementary Materials: The following are available online at https://www.mdpi.com/article/10.3390/horticulturae7060139/s1.

Author Contributions: Conceptualization, H.L.; data curation, J.C. and S.S.; formal analysis, S.S.; funding acquisition, H.L.; investigation, J.C.; methodology, J.Y. and H.L.; writing—original draft, J.C. and S.S.; writing—review and editing, J.Y. and H.L. All authors have read and agreed to the published version of the manuscript.

Funding: This research was funded by National Key Research and Development Program of China (2017YFE0131000, 2019YFD1001900).

Institutional Review Board Statement: Not applicable.

Informed Consent Statement: Not applicable.

Data Availability Statement: Raw data is available on request.

Conflicts of Interest: The authors declare no conflict of interest.

References

1. Faust, J.E.; Logan, J. Daily Light Integral: A research review and high-resolution maps of the United States. *HortScience* **2018**, *53*, 1250–1257. [CrossRef]
2. Christiaens, A.; Lootens, P.; Roldán-Ruiz, I.; Pauwels, E.; Gobin, B.; Van Labeke, M.C. Determining the minimum daily light integral for forcing of azalea (*Rhododendron simsii*). *Sci. Hortic.* **2014**, *177*, 1–9. [CrossRef]
3. Lee, J.H. *Vegetative Growth and Inflorescence Emergence of Phalaenopsis 'Mantefon' as Affected by Photoperiod, Light Intensity, and Daily Light Integral*; Seoul National University: Seoul, Korea, 2018.
4. Lee, H.B.; Lee, J.H.; An, S.K.; Park, J.H.; Kim, K.S. Growth characteristics and flowering initiation of *Phalaenopsis* Queen Beer 'Mantefon' as affected by the daily light integral. *Hortic. Environ. Biote.* **2019**, *60*, 637–645. [CrossRef]
5. Zhang, S.; Pan, X.; Ma, L.; Zhang, Y. Influence of light on germination and seedling status of *Centipeda herba*. *Seed* **2016**, *35*, 91–93.
6. Walters, K.J.; Curry, C.J. Effects of nutrient solution concentration and daily light integral on growth and nutrient concentration of several basil species in hydroponic production. *HortScience* **2018**, *53*, 1319–1325. [CrossRef]
7. Currey, C.J.; Hutchinson, V.A.; Lopez, R.G. Growth, morphology, and quality of rooted cuttings of several herbaceous annual bedding plants are influenced by photosynthetic daily light integral during root development. *HortScience* **2012**, *47*, 25–30. [CrossRef]
8. Warner, R.M.; Erwin, J.E. Effect of photoperiod and daily light integral on flowering of five *Hibiscus* sp. *Sci. Hortic.* **2003**, *97*, 341–351. [CrossRef]
9. Solis-Toapanta, E.; Gómez, C. Growth and photosynthetic capacity of basil grown for indoor gardening under constant or increasing daily light integrals. *HortTechnology* **2019**, *29*, 880–888. [CrossRef]
10. Baumbauer, D.A.; Schmidt, C.B.; Burgess, M.H. Leaf lettuce yield is more sensitive to low daily light integral than kale and spinach. *HortScience* **2019**, *54*, 2159–2162. [CrossRef]

11. Currey, C.J.; Lopez, R.G. Biomass Accumulation and allocation, photosynthesis, and carbohydrate status of new guinea impatiens, geranium, and petunia cuttings are affected by photosynthetic daily light integral during root development. *J. Amer. Soc. Hort. Sci.* **2015**, *140*, 542–549. [CrossRef]
12. Kelly, N.; Choe, D.; Meng, Q.; Runkle, E.S. Promotion of lettuce growth under an increasing daily light integral depends on the combination of the photosynthetic photon flux density and photoperiod. *Sci. Hortic.* **2020**, *272*, 1–8. [CrossRef]
13. Graver, J.K.; Boldt, J.K.; Lopez, R.G. Radiation intensity and quality from sole-source light-emitting diodes affect seedling quality and subsequent flowering of long-day bedding plant species. *HortScience* **2018**, *53*, 1407–1415.
14. Owen, W.G.; Meng, Q.W.; Lopez, R.G. Promotion of flowering from far red radiation depends on the photosynthetic daily light integral. *HortScience* **2018**, *53*, 1–7. [CrossRef]
15. Randall, W.C.; Lopez, R.G. Comparison of supplemental lighting from high-pressure sodiumlamps and light-emitting diodes during bedding plant seedling production. *HortScience* **2014**, *49*, 589–595. [CrossRef]
16. Torres, A.P.; Lopez, R.G. Photosynthetic daily light integral during propagation of *Tecoma stans* influences seedling rooting and growth. *HortScience* **2011**, *46*, 282–286. [CrossRef]
17. Zhang, Y.; Dong, H.; Song, S.; Su, W.; Liu, H. Morphological and physiological responses of cucumber seedlings to supplemental led light under extremely low irradiance. *Agronomy* **2020**, *10*, 1698. [CrossRef]
18. Elkins, C.; van Iersel, M.W. Longer photoperiods with the same daily light integral improve growth of rudbeckia seedlings in a greenhouse. *HortScience* **2020**, *55*, 1676–1682. [CrossRef]
19. Hernández, R.; Kubota, C. Growth and morphological response of cucumber seedlings to supplemental red and blue photon flux ratios under varied solar daily light integrals. *Sci. Hortic.* **2014**, *173*, 92–99. [CrossRef]
20. Ji, F.; Wei, S.; Liu, N.; Xu, L.; Yang, P. Growth of cucumber seedlings in different varieties as affected by light environment. *Int. J. Agric. Biol. Eng.* **2020**, *13*, 73–78. [CrossRef]
21. Oh, W.; Runkle, E.S.; Warner, R.M. Timing and duration of supplemental lighting during the seedling stage influence quality and flowering in petunia and pansy. *HortScience* **2010**, *45*, 1332–1337. [CrossRef]
22. Pramuk, L.A.; Runkle, E.S. Photosynthetic daily light integral during the seedling stage influences subsequent growth and flowering of *Celosia, Impatiens, Salvia, Tagetes,* and *Viola. HortScience* **2005**, *40*, 1336–1339. [CrossRef]
23. Wang, R.; Gui, Y.; Zhao, T.; Ishii, M.; Eguchi, M.; Xu, H.; Li, T.; Iwasaki, Y. Determining the relationship between floral initiation and source–sink dynamics of tomato seedlings affected by changes in shading and nutrients. *HortScience* **2020**, *55*, 457–464. [CrossRef]
24. Velez-Ramirez, A.I.; Dünner-Planella, G.; Vreugdenhil, D.; Millenaar, F.F.; van Ieperen, W. On the induction of injury in tomato under continuous light: Circadian asynchrony as the main triggering factor. *Funct. Plant Biol.* **2017**, *44*, 597–611. [CrossRef]
25. Fan, X.; Xu, Z.; Liu, X.; Tang, C.; Wang, L.; Han, X. Effects of light intensity on the growth and leaf development of young tomato plants grown under a combination of red and blue light. *Sci. Hortic.* **2013**, *153*, 50–55. [CrossRef]
26. Fu, W.; Li, P.; Wu, Y. Effects of different light intensities on chlorophyll fluorescence characteristics and yield in lettuce. *Sci. Hortic.* **2012**, *135*, 45–51. [CrossRef]
27. Gao, W.; He, D.; Ji, F.; Zhang, S.; Zheng, J. Effects of daily light integral and LED spectrum on growth and nutritional quality of hydroponic spinach. *Agronomy* **2020**, *10*, 1082. [CrossRef]
28. Lopez, R.G.; Runkle, E.S. Photosynthetic daily light integral during propagation influences rooting and growth of cuttings and subsequent development of new guinea impatiens and petunia. *HortScience* **2008**, *43*, 2052–2059. [CrossRef]
29. Blanchard M, G. *Manipulating Light and Temperature for Energy-Efficient Greenhouse Production of Ornamental Crops*; Michigan State University: East Lansing, MI, USA, 2009.
30. Hutchinson, V.A.; Currey, C.J.; Lopez, R.G. Photosynthetic daily light integral during root development influences subsequent growth and development of several herbaceous annual bedding plants. *HortScience* **2012**, *47*, 856–860. [CrossRef]
31. Palmer, S.; van Iersel, M.W. Increasing growth of lettuce and mizuna under sole-source LED lighting using longer photoperiods with the same daily light integral. *Agronomy* **2020**, *10*, 1659. [CrossRef]
32. Mao, H.; Hang, T.; Zhang, X.; Lu, N. Both multi-segment light intensity and extended photoperiod lighting strategies, with the same daily light integral, promoted *Lactuca sativa* L. growth and photosynthesis. *Agronomy* **2019**, *9*, 857. [CrossRef]
33. Bai, Y.; Shi, W.; Xing, X.; Wang, Y.; Jin, Y.; Zhang, L.; Song, Y.; Dong, L.; Liu, H. Study on tobacco vigorous seedling indexes model. *Sci. Agri. Sin.* **2014**, *47*, 1086–1098.
34. Faust, J.E.; Holcombe, V.; Rajapakse, N.C.; Layne, D.R. The effect of daily light integral on bedding plant growth and flowering. *HortScience* **2005**, *40*, 645–649. [CrossRef]
35. Li, H.S. *Principles and Techniques of Plant Physiological and Biochemical Experiments*; Higher Education Press: Beijing, China, 2000.
36. Hao, J.J.; Kang, Z.L.; Yu, Y. *Experimental Techniques for Plant Physiology*; Chemical Industrial Press: Beijing, China, 2007.
37. Currey, C.J.; Lopez, R.G. Biomass accumulation, allocation and leaf morphology of *impatiens hawker* 'Magnum Salmon' cuttings is affected by photosynthetic daily light integral in propagation. *Acta Hortic.* **2012**, *956*, 349–356. [CrossRef]
38. Hurt, A.; Lopez, R.G.; Craver, J.K. Supplemental but not photoperiodic lighting increased seedling quality and reduced production time of annual bedding Plants. *HortScience* **2019**, *54*, 289–296. [CrossRef]
39. Garland, K.F.; Burnett, S.E.; Day, M.E.; van Iersel, M.W. Influence of substrate water content and daily light integral on photosynthesis, water use efficiency, and morphology of *Heuchera americana. J. Amer. Soc. Hort. Sci.* **2012**, *137*, 57–67. [CrossRef]
40. Gent, M.P.N. Effect of daily light integral on composition of hydroponic lettuce. *HortScience* **2014**, *49*, 173–179. [CrossRef]

41. Gerovac, J.R.; Craver, J.K.; Boldt, J.K.; Lopez, R.G. Light intensity and quality from sole-source light-emitting diodes impact growth, morphology, and nutrient content of *Brassica* microgreens. *HortScience* **2016**, *51*, 497–503. [CrossRef]
42. Kjaer, K.H.; Ottosen, C.; Jørgensen, B.N. Timing growth and development of *Campanula* by daily light integral and supplemental light level in a cost-efficient light control system. *Sci. Hortic.* **2012**, *143*, 189–196. [CrossRef]
43. Hutchinson, V.A. *Photosynthetic Daily Light Integral During Vegetative Propagation and Finish Stages Influence Growth and Development of Annual Bedding Plant Species*; Purdue University: West Lafayette, IN, USA, 2012.
44. Vaid, T.M. *Improving the Scheduling and Profitability of Annual Bedding Plant Production by Manipulating Temperature, Daily Light Integral, Photoperiod, and Transplant Size*; Michigan State University: East Lansing, MI, USA, 2012.
45. Zhen, S.; van Iersel, M.W. Photochemical acclimation of three contrasting species to different light levels: Implications for optimizing supplemental lighting. *J. Amer. Soc. Hort. Sci.* **2017**, *142*, 346–354. [CrossRef]
46. Elkins, C.; Van Iersel, M.W. Longer photoperiods with the same daily light integral increase daily electron transport through photosystem II in lettuce. *Plants* **2020**, *9*, 1172. [CrossRef]
47. Murchie, E.H.; Lawson, T. Chlorophyll fluorescence analysis: A guide to good practice and understanding some new applications. *J. Exp. Bot.* **2013**, *64*, 3983–3998. [CrossRef]
48. Dou, H.; Niu, G.; Gu, M.; Masabni, J.G. Responses of sweet basil to different daily light integrals in photosynthesis, morphology, yield, and nutritional quality. *HortScience* **2018**, *53*, 496–503. [CrossRef]
49. Dai, Y.; Shen, Z.; Liu, Y.; Wang, L.; Hannaway, D.; Lu, H. Effects of shade treatments on the photosynthetic capacity, chlorophyll fluorescence, and chlorophyll content of *Tetrastigma hemsleyanum* Diels et Gilg. *Enviro. Exp. Bot.* **2009**, *65*, 177–182. [CrossRef]
50. Pettersen, R.I.; Torreand, S.; Gislerød, H.R. Effects of leaf aging and light duration on photosynthetic characteristics in a cucumber canopy. *Sci. Hortic.* **2010**, *125*, 82–87. [CrossRef]
51. Du, X.; Huang, P.; Yang, M. Phosphorus fertilizers on the growth and resistance physiology of Eucalyptus *urophylla* × *Eucalyptus* grandis DH3229 seedlings. *J. For. Environ.* **2020**, *40*, 526–533.

Article

Effects of Application Methods of Boron on Tomato Growth, Fruit Quality and Flavor

Weinan Xu [1,2,3], Pengju Wang [1,4,5], Luqiao Yuan [1,4,5], Xin Chen [2] and Xiaohui Hu [1,4,5,*]

1   College of Horticulture, Northwest A&F University, Yangling 712100, China; xuweinan0817@163.com (W.X.); pjwang98@163.com (P.W.); wjswylq@163.com (L.Y.)
2   Institute of Industrial Crops, Jiangsu Academy of Agricultural Sciences, Nanjing 210014, China; cx@jaas.ac.cn
3   Plant Biotechnology Research Group, Western Australian State Agricultural Biotechnology Centre, Murdoch University, Perth 6150, Australia
4   Key Laboratory of Protected Horticultural Engineering in Northwest, Ministry of Agriculture, Yangling 712100, China
5   Shaanxi Protected Agriculture Research Centre, Yangling 712100, China
*   Correspondence: hxh1977@163.com; Tel.: +86-29-870-811-92

Abstract: The effect of application methods with different boron levels on the growth, fruit quality and flavor of tomato (*Solanum lycopersicum* L., cv. 'Jinpeng No.1') were investigated under greenhouse conditions. Seven treatments used included two application methods (leaf and root application) with four boron levels (0, 1.9, 3.8 and 5.7 mg·L^{-1} H$_3$BO$_3$). Experimental outcomes revealed that both application methods significantly increased net photosynthetic rate and chlorophyll content, and stabilized leaf structure of tomato. Leaf spray of 1.9 mg·L^{-1} H$_3$BO$_3$ was more effective at improving plant growth and photosynthetic indices in tomato compared to other treatments. Additionally, root application of 3.8 mg·L^{-1} H$_3$BO$_3$ resulted in better comprehensive attributes of fruit quality and flavor than other treatments in terms of amounts of lycopene, β-carotene, soluble protein, the sugar/acid ratio and characteristic aromatic compounds in fruit. The appropriate application of boron can effectively improve the growth and development of tomato, and change the quality and flavor of fruit, two application methods with four boron levels had different effects on tomato.

Keywords: boron; application methods; growth; fruit quality; tomato

Citation: Xu, W.; Wang, P.; Yuan, L.; Chen, X.; Hu, X. Effects of Application Methods of Boron on Tomato Growth, Fruit Quality and Flavor. *Horticulturae* **2021**, *7*, 223. https://doi.org/10.3390/horticulturae7080223

Academic Editor: Francisco Garcia-Sanchez

Received: 11 July 2021
Accepted: 3 August 2021
Published: 5 August 2021

Publisher's Note: MDPI stays neutral with regard to jurisdictional claims in published maps and institutional affiliations.

1. Introduction

Tomato (*Solanum lycopersicum* L.) is one of the most important horticultural crops in the world [1,2]. Tomato fruit quality indices comprise appearance, taste, nutrient and characteristic aroma compounds [3]. Flavor is one of the most important factors of fruit quality, and levels of micronutrients are closely related to the flavor of the fruit [4,5].

Micronutrients can play direct or indirect roles in plant growth and quality [6,7]. Boron is one of the most important micronutrients, involved in carbohydrate metabolism, cell division and maintaining cell wall structure, flowering and fruit setting [8]. Boron has extremely low mobility compared to other micronutrients in plants [9], and it is transported in the xylem through the root pressure generated by leaf transpiration and in the phloem through other mechanisms [10,11]. Boron is absorbed by plants in the form of BO$_3$$^{3-}$ [9]. Although boron is an essential micronutrient for plants [12], it has a narrow optimal concentration. In excess it causes biochemical, physiological and anatomical aberrations in plants [9,13,14]. When in boron-deficient conditions, plants have reduced photosynthetic capacity, thereby preventing cell division and development in chickpea plants [15]. Boron promotes the absorption of calcium and increases the content of vitamin C in the fruit by improving membrane integrity, slowing biosynthesis and reducing respiration in cherry tomato [16]. Davis et al. [17] found leaf sprays of boron increased shelf life of tomato fruit. Sotiropoulos et al. [18] reported that application of boron increased photosynthetic rate

(Pn), stomatal conductance (Cs) and intracellular carbon dioxide concentrations (Ci) in kiwifruit, and boron synergistically promoted the absorption of other plant nutrients and improved quality of kiwifruit [19].

Most studies have focused on the effect of boron stress on tomato growth, but research about different application methods with different boron levels on tomato fruit quality and flavor components is not well studied. The aim of this study was to explore effects of different application methods and boron levels on plant growth and fruit quality in tomato to select optimal application method.

2. Materials and Methods

2.1. Plant Material and Experimentation

This experiment was conducted in the Horticulture Research Greenhouse (College of Horticulture, Northwest A&F University, China). Tomato cultivar 'Jinpeng No.1' was used for experiments. At the five true leaf stage, seedlings were transplanted into 15 L pots containing perlite and vermiculite in a 3:1 v/v ratio. There were seven treatments in this study, and each treatment consisted of 20 pots placed randomly with a row spacing of 30 cm $\times$ 60 cm area. The plants were treated with 300 mL nutrient solution once every six days after the seedlings were transplanted until the first truss of fruit ripened. Then, 500 mL of nutrient solution was applied per plant every four days until to the second truss was harvested. The nutrient solution (control, CK) was modified slightly by Yamazaki nutrient solution formula without H_3BO_3. According to our previous results and other reports [17,20,21], three levels of boron were used in this experiment. The factorial experiment was performed in a completely random design using leaf and root application methods: CK; L1: control + leaf spray 1.9 mg·L^{-1} H_3BO_3; L2: control + leaf spray 3.8 mg·L^{-1} H_3BO_3; L3: control + leaf spray 5.7 mg·L^{-1} H_3BO_3; W1: control + root application 1.9 mg·L^{-1} H_3BO_3; W2: control + root application 3.8 mg·L^{-1} H_3BO_3; W3: control + root application 5.7 mg·L^{-1} H_3BO_3. To increase the absorption of nutrient elements, the pH of the solution was adjusted to 6.5.

2.2. Plant Growth Parameters and Photosynthetic Indices

In this experiment, three plants were randomly collected in each treatment for plant growth parameters and photosynthetic indices measurement.

Plant growth measurements: After the second fruit truss was harvested, each part of the plant was washed and the fresh weight (FW) was measured, and then plants were put into an oven to determine the dry weight (DW).

Leaf physiological indices: 2–3 leaves of each plant were collected at the fifth leaf below the growth point in the period of the first fruit truss ripening in each experiment. Chlorophyll content was measured using the method of Arnon [22]. Ascorbic acid content was measured using the method of Gao et al. [23]. Soluble protein content was measured using the method of Bradford et al. [24].

Photosynthetic parameters: On a sunny day in the morning at the stage of the first truss ripening stage, the fifth leaf below the growth point was selected to measure the net photosynthetic rate (Pn), intercellular carbon dioxide concentration (Ci), stomatal conductance (Cs), and transpiration rate (Tr) by Li-6400 platform (LI-COR, USA). The determination of each index was repeated four times in each experiment.

2.3. Tomato Fruit Quality Parameters

Two kilograms of fruit at similar ripening stage from the second truss (about 45 days after flowering) were collected from each treatment for fruit quality index determination. Soluble solid content was measured using a PAl-1 digital refractometer (Atago, Japan). Titratable acid content was measured by GMK-835f fruit acidity tester (G-WON, Korea). Sugar/acid rate was calculated by soluble solids/titratable acid. Vitamin C and soluble protein content were described as above. Nitrate nitrogen content was measured using the method of Gao et al. [19]. Lycopene, β-Carotene and Carotenoid content were measured by the methods of De Nardo et al. [25] and Radu et al. [26], with slight modification. Single fruit weight measured from three plants with similar growth trends were selected in each treatment, and fruits of the second truss were weighed. The determination of each index was repeated four times.

2.4. Analysis of Characteristic Aromatic Compounds

Tomato fruit characteristic aromatic compounds analysis was measured as described by Tikunov et al. [27]. Five fruits at similar ripening stage were selected and homogenized in each treatment, 15 g of the finely powdered tomato sample from each treatment was transferred into a 40 mL Teflon cap vial (Thermo Fisher Scientific, USA) with 5 g of NaCl. The vials were sealed using a silicone/PTFE septum and a magnetic cap, and each sample was agitated (500 rpm) and sonicated for 10 min, then incubated at 50 °C for 10 min prior to HS-SPME-GC-MS analysis. Headspace volatiles were extracted by exposing each sample to a 75 μm carboxen-polydimethylsiloxane SPME fiber (Supelco, USA) for 30 min under continuous agitation (500 rpm) and heating at 40 °C. The fiber was then inserted into an ISQ GC-MS (Thermo Scientific instruments, USA) injection port and the volatiles were desorbed for 3 min at 250 °C. The chromatography was analyzed on an HP-INNOWAX column (60 m × 0.25 mm × 0.25 μm) with helium as the carrier gas at a constant flow of 1 mL·min^{-1}. The temperature of both the GC interface and MS source was 230 °C. The GC temperature program began at 40 °C (2.5 min), and then was raised to 160 °C (5 °C min^{-1}) and to 230 °C (10 °C min^{-1}) before being held at 230 °C for 5 min. The raw data of each sample were obtained from GC-MS with Xcalibur and AMDIS software and identified the volatile compounds on the basis of the NIST/EPA/NIH Mass Spectral Library (NIST, 2008) and Wiley Registry of Mass Spectral Data 8th edition. The characteristic aromatic compounds were measured according to the Baek and Cadwallader [28] and Wang et al. [29] modified slightly. The characteristic aromatic compounds (%) = (Peak area of particular compound/Total peak area) × 100%. Three independent reactions for each sample of each experiment were performed for the analysis of tomato fruits characteristic aromatic compounds.

2.5. Statistical Analysis

The statistical analysis was performed using SPSS software, version 16 (SPSS Inc., Chicago, IL, USA), and the means were compared using Duncan's multiple range test method at a significance level of 0.05.

3. Results

3.1. Effects of Boron on Plant Growth Parameters

There were different effects on growth parameters for different boron treatments (Table 1). Both leaf spray and root application methods significantly increased FW and DW of shoots. Root application of boron significantly increased the shoot weight (FW) compared to leaf spray, except for the level of 1.9 mg·L^{-1} H$_3$BO$_3$. Leaf spray of boron decreased root FW with the increase of boron level, while the trend was opposite when boron was applied to the root. Boron treatment significantly increased the root DW in each treatment except L3, and W3 gave the largest root DW. The control and W1 had the highest (0.21) and lowest (0.11) root/shoot FW, respectively, and the root/shoot FW decreased after leaf spray treatment as boron levels increased. The root/shoot DW, except L1, it was better than W1 in the root/shoot DW, the root application boron had the higher rate of root/shoot DW than leaf spray with same boron level.

Table 1. Effects of boron on tomato growth parameters.

Treatment	Shoot FW (g)	Shoot DW (g)	Root FW (g)	Root DW(g)	R/S (FW)	R/S (DW)
CK	231.63 ± 1.09 e	34.43 ± 1.023 c	48.81 ± 1.26 c	5.45 ± 0.20 d	0.210 ± 0.006 a	0.160 ± 0.006 bc
L1	276.70 ± 6.07 d	41.55 ± 0.881 b	47.55 ± 2.35 c	6.79 ± 0.53 c	0.173 ± 0.009 b	0.160 ± 0.010 bc
L2	281.86 ± 0.70 cd	37.95 ± 0.731 c	45.66 ± 0.80 c	6.71 ± 0.55 c	0.163 ± 0.003 b	0.177 ± 0.012 b
L3	292.30 ± 0.60 c	34.64 ± 1.060 c	37.68 ± 1.15 d	4.77 ± 0.26 d	0.130 ± 0.006 c	0.140 ± 0.010 cd
W1	324.17 ± 7.42 b	43.58 ± 0.281 b	35.50 ± 1.62 d	5.50 ± 0.13 d	0.110 ± 0.006 d	0.127 ± 0.003 d
W2	326.50 ± 5.03 b	42.87 ± 0.925 b	64.52 ± 1.49 b	7.98 ± 0.36 b	0.197 ± 0.007 a	0.187 ± 0.003 b
W3	352.14 ± 4.50 a	48.88 ± 2.159 a	70.44 ± 1.50 a	10.63 ± 0.45 a	0.197 ± 0.003 a	0.220 ± 0.015 a

FW: Fresh weight; DW: Dry weight; Data are expressed as the mean ± standard error of three independent biological replicates. Different letters indicate significant differences at $p < 0.05$; CK: Yamazaki nutrient solution without H_3BO_3; L1: CK + leaf spray 1.9 mg·L^{-1} H_3BO_3; L2: CK + leaf spray 3.8 mg·L^{-1} H_3BO_3; L3: CK + leaf spray 5.7 mg·L^{-1} H_3BO_3; W1: CK + root application 1.9 mg·L^{-1} H_3BO_3; W2: CK + root application 3.8 mg·L^{-1} H_3BO_3; W3: CK + root application 5.7 mg·L^{-1} H_3BO_3.

3.2. Effects of Boron on Soluble Protein and Ascorbic Acid Content in Leaves

Boron treatment significantly increased soluble protein and ascorbic acid content in leaves. Each application method with different boron levels had different effects (Figure 1). Soluble protein in leaves decreased with the increased level of boron by leaf spray, and there was no difference in root application with different boron levels. L1 and W2 treatments increased the soluble protein by 177.6% and 192.8%, respectively. Both leaf spray and root application of boron increased the content of ascorbic acid with increasing boron. Root application of boron was better than leaf spray at the same boron levels. W3 treatment resulted in the highest content of ascorbic acid in leaves, increasing ascorbic acid by 185.2% compared to the control.

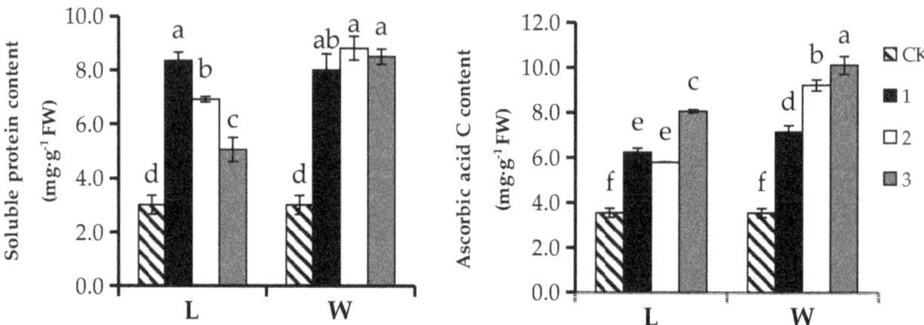

Figure 1. Effects of boron on soluble protein and ascorbic acid content in leaves. Different letters indicate significant differences at $p < 0.05$; L: Leaf spray; W: Root application; CK: Yamazaki nutrient solution without H_3BO_3; 1: 1.9 mg·L^{-1} H_3BO_3; 2: 3.8 mg·L^{-1} H_3BO_3; 3: 5.7 mg·L^{-1} H_3BO_3.

3.3. Effects of Boron on Chlorophyll and Carotenoid Contents and Photosynthetic Parameters

Boron treatment significantly increased Chl a, Chl b, total Chl and carotenoid contents and Chl a/b ratios in leaves (Table 2). Both treatment of leaf spray and root application boron had no significant difference in Chl a, Chl b, total Chl and carotenoid contents using same boron level. L2 treatment had the highest content of Chl a, Chl b, total Chl and carotenoid in the leaf spray group, while W3 had the highest Chl a, Chl b, total Chl and carotenoid content in root application group. Boron treatments had no obvious effects on Chl a/b, which ranged from 1.46 (W2) to 1.54 (L1), while the control had the lowest Chl a/b with only 1.35.

Table 2. Effects of boron on content of Chlorophyll a, Chlorophyll b, total Chlorophyll and Carotenoid in tomato leaves.

Treatment	Chl a (mg·g^{-1})	Chl b (mg·g^{-1})	Total Chl (mg·g^{-1})	Chl a/b	Carotenoid (mg·g^{-1})
CK	0.76 ± 0.07 c	0.56 ± 0.05 c	1.33 ± 0.12 c	1.35 ± 0.01 d	0.29 ± 0.03 c
L1	1.75 ± 0.06 b	1.14 ± 0.04 b	2.89 ± 0.10 b	1.54 ± 0.01 a	0.59 ± 0.02 b
L2	2.16 ± 0.03 a	1.42 ± 0.03 a	3.58 ± 0.07 a	1.53 ± 0.02 ab	0.74 ± 0.02 a
L3	2.07 ± 0.13 a	1.39 ± 0.10 a	3.46 ± 0.23 a	1.48 ± 0.01 c	0.68 ± 0.05 ab
W1	1.75 ± 0.06 b	1.17 ± 0.03 b	2.93 ± 0.09 b	1.49 ± 0.02 bc	0.58 ± 0.02 b
W2	2.07 ± 0.09 a	1.42 ± 0.07 a	3.49 ± 0.16 a	1.46 ± 0.01 c	0.69 ± 0.03 a
W3	2.13 ± 0.02 a	1.42 ± 0.01 a	3.56 ± 0.03 a	1.50 ± 0.01 bc	0.72 ± 0.01 a

Chl: Chlorophyll; Data are expressed as the mean ± standard error of three independent biological replicates. Different letters indicate significant differences at $p < 0.05$; CK: Yamazaki nutrient solution without H_3BO_3; L1: CK + leaf spray 1.9 mg·L^{-1} H_3BO_3; L2: CK + leaf spray 3.8 mg·L^{-1} H_3BO_3; L3: CK + leaf spray 5.7 mg·L^{-1} H_3BO_3; W1: CK + root application 1.9 mg·L^{-1} H_3BO_3; W2: CK + root application 3.8 mg·L^{-1} H_3BO_3; W3: CK + root application 5.7 mg·L^{-1} H_3BO_3.

Boron treatments significantly increased net photosynthetic rate (Pn), intercellular carbon dioxide concentration (Ci), stomatal conductance (Cs) and transpiration rate (Tr) in plants compared with the control (Figure 2). Different application methods showed no significant difference in Pn using the same boron level, except L1, Pn in L1 was significantly higher than other treatments. For Cs, L1 had the highest Cs in leaf spray boron group, and root application significantly increased Cs compared to leaf spray using the higher boron levels. W2 had the highest Cs, which increased by 27.7% compared with the control. The effects on Tr were similar to that in Pn by using different application methods of boron. L1 and W2 had the highest Tr in each treatment group, increasing significantly by 51.1% and 40.8%, respectively. There was no difference in Ci by leaf spray with different boron levels, while in root application treatments, W2 showed significantly increased Ci compared to others. W2 had the highest Ci increasing significantly by 72.6%.

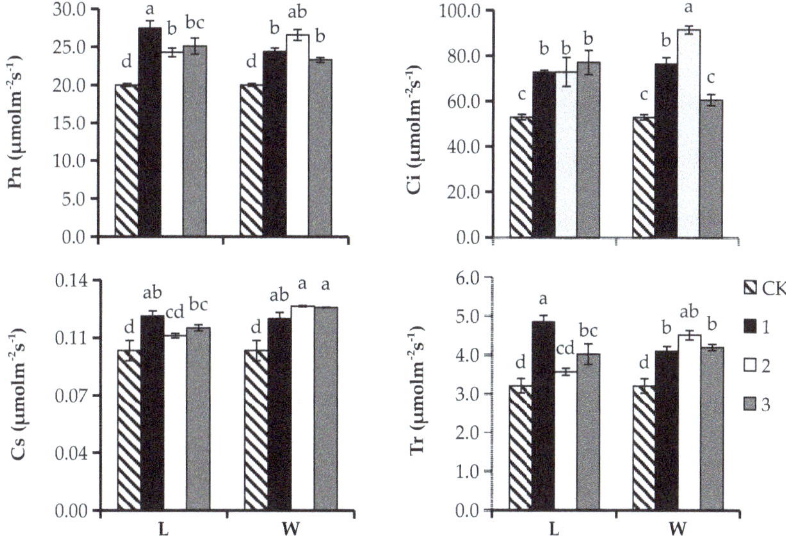

Figure 2. Effect of boron on net photosynthetic rate (Pn), intercellular carbon dioxide concentration (Ci), stomatal conductance (Cs) and transpiration rate (Tr) in tomato. Different letters indicate significant differences at $p < 0.05$; L: Leaf spray; W: Root application; CK: Yamazaki nutrient solution without H_3BO_3; 1: 1.9 mg·L^{-1} H_3BO_3; 2: 3.8 mg·L^{-1} H_3BO_3; 3: 5.7 mg·L^{-1} H_3BO_3.

3.4. Effects of Boron on Quality Indices of Tomato Fruit

In this experiment, two application methods with different boron levels had different effects on soluble solids, titratable acid and sugar/acid rate in fruit (Table 3). For soluble solids, W3 treatment significantly increased the percentage of soluble solids, but there were no obvious differences in other treatments. Boron treatments significantly improved the percentage of titratable acid in fruit. Leaf spray boron gave a higher sugar/acid rate than root application using same boron level, except W3, it had the highest sugar/acid rate with 7.10 compared to other treatments. Additionally, both leaf spray and root application of boron significantly increased the content of soluble protein and vitamin C in fruit. Root application was better at improving the soluble protein and vitamin C content in fruit than leaf spray, except W2 treatment. Boron treatments significantly increased the nitrate content in fruit compared to the control, except L3 treatment. W3 had the highest nitrate nitrogen content with 403.02 $\mu g \cdot g^{-1}$. Both leaf spray and root application increased the single fruit weight. Root application gave heavier single fruit than leaf spray treatments using same boron level. W1 had the heaviest single fruit with 193.81 g on average.

Table 3. Effects of boron on quality indices of tomato fruit.

Treatment	Soluble Solids (%)	Titratable Acid (%)	Sugar/Acid Rate	Soluble Protein ($mg \cdot g^{-1}$)	Vitamin C ($mg \cdot kg^{-1}$)	Nitrate Nitrogen ($\mu g \cdot g^{-1}$)	Single Fruit Weight (g)
CK	4.63 ± 0.20 bc	0.69 ± 0.02 c	6.75 ± 0.26 ab	3.01 ± 0.35 d	69.82 ± 0.52 e	177.58 ± 9.46 d	110.96 ± 1.11 e
L1	4.40 ± 0.15 c	0.76 ± 0.01 a	5.81 ± 0.17 d	8.36 ± 0.30 a	72.94 ± 1.50 de	213.19 ± 5.04 c	117.78 ± 1.72 e
L2	5.03 ± 0.09 ab	0.77 ± 0.01 a	6.57 ± 0.11 abc	6.92 ± 0.10 b	82.67 ± 0.90 b	231.98 ± 3.83 c	130.16 ± 3.94 d
L3	4.63 ± 0.19 bc	0.77 ± 0.01 a	6.05 ± 0.27 cd	5.06 ± 0.45 c	76.54 ± 0.44 cd	172.48 ± 4.26 d	130.78 ± 5.28 d
W1	4.47 ± 0.12 c	0.79 ± 0.01 a	5.68 ± 0.10 d	8.00 ± 0.60 ab	79.64 ± 1.98 bc	253.74 ± 7.10 b	193.81 ± 4.59 a
W2	4.53 ± 0.15 c	0.70 ± 0.01 bc	6.50 ± 0.15 bc	8.82 ± 0.44 a	75.58 ± 1.37 cde	220.11 ± 10.02 c	149.48 ± 2.17 c
W3	5.13 ± 0.03 a	0.72 ± 0.01 b	7.10 ± 0.05 a	8.51 ± 0.27 a	92.37 ± 4.00 a	403.02 ± 0.72 a	168.57 ± 3.41 b

Data are expressed as the mean ± standard error of three independent biological replicates. Different letters indicate significant differences at $p < 0.05$; CK: Yamazaki nutrient solution without H_3BO_3; L1: CK + leaf spray 1.9 $mg \cdot L^{-1}$ H_3BO_3; L2: CK + leaf spray 3.8 $mg \cdot L^{-1}$ H_3BO_3; L3: CK + leaf spray 5.7 $mg \cdot L^{-1}$ H_3BO_3; W1: CK + root application 1.9 $mg \cdot L^{-1}$ H_3BO_3; W2: CK + root application 3.8 $mg \cdot L^{-1}$ H_3BO_3; W3: CK + root application 5.7 $mg \cdot L^{-1}$ H_3BO_3.

3.5. Effects of Boron on Lycopene, β-Carotene and Carotenoid Content in Fruit

Both leaf spray and root application had different effects on the content of lycopene, β-carotene and carotenoid in fruit (Figure 3). Leaf spray significantly increased the lycopene content with increasing boron, while in root application treatment, W1 treatment had the highest content of lycopene compared to other levels applied the same way. Lycopene content decreased with the increase of boron level. Both leaf spray and root application treatments significantly increased β-carotene content in fruit, and the effects on the content of β-carotene were similar to that of lycopene in both treatments. W1 had the highest β-carotene content with 5.6 $\mu g \cdot g^{-1}$ compared to others. Moreover, boron treatments significantly increased the carotenoid content in fruit. Leaf spray and root application treatments with same boron level (3.8 $mg \cdot L^{-1}$ H_3BO_3) resulted in the highest carotenoid content in each group, respectively.

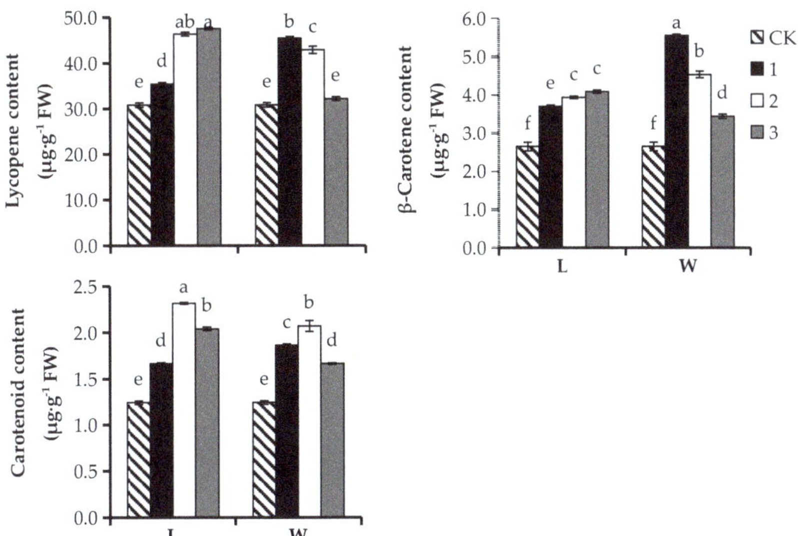

Figure 3. Effects of boron on lycopene, β-carotene and carotenoid in fruits. Different letters indicate significant differences at $p < 0.05$; L: Leaf spray; W: Root application; CK: Yamazaki nutrient solution without H_3BO_3; 1: 1.9 mg·L^{-1} H_3BO_3; 2: 3.8 mg·L^{-1} H_3BO_3; 3: 5.7 mg·L^{-1} H_3BO_3.

3.6. Changes in Volatile Substances and Characteristic Aromatic Compounds in Fruit

In this experiment, six kinds of volatile substances were measured by GC-MS: aldehydes, hydrocarbons, alcohols, ketones, esters and others. The total numbers of volatile substances in each treatment were 58 (control), 65 (L1), 64 (L2), 64 (L3), 66 (W1), 63 (W2) and 68 (W3) (Table 4). There were different degrees of decline in percentage of aldehydes, eaters, alcohols (except 1.9 mg·L^{-1} H_3BO_3 treatment) and ketones (except L2) in both leaf spray and root application treatments, and both application methods had higher percentage in hydrocarbons (except W1) and other volatile substances. Additionally, there were 14 characteristic aromatic compounds of tomato selected from the total volatile substances: 11 (control), 13 (L1), 12 (L2), 13 (L3), 12 (W1), 14 (W2) and 14 (W3) in each treatment (Table 5). Four categories of the tomato specific aroma were described: fruit aroma, fresh scent, floral fragrance and pungent smell (Table 5). Both leaf spray and root application treatments increased the numbers and the percentage of characteristic aromatic compounds compared with the control. In leaf spray treatments, the percentage of characteristic aromatic compounds increased with increase of boron level, while that in root application was variable. W2 had the highest percentage at 54.18%.

Table 4. Effects of boron on the quantity and the percentage of different volatile substances in fruits.

Treatment	CK		L1		L2		L3		W1		W2		W3	
	Qty.	Pct. (%)	Qty.	Pct. (%)	Qty.	Pct. (%)	Qty.	Pct. (%)	Qty.	Pct. (%)	Qty.	Pct. (%)	Qty.	Pct. (%)
Aldehydes	23	14.50	28	11.10	27	8.28	26	8.66	26	11.69	29	11.88	29	11.19
Hydrocarbons	5	49.00	7	54.05	6	59.39	6	59.52	7	46.38	4	64.37	8	60.38
Alcohols	13	24.58	13	24.61	13	20.15	14	17.34	14	30.21	12	13.32	13	17.93
Ketones	9	2.00	8	1.98	8	3.33	8	1.03	8	1.33	8	0.43	8	1.62
Esters	2	6.49	1	2.59	3	3.87	3	6.21	3	3.49	2	4.94	2	3.16
Others	6	3.43	8	5.67	7	4.98	7	7.24	8	6.90	8	5.06	8	5.71
Total	58	100	65	100	64	100	64	100	66	100	63	100	68	100

Qty.: quantity; Pct.: percentage; CK: Yamazaki nutrient solution without H_3BO_3; L1: CK + leaf spray 1.9 mg·L^{-1} H_3BO_3; L2: CK + leaf spray 3.8 mg·L^{-1} H_3BO_3; L3: CK + leaf spray 5.7 mg·L^{-1} H_3BO_3; W1: CK + root application 1.9 mg·L^{-1} H_3BO_3; W2: CK + root application 3.8 mg·L^{-1} H_3BO_3; W3: CK + root application 5.7 mg·L^{-1} H_3BO_3.

Table 5. Effects of boron on the percentage of characteristic aroma in the total volatile substances in fruits.

Volatile Components	Flavor Description	Treatment (%)						
		CK	L1	L2	L3	W1	W2	W3
3-methyl-butanal	Fruit aroma	-	0.021%	-	-	-	0.037%	0.025%
3-methyl-1-butanol	Fruit aroma	0.136%	0.147%	0.108%	0.262%	0.135%	0.050%	0.197%
3-hexenal	Fruit aroma	0.063%	0.039%	0.036%	0.060%	0.028%	0.045%	0.012%
6-methyl-5-hepten-2-one	Fruit aroma	7.647%	5.862%	4.548%	5.148%	5.813%	7.004%	6.223%
(*E*)-2-hexenal	Fruit aroma	0.447%	-	0.472%	0.855%	0.440%	0.816%	0.548%
(*Z*)-3-hexen-1-ol	Fresh scent	4.590%	5.935%	5.681%	5.142%	4.782%	3.056%	3.725%
(*E*)-2-heptenal	Fresh scent	5.197%	5.120%	4.090%	3.339%	3.319%	3.288%	3.823%
Hexanal	Fresh scent	13.830%	23.093%	28.540%	26.790%	18.012%	30.590%	26.426%
2-isobutylthiazole	Fresh scent	-	3.324%	3.036%	5.886%	4.424%	3.266%	3.185%
Methylsalicylate	Fresh scent	6.428%	2.593%	3.653%	5.887%	3.392%	4.922%	3.110%
Benzeneacetaldehyde	Floral fragrance	-	0.166%	0.062%	0.225%	-	0.146%	0.182%
Phenylethyl Alcohol	Floral fragrance	0.347%	0.186%	-	0.449%	0.231%	0.220%	0.255%
Trans-á-ionone	Floral fragrance	0.355%	0.253%	0.193%	0.125%	0.202%	0.147%	0.155%
1-penten-3-one	Pungent smell	0.735%	0.646%	1.009%	0.808%	0.600%	0.590%	0.838%
Total		39.775%	47.385%	51.428%	54.975%	41.376%	54.176%	48.704%

'-' means Not detected; CK: Yamazaki nutrient solution without H_3BO_3; L1: CK + leaf spray 1.9 mg·L^{-1} H_3BO_3; L2: CK + leaf spray 3.8 mg·L^{-1} H_3BO_3; L3: CK + leaf spray 5.7 mg·L^{-1} H_3BO_3; W1: CK + root application 1.9 mg·L^{-1} H_3BO_3; W2: CK + root application 3.8 mg·L^{-1} H_3BO_3; W3: CK + root application 5.7 mg·L^{-1} H_3BO_3.

4. Discussion

Boron, as one of the essential micronutrients in plants, directly or indirectly participates in a variety of physiological and biochemical processes in plant growth and development [30]. Boron has very low mobility in tomato plants [8,31], and the optimum boron level varies with different tomato varieties, ranging from ~0.2 mg·L^{-1} (1.2 mg·L^{-1} H_3BO_3) to ~1 mg·L^{-1} (6 mg·L^{-1} H_3BO_3) [17,20,21]. Both deficiency and toxicity lead to irreversible effects on the growth and development in plants [32,33]. We compared the effects of two application methods with three boron levels on tomato plants. As expected, we found boron treatments significantly affected the biomass accumulation in tomato plants. Both leaf spray and root application of boron increased the FW and DW of shoots, and the root application resulted in higher ratios of root/shoot DW than leaf spray. Davis et al. [17] reported boron has an important function in cell wall metabolism and the stability of root structure, and Loomis and Durst [34] reported the appropriate boron level can promote the absorption of calcium (Ca) by tomato roots and increase the biomass accumulation of tomato plants, which is consistent with the results in this study.

Boron treatment significantly increased the soluble protein and ascorbic acid content compared to the control, and the root application boron had higher content than that of leaf spray using the same boron level. Chermsiri et al. [35] reported spraying boron significantly increased soluble protein and ascorbic acid in garlic. Ali et al. [36] reported boron application decreased the percentage of fruit drop and increased the total chlorophyll concentration in fruit peel of mango. We speculate that boron can improve the antioxidant capacity in tomato leaves. Our results were different between the two application methods with same boron level, possibly due to differences in absorption efficiency regulated by sink–source distribution of boron in leaves and roots.

Photosynthesis is one of the most important indices of plant growth and development, and chlorophyll content is a measure of photosynthetic efficiency [37]. Han et al. [38] reported photosynthesis was reduced under boron deficiency that damaged the structure of thylakoids, which affects electron transmission and decrease the optimal/maximal quantum yield of PSII (Fv/Fm). Excessive boron leads to the production of reactive oxygen species (ROS) in leaves, causing photo-oxidative damage to organic molecules and chloroplast structure and consequently decrease in the content of chlorophyll [39]. In this study, both methods significantly increased the content of chlorophyll a, chlorophyll b and carotenoids, and there were no obvious differences between the two application methods at the same boron level. We speculate boron could not the direct factor for photosynthetic efficiency of leaves that closely related to chlorophyll content and leaf structure affected by different boron levels. Kastori et al. [40] found boron deficiency led to a decrease of chlorophyll content in sunflower leaves and the chloroplast in mesophyll cells, which damage the structure of chloroplast membrane, grana lamellae and grana, but the symptom was effectively alleviated by spraying boron, which is consistent with our results. Moreover, boron treatments effectively improved the ability of photosynthesis in terms of Pn, Ci, Cs and Tr in tomato plants, and two application methods had little effect with different boron levels, for example, L1 and W2 had the highest photosynthesis ability in each group, respectively (Figure 2). We think appropriate boron can stable leaf structure, which can effectively promote the synthesis of photosynthetic pigments that affect directly the photosynthetic capacity in tomato plants.

How to improve fruit flavor is an important area of research [41,42]. Micronutrients directly affect fruit quality. Boron plays an important role to improve fruit quality. For example, Niu et al. [43] found the flavor quality in konjac (*Amorphophallus rivieri*) was improved by spraying boron, which promoted the transportation of carbohydrates, increase glucomannan and soluble sugar content. Islam et al. [44] reported leaf spray of boron increased vitamin C and Ca in tomato fruit, and increased shelf life and single fruit weight [17]. We found that both leaf spray and root application boron treatments significantly increased the soluble protein and vitamin C content in fruit, and except L3, boron treatments significantly increased nitrate accumulation in fruit compared to the control, which could result from the absorption and transportation of boron always cooperate with other nutrients transportation in plants that lead to higher accumulation of nitrate, and the efficiency could be affected by different application methods with different boron levels [45]. We found boron-treated plants had a higher percentage of soluble solids and titratable acid in fruit, and the sugar-acid rate was affected by different boron levels (Table 3), which is consistent with Wójcik and Lewandowski [46], who found spraying Ca and boron on strawberry changed the ratio of soluble solids/titratable acids, inducing better flavor. Additionally, we found both application methods increased the lycopene, β-carotene and carotenoid content in fruit. Leaf spray of boron increased lycopene and β-carotene content with the increase of boron level, while root application of boron had a decreasing trend using same boron level (Figure 3). We speculate the efficiency of boron uptake by plants is affected by application methods and boron levels. Lycopene, β-Carotene and carotenoid are natural antioxidants in fruit, which is closely related to the quality of tomato fruit [47]. Lahoz et al. [48] reported lycopene is one of the most important quality indices of tomato, affecting flavor.

Over 400 volatile substances have been reported in tomato, including hydrocarbons, alcohols, aldehydes, ketones and esters [49], but only 30 volatile substances of levels of more than $1 \text{ nL} \cdot \text{L}^{-1}$ [50]. Buttery [51] reported there were just 29 volatile substances of concentration more than $1 \text{ nL} \cdot \text{L}^{-1}$ in tomato fruit, and among these substances, 16 substances with logarithmic threshold units more than 0 were identified. Baldwin et al. [52] claimed these 16 volatile substances are the main characteristic aroma compounds in tomato fruit. In this study, 68 volatile substances were detected by using GC-MS, including 14 characteristic aromatic compounds, but some characteristic aromatic compounds such as β-damascone and 1-nitro-2-ethylbenzene were not detected, which may be caused by different tomato varieties and cultivation conditions (Tables 4 and 5). Both application methods of boron increased the numbers of volatile substances in fruit, and leaf and root application of boron decreased the percentage of aldehydes and esters, but significantly increased the percentage of hydrocarbons (except W1) in fruit. We speculate that the efficiency of some substances synthesis pathways was affected by different application methods with different boron levels. Liu et al. [53,54] reported leaf spray and root application of melatonin with different concentrations had effects on the composition and quantity of tomato volatile substances, speculating that exogenous nutrients may directly or indirectly participate in synthesis of volatile substances in tomato fruit. Buttery et al. [51] reported the unique aroma of tomato is mainly composed of its characteristic aroma compounds. In this study, both leaf spray and root application boron treatment increased the percentage of characteristic aroma compounds in fruit, leaf spray boron resulted in a higher percentage in characteristic aroma compounds than root application using same boron level (Table 5). Krumbein and Auerswald [55] and Lecomte et al. [56] reported the characteristic aroma of tomato can be divided into four categories: fruit aroma, fresh scent, floral fragrance and pungent smell, but it is still unknown that how the characteristic aroma compounds of tomato are directly regulated. In this study, L3 and W2 had the highest percentage of characteristic aromatic compounds in fruit in each application method group, which could have much richer tomato flavor.

5. Conclusions

The appropriate application of boron can effectively improve the growth and development of tomato, and change the quality and flavor of tomato fruit, two application methods with three boron levels had different effects on these indices of tomato. Based on the results of the tomato cultivar used in this study, leaf spray of 1.9 mg·L^{-1} H$_3$BO$_3$ and root application of 3.8 mg·L^{-1} H$_3$BO$_3$ gave better performance for each application method.

Author Contributions: W.X. and X.H. designed the experiments. W.X., P.W., L.Y. and X.C. performed the experiments and analysis the data. W.X. and X.H. wrote the manuscript. All authors have read and agreed to the published version of the manuscript.

Funding: This research was funded by Technological Innovative Research Team of Shaanxi Province (Projects No. 2021TD-34) and Technical Innovation Guidance Special Project of Shaanxi Province, China (Projects No. 2021QFY08-03).

Institutional Review Board Statement: The study did not involve humans or animals.

Informed Consent Statement: The study did not involve humans.

Data Availability Statement: The data are available upon request from the corresponding author.

Acknowledgments: The authors sincerely thank Stephen J. Wylie (S.W.) from Western Australian State Agricultural Biotechnology Centre, Murdoch University, for manuscript revision and suggestions.

Conflicts of Interest: The authors declare no conflict of interest.

References

1. Wang, X.; Xing, Y. Evaluation of the effects of irrigation and fertilization on tomato fruit yield and quality: A principal component analysis. *Sci. Rep.* **2017**, *7*, 350. [CrossRef]
2. Tahamolkonan, M.; Ghahsareh, A.M.; Ashtari, M.K.; Honarjoo, N. Tomato (*Solanum lycopersicum*) growth and fruit quality affected by organic fertilization and ozonated water. *Protoplasma* **2021**, 1–9. [CrossRef]
3. Ripoll, J.; Urban, L.; Staudt, M.; Lopez-Lauri, F.; Bidel, L.P.; Bertin, N. Water shortage and quality of fleshy fruits-making the most of the unavoidable. *J. Exp. Bot.* **2014**, *65*, 4097–4117. [CrossRef]
4. Davies, J.N.; Hobson, G.E.; McGlasson, W.B. The constituents of tomato fruit-the influence of environment, nutrition, and genotype. *Crit. Rev. Food Sci. Nutr.* **1981**, *15*, 205–280. [CrossRef] [PubMed]
5. Klee, H.J. Improving the flavor of fresh fruits: Genomics, biochemistry, and biotechnology. *New Phytol.* **2010**, *187*, 44–56. [CrossRef]
6. Goldbach, H.E.; Wimmer, M.A. Boron in plants and animals: Is there a role beyond cell-wall structure? *J. Plant Nutr. Soil Sci.* **2007**, *170*, 39–48. [CrossRef]
7. Camacho-Cristóbal, J.J.; Rexach, J.; González-Fontes, A. Boron in plants: Deficiency and toxicity. *J. Integr. Plant Biol.* **2008**, *50*, 1247–1255. [CrossRef]
8. Gupta, U.; Solanki, H. Impact of boron deficiency on plant growth. *Int. J. Bioassays.* **2013**, *2*, 1048–1050.
9. Shelp, B.J.; Marentes, E.; Kitheka, A.M.; Vivekanandan, P. Boron mobility in plants. *Physiol. Plant.* **1995**, *94*, 356–361. [CrossRef]
10. Ahmad, W.; Zia, M.H.; Malhi, S.S.; Niaz, A.; Ullah, S. Boron Deficiency in soils and crops: A review. *Crop Plant.* **2012**, *2012*, 65–97. [CrossRef]
11. Li, W.L.; Chen, X.; Jian, T.; Chen, T.T.; Li, J.; Wang, L.S. From planar boron clusters to borophenes and metalloborophenes. *Nat. Rev. Chem.* **2017**, *1*, 1–9. [CrossRef]
12. Nable, R.O.; Bañuelos, G.S.; Paull, J.G. Boron toxicity. *Plant Soil.* **1997**, *193*, 181–198. [CrossRef]
13. Rees, R.; Robinson, B.H.; Menon, M.; Lehmann, E.; Günthardt-Goerg, M.S.; Schulin, R. Boron accumulation and toxicity in hybrid poplar (*Populus nigra × euramericana*). *Environ. Sci. Technol.* **2011**, *45*, 10538–10543. [CrossRef]

14. Kelly, J.; Unkovich, M.; Stevens, D. Crop nutrition considerations in reclaimed water irrigation systems. In *Growing Crops with Reclaimed Wastewater*; Stevens, D., Kelly, J., McLaughlin, M., Unkovich, M., Eds.; CSIRO Publishing: Collingwood, Australia, 2006; pp. 91–105.

15. Ardic, M.; Sekmen, A.H.; Tokur, S.; Ozdemir, F.; Turkan, I. Antioxidant responses of chickpea plants subjected to boron toxicity. *Plant Biol.* **2009**, *11*, 328–338. [CrossRef] [PubMed]

16. Milagres, C.d.C.; Maia, J.T.L.S.; Ventrella, M.C.; Martinez, H.E.P. Anatomical changes in cherry tomato plants caused by boron deficiency. *Braz. J. Bot.* **2019**, *42*, 319–328. [CrossRef]

17. Davis, J.M.; Sanders, D.C.; Nelson, P.V.; Lengnick, L.; Sperry, W.J. Boron improves growth, yield, quality, and nutrient content of tomato. *J. Am. Soc. Hortic. Sci.* **2003**, *128*, 441–446. [CrossRef]

18. Sotiropoulos, T.E.; Therios, I.N.; Dimassi, K.N.; Bosabalidis, A.; Kofidis, G. Nutritional status, growth, CO_2 assimilation, and leaf anatomical responses in two kiwifruit species under boron toxicity. *J. Plant Nutr.* **2002**, *25*, 1249–1261. [CrossRef]

19. Xu, W.P.; Wang, L.; Yang, Q.; Wei, Y.H.; Zhang, C.X.; Wang, S.P. Effect of Calcium and Boron on the Quality of Kiwifruit. *VIII Int. Symp. Kiwifruit* **2014**, *1096*, 317–320. [CrossRef]

20. Cervilla, L.M.; Blasco, B.; Ríos, J.J.; Rosales, M.A.; Rubio-Wilhelmi, M.M.; Sánchez-Rodríguez, E.; Ruiz, J.M. Response of nitrogen metabolism to boron toxicity in tomato plants. *Plant Biol.* **2009**, *11*, 671–677. [CrossRef] [PubMed]

21. Choi, E.Y.; Park, H.I.; Ju, J.H.; Yoon, Y.H. Boron availability alters its distribution in plant parts of tomato. *Hortic. Environ. Biotechnol.* **2015**, *56*, 145–151. [CrossRef]

22. Arnon, D.I. Copper enzymes in isolated chloroplasts. Polyphenoloxidase in Beta vulgaris. *Plant Physiol.* **1949**, *24*, 1–15. [CrossRef]

23. Gao, J.F. *Plant Physiology Experiment Guidance*; World Publishing Corporation: Xi'an, China, 2006.

24. Bradford, M.M. A rapid and sensitive method for the quantitation of microgram quantities of protein utilizing the principle of protein-dye binding. *Anal. Biochem.* **1976**, *72*, 248–254. [CrossRef]

25. De Nardo, T.; Shiroma-Kian, C.; Halim, Y.; Francis, D.; Rodriguez-Saona, L.E. Rapid and simultaneous determination of lycopene and β-carotene contents in tomato juice by infrared spectroscopy. *J. Agric. Food Chem.* **2009**, *57*, 1105–1112. [CrossRef] [PubMed]

26. Radu, A.I.; Ryabchykov, O.; Bocklitz, T.W.; Huebner, U.; Weber, K.; Cialla-May, D.; Popp, J. Toward food analytics: Fast estimation of lycopene and β-carotene content in tomatoes based on surface enhanced Raman spectroscopy (SERS). *Analyst* **2016**, *141*, 4447–4455. [CrossRef] [PubMed]

27. Tikunov, Y.; Lommen, A.; De Vos, C.R.; Verhoeven, H.A.; Bino, R.J.; Hall, R.D.; Bovy, A.G. A novel approach for nontargeted data analysis for metabolomics. Large-scale profiling of tomato fruit volatiles. *Plant Physiol.* **2005**, *139*, 1125–1137. [CrossRef]

28. Baek, H.H.; Cadwallader, K.R. Volatile compounds in flavor concentrates produced from crayfish-processing byproducts with and without protease treatment. *J. Agric. Food Chem.* **1996**, *44*, 3262–3267. [CrossRef]

29. Wang, L.; Baldwin, E.A.; Bai, J. Recent advance in aromatic volatile research in tomato fruit: The metabolisms and regulations. *Food Bioprocess Technol.* **2016**, *9*, 203–216. [CrossRef]

30. Schuman, G.E.; Stanley, M.A.; Knudsen, D. Automated total nitrogen analysis of soil and plant samples. *Soil Sci. Soc. Am. J.* **1973**, *37*, 480–481. [CrossRef]

31. Nejad, S.A.G.; Etesami, H. The Importance of Boron in Plant Nutrition. *Met. Plants Adv. Future Prospect.* **2020**, 433–449. [CrossRef]

32. Haleema, B.; Rab, A.; Hussain, S.A. Effect of Calcium, Boron and Zinc Foliar Application on Growth and Fruit Production of Tomato. *Sarhad J. Agric.* **2018**, *34*. [CrossRef]

33. Gündeş, F.A.; Sönmez, İ. The effects of different doses of nitrogen on tomato plant mineral contents under boron toxicity. *Acta Sci. Pol. Hortorum Cultus.* **2020**, *19*, 97–104. [CrossRef]

34. Loomis, W.D.; Durst, R.W. Chemistry and biology of boron. *BioFactors.* **1992**, *3*, 229–239.

35. Chermsiri, C.; Watanabe, H.; Attajarusit, S.; Tuntiwarawit, J.; Kaewroj, S. Effect of boron sources on garlic (*Allium sativum* L.) productivity. *Biol. Fertil. Soils.* **1995**, *20*, 125–129. [CrossRef]

36. Ali, M.S.; Elhamahmy, M.A.; El-Shiekh, A.F. Mango trees productivity and quality as affected by boron and putrescine. *Sci. Hortic.* **2017**, *216*, 248–255. [CrossRef]

37. Boardman, N.T. Comparative photosynthesis of sun and shade plants. *Annu. Rev. Plant Physiol.* **1977**, *28*, 355–377. [CrossRef]

38. Han, S.; Chen, L.S.; Jiang, H.X.; Smith, B.R.; Yang, L.T.; Xie, C.Y. Boron deficiency decreases growth and photosynthesis, and increases starch and hexoses in leaves of citrus seedlings. *J. Plant Physiol.* **2008**, *165*, 1331–1341. [CrossRef] [PubMed]

39. Khaleghi, E.; Arzani, K.; Moallemi, N.; Barzegar, M. Evaluation of chlorophyll content and chlorophyll fluorescence parameters and relationships between chlorophyll a, b and chlorophyll content index under water stress in *Olea europaea* cv. Dezful. *World Acad. Sci. Eng. Technol.* **2012**, *6*, 1154–1157.

40. Kastori, R.; Plesničar, M.; Panković, D.; Sakač, Z. Photosynthesis, chlorophyll fluorescence and soluble carbohydrates in sunflower leaves as affected by boron deficiency. *J. Plant Nutr.* **1995**, *18*, 1751–1763. [CrossRef]

41. Gautier, H.; Diakou-Verdin, V.; Bénard, C.; Reich, M.; Buret, M.; Bourgaud, F.; Génard, M. How does tomato quality (sugar, acid, and nutritional quality) vary with ripening stage, temperature, and irradiance? *J. Agric. Food Chem.* **2008**, *56*, 1241–1250. [CrossRef] [PubMed]

42. Mavlyanova, R.F.; Lyan, E.E.; Karimov, B.A.; Dubinin, B.V. The vegetative grafting effect on increasing tomato fruit quality. In *IOP Conference Series: Earth and Environmental Science*; IOP Publishing Ltd.: Redcliffe Way, Bristol, UK, 2020; p. 12077. [CrossRef]

43. Niu, Y.; Wang, Q.; Zhang, S.; Wang, Z. Effect of boron nutrition on quality and enzyme activity of Amorphophallus. *China Veg.* **2011**, *20*, 63–68.

44. Islam, M.Z.; Mele, M.A.; Baek, J.P.; Kang, H.M. Cherry tomato qualities affected by foliar spraying with boron and calcium. *Hortic. Environ. Biotechnol.* **2016**, *57*, 46–52. [CrossRef]

45. Eraslan, F.; Inal, A.; Gunes, A.; Alpaslan, M. Boron toxicity alters nitrate reductase activity, proline accumulation, membrane permeability, and mineral constituents of tomato and pepper plants. *J. Plant Nutr.* **2007**, *30*, 981–994. [CrossRef]

46. Wójcik, P.; Lewandowski, M. Effect of calcium and boron sprays on yield and quality of 'Elsanta' strawberry. *J. Plant Nutr.* **2003**, *26*, 671–682. [CrossRef]

47. Shi, J.; Maguer, M.L. Lycopene in tomatoes: Chemical and physical properties affected by food processing. *Crit. Rev. Food Sci. Nutr.* **2000**, *40*, 1–42. [CrossRef] [PubMed]

48. Lahoz, I.; Leiva-Brondo, M.; Martí, R.; Macua, J.I.; Campillo, C.; Roselló, S.; Cebolla-Cornejo, J. Influence of high lycopene varieties and organic farming on the production and quality of processing tomato. *Sci. Hortic.* **2016**, *204*, 128–137. [CrossRef]

49. Buttery, R.G.; Takeoka, G.R. Some unusual minor volatile components of tomato. *J. Agric. Food Chem.* **2004**, *52*, 6264–6266. [CrossRef] [PubMed]

50. Azodanlou, R.; Darbellay, C.; Luisier, J.L.; Villettaz, J.C.; Amadò, R. Development of a model for quality assessment of tomatoes and apricots. *LWT Food Sci. Technol.* **2003**, *36*, 223–233. [CrossRef]

51. Buttery, R.G. Quantitative and sensory aspects of flavor of tomato and other vegetables and fruits. In *Flavor Sci. Sensib. Princ. Tech*; Acree, T., Teranishi, R., Eds.; ACS: Washington, DC, USA, 1993; pp. 259–286.

52. Baldwin, E.A.; Scott, J.W.; Shewmaker, C.K.; Schuch, W. Flavor trivia and tomato aroma: Biochemistry and possible mechanisms for control of important aroma components. *HortScience* **2000**, *35*, 1013–1022. [CrossRef]

53. Liu, J.; Zhang, R.; Sun, Y.; Liu, Z.; Jin, W.; Sun, Y. The beneficial effects of exogenous melatonin on tomato fruit properties. *Sci. Hortic.* **2016**, *207*, 14–20. [CrossRef]

54. Liu, J.; Liu, H.; Wu, T.; Zhai, R.; Yang, C.; Wang, Z.; Xu, L. Effects of melatonin treatment of postharvest pear fruit on aromatic volatile biosynthesis. *Molecules* **2019**, *24*, 4233. [CrossRef]

55. Krumbein, A.; Auerswald, H. Characterization of aroma volatiles in tomatoes by sensory analyses. *Food Nahr.* **1998**, *42*, 395–399. [CrossRef]

56. Lecomte, L.; Gautier, A.; Luciani, A.; Duffé, P.; Hospital, F.; Buret, M.; Causse, M. Recent advances in molecular breeding: The example of tomato breeding for flavor traits. *Acta Hortic.* **2004**, *637*, 231–242. [CrossRef]

horticulturae

Article

Productivity Enhancement of Cucumber (*Cucumis sativus* L.) through Optimized Use of Poultry Manure and Mineral Fertilizers under Greenhouse Cultivation

Basheer Noman Sallam [1,2,†], Tao Lu [1,†], Hongjun Yu [1,*], Qiang Li [1,†], Zareen Sarfraz [3], Muhammad Shahid Iqbal [3], Shumaila Khan [1], Heng Wang [1], Peng Liu [1] and Weijie Jiang [1,*]

[1] Key Laboratory of Horticultural Crops Genetic Improvement (Ministry of Agriculture), Institute of Vegetables and Flowers, Chinese Academy of Agricultural Sciences, No.12 Zhongguancun, South Street, Haidian District, Beijing 100081, China; bns.alhakimi@gmail.com (B.N.S.); lutao@caas.cn (T.L.); liqiang05@caas.cn (Q.L.); shumaila.khan@kfueit.edu.pk (S.K.); 82101181122@caas.cn (H.W.); 82101179107@caas.cn (P.L.)

[2] Department of Horticulture and Its Technologies, Faculty of Agriculture, Foods and Enviromant, Sana'a University, Sana'a P.O. Box 1247, Yemen

[3] Cotton Research Institute, Chinese Academy of Agricultural Sciences, Anyang 455000, China; zskpbg@gmail.com (Z.S.); shahidkooria@gmail.com (M.S.I.)

* Correspondence: yuhongjun@caas.cn (H.Y.); jiangweijie@caas.cn (W.J.); Tel.: +86-010-8210-8797 (W.J.)

† These authors contributed equally to this work.

Citation: Sallam, B.N.; Lu, T.; Yu, H.; Li, Q.; Sarfraz, Z.; Iqbal, M.S.; Khan, S.; Wang, H.; Liu, P.; Jiang, W. Productivity Enhancement of Cucumber (*Cucumis sativus* L.) through Optimized Use of Poultry Manure and Mineral Fertilizers under Greenhouse Cultivation. *Horticulturae* **2021**, *7*, 256. https://doi.org/ 10.3390/horticulturae7080256

Academic Editors: Xun Li, Xiaohui Hu and Shiwei Song

Received: 9 July 2021
Accepted: 9 August 2021
Published: 20 August 2021

Publisher's Note: MDPI stays neutral with regard to jurisdictional claims in published maps and institutional affiliations.

Abstract: Cucumber, a widely cultivated vegetable, is mostly grown under greenhouse conditions. In recent years, the overuse of inorganic fertilizers for higher yield attainment adversely has affected human health and the environment. Therefore, a greenhouse experiment was designed to evaluate the effects of different nutrient sources (poultry manure (PM) and mineral fertilizer (MF)) on productivity-enhancing parameters of cucumber via univariate and multivariate analyses. Amounts of PM and MF (NPK15:15:15) were added to coco-peat per cubic meter by weight/volume (w/v) ratios as follows: T_1 (control), 60 kg PM; T_2, 30 kg PM + 3 kg MF; T_3, 30 kg PM + 5 kg MF, and T_4, 30 kg PM + 7 kg MF. The univariate analysis performed on the collected data illustrated the significant enhancement in growth and productivity for the integrated use of PM and MF. Multivariate analyses (correlation, clustering, and Principal Component Analysis) validated the results of univariate analysis by differentiating treatments into two groups. The three treatments obtained a distinguished group from T_1 (Control) and did not show significant differences among each other, with a maximum yield increase by T_2 (74.6%). According to these results, T_2 could improve cucumber productivity under greenhouse conditions. It can be taken as recommendations for better quality and yield enhancement in future improvement programs and cucumber-related farming communities.

Keywords: protected greenhouse; soilless culture; poultry manure; yield enhancement; cucumber; univariate analysis; multivariate analysis

1. Introduction

Cucumber (*Cucumis sativa* L.), a member of the family Cucurbitaceae, is regarded as an essential vegetable for fresh consumption crops worldwide and is a rich source of vitamins, minerals, and antioxidants [1,2]. Cucumber is a low-calorie food, consisting of 90% water, which is why it provides superior hydration. Its eminent texture and flavor are the main reasons for its use in salads in fresh form and pickles in the processed form [3]. Its medicinal value is another distinguished property, which includes its antioxidant ability, ability to lower glycemic and antimicrobial activity, etc. Its intake regularly helps to boost metabolism and improve immunity [3]. Due to its high yields and economic value, cucumber is extensively cultivated in greenhouses in China throughout the year. Its global production in 2019 was 87,805,086 tons on 2,231,402 hectares of cultivated area. It is ranked

10th among the most important vegetable crops worldwide. China shared 70,338,971 tons (80.11%) in global production in 2019 from 1,258,370 ha (56.39%) of cultivation area [4].

To cover rising demands, it is essential to produce a sufficient quantity of excellent-quality cucumber. Several cropping obstacles prevent productivity enhancement at the desired pace, and these are responsible for restricting the healthy development of industry [5]. Finding the possibilities and increasing the pool of knowledge regarding continuous cropping obstacles are among the possibilities to develop practical approaches to overcome the issues present in long-term intensive cucumber production [6]. Cucumber is assumed to perform better with enhanced productivity under early-season crop growing and during the summer [7,8]. The optimum growth occurs between 20 °C and 25 °C and with growth reduction below 16 °C or above 30 °C. Recently, cucumber cultivation in soilless mediums under controlled greenhouse conditions has become more developed and trendier around the globe than in the past [7–10]. Fertilization plays a vital role in improving soil fertility and crop yield [11,12]. Cucumber requires many nutrients for proper growth and yield.

Nitrogen, phosphorous, and potassium (NPK) are the essential nutrients for plant growth when used in optimum amounts of 15:15:15, respectively [13]. However, the improper use of fertilizers is causing severe soil degradation and limited crop productivity. Poor management of fertilizers leads to the accumulation of salt in soil aquifers. Fertilizer management should be applied in an integrated manner according to soil type, climatic factors, and crop requirements [14,15]. The integrated utilization of mineral and organic fertilizers can enhance soil fertility. Previous studies have reported that the integration of mineral and organic fertilizers enhances plant growth, yield, and quality. It has been reported that both organic and mineral fertilizers, when used together, led to higher nutrient uptake and increased fruit production [16–18]. It is also reported that optimum poultry manure significantly influenced the growth, yield, and nutritional quality of lettuce [19].

Due to the higher cost of inputs, commercial vegetable production has become an expensive business. One prominent cause is the availability of arable land that is declining due to extensive urbanization. Furthermore, accelerated poverty and unemployment are faced mainly in all metropolitan cities across emerging nations, which can be prevented by opting for soilless culture to grow fresh vegetables in the adjoined outskirts. Soilless cultivation is defined as the technique of growing various crops in the absence of soil used as rooting media [20,21]. Numerous nations have extensively adopted it during the last five decades to sustain their crop production efficiently [22,23]. Several research reports have been published regarding cucumber cultivation in soilless media under greenhouse conditions [24–30]. The current climate change situation has accelerated interest in and choice of this approach, mainly due to environmental pollution, water shortage, abrupt changes in the surrounding environment, etc. Recently, China has widely adopted this technique as the eco-organic soilless culture at the commercial level, enabling organic manures as the substrate for plant production [31,32].

This economic technique has been recommended for developing countries due to its cost-effectiveness compared to mineral solution systems [21,33]. Usage of organic materials, such as crop residues, sewage sludge, compost, and poultry manure, is well recognized as beneficial in soil resumption [18,34–36]. Keeping in view the health benefits and nutritional significance, organically produced vegetables are in high demand these days. However, there is a lack of information regarding suitable proportions of mineral and organic fertilizers required for vegetable production: excessive nitrogen fertilizers in agriculture lead to the accumulation of nitrates in plants and groundwater [15,16,37–41], thus reducing chemical fertilizers required.

The assessment and description of trait variation are essential in selecting and devising best practices to achieve productivity and quality enhancement. Furthermore, studies on variation under safeguard and decision-making are valuable in performing efficient conservation phases and avoiding redundant practices that increase production [42]. Therefore, developing procedures for both characterizing variability and reducing data size to manageable and accessible levels is essential in such studies. It is critical for agronomic characterization to be carried out using appropriate statistical methods. The widely used techniques include various univariate and multivariate methods such as ANOVA, mean comparison, regression, PCA, and AHC [43–45].

The main objective of the research was to assess and describe the impact of different ratios of PM and MF on the growth and productivity enhancement of cucumber crops by using univariate and multivariate analyses. The investigated fertilizer ratios or amounts can be recommended for better crop quality and yield attainment in future productivity enrichment programs and cucumber-related farming communities.

2. Materials and Methods

The experiment was conducted in a protected greenhouse at the Institute of Vegetables and Flowers (IVF), Soilless Culture Department, Chinese Academy of Agricultural Sciences (CAAS), during spring 2017 from February to the end of February May. The experiment followed a triplicated, completely randomized design CRD with four treatments with different substrate compositions as given Table 1. The experiment was conducted under a protected greenhouse using troughs built of red bricks. To execute the experiment, the sand, substrate (coco-peat), poultry manure, and MF were purchased from the Hebei Lingshousi County market to execute the experiment.

Table 1. Details of four investigated treatments with their respective substrate compositions.

Treatment	Substrate Composition
T_1 (Control)	60 kg m^{-3} PM
T_2	30 kg m^{-3} PM + 3 kg m^{-3} MF
T_3	30 kg m^{-3} PM + 5 kg m^{-3} MF
T_4	30 kg m^{-3} PM + 7 kg m^{-3} MF

The treatments of the experiment were in each cubic substrate with 60 kg m^{-3} PM (T_1) used as a control, 30 kg m^{-3} PM plus 3 kg m^{-3} MF (T_2), 30 kg m^{-3} PM plus 5 kg m^{-3} MF (T_3), and 30 kg m^{-3} PM plus 7 kg MF (T_4).

The physicochemical properties of all substrates consisting of bulk density, air space, and porosity were evaluated according to [2,46]. The water-holding capacity (WHC), electrical conductivity (EC), and pH were measured carefully according to the methods of [47–49]. The aforementioned physico-chemical properties of sand and coco-peat utilized in this experiment are presented in Table 2.

Table 2. Some physicochemical characteristics of substrates.

Properties	Coco-Peat	Sand
Bulk Density (g cm^{-3})	0.499	1.487
Air Space (%)	20.87	25.01
Water holding capacity (%)	75.7	27.14
Total Porosity (%)	94.9	51.99
Electrical Conductivity (dS m^{-1})	0.207	0.122
pH	7.3	7.9

The seedlings of cucumber (*Cucumis Sativus* L. cv Zhong Nong. No. 26. F1) were transplanted from the nursery to troughs at the four-leaf stage (45 days in age). Each treatment was replicated three times, and eight plants were grown as a plot. The irrigation practices were conducted daily, and fruit harvesting was carried out twice a week in the early morning. Sticky and yellow bands were used as insect control. The average relative humidity was maintained at 65%. However, the temperature was maintained on an average basis at 26 °C during daytime and 16 °C during nighttime. Cooling panels and fans were installed in the greenhouse at intervals working together to keep the temperature, and relative humidity maintained and kept the environment ventilated. An automatic photoperiod adjustment system was used to adjust the optimal amount of supplemental light needed to maintain plant growth. The system automatically directs LED lights to be dim, brighten, and turn off throughout the day while targeting a specific daily light integral (DLI). In the current study, the photoperiod ranged between 10–12 h, and light intensity oscillated between 95–155 W m^{-2}. Irrigation water was scheduled every week as per requirements and when needed.

The total depth of the root zone (30 cm) was divided into three layers, and the thickness of each layer was 10 cm (Figure 1). The bottom layer was coco-peat, the top layer was sand, and the middle layer was used to treat the combination of coco-peat, MF, and PM (Table 1). The poultry manure and MF (N-P$_2$O$_5$-K$_2$O 15-15-15) were added well to the coco-peat based on the weight/volume (w/v).

Figure 1. The schematic design three layers of the trough. The bottom layer contained coco-peat, the top layer contained sand, and the middle layer was used for different treatments of the combination of coco-peat, MF, and poultry manure (PM) except T1, which was used as control comprising only 60 kg of PM.

The reference level of the poultry manure (60 kg) was chosen based on the preliminary experiment investigating the effect of different manure levels 15, 30, 45, 60, and 75 kg m^{-3} on the yield, quality, and some growth parameters of the cucumber. According to the nitrogen (N) content in the MF and PM, the concentration of N in 30 kg m^{-3} of poultry manure was equivalent to its concentration in 5 kg MF. Accordingly, 5 kg of MF was chosen to be combined with 30 kg m^{-3} PM to be similar to the control of nitrogen content and different fertilizer forms. In addition, ± 2 kg of MF were added to 5 kg m^{-3} of MF and used in combination with PM.

Ten productivity-enhancing parameters were measured, including yield and quality parameters. Fruit yield per plant, fruit weight, leaf area, vine length, vine girth, and the number of leaves are yield traits, whereas dry matter, total soluble solids (TSS), leaf nitrogen contents, and fruit are quality traits. The data were collected from each plant in each replication, and then an average of eight plants per plot was considered for further analysis. The vine length was calculated starting from ground level until the tip of the vine with a measuring tape calibrated in centimeters, and the mean was determined for computational analysis. The vine girth was taken at 5 cm above the base on each sample plant using a Vernier caliper. The number of leaves from each plant was counted, and the mean value was estimated for further analysis. Leaf area was determined from three leaflets of each of the eight sampled plants based on the rectangular method. The leaflet's length and maximum width were determined from the plants' top, middle, and bottom portions.

The yield was calculated by weighing harvested fruits per plant. Fruit weight was measured at the end of experiments by analytical balance at a precision level of 0.001 (Mettler Toledo, LLC 1900 Polaris Parkway Columbus, OH 43240, USA). Dry matter and total soluble solids (TSS) were measured as indicators for fruit quality. Dry matter of the fruit (%) was obtained after the samples were dried at 60 °C until they had a constant weight. The TSS was measured by a refractometer.

Scored data were statistically analyzed for analysis of variance ANOVA to determine significant differences ($p < 0.05$) for traits under study. Treatment means were further compared using the Tuckey (HSD) test to determine the mean differences of different treatments.

A multivariate scatterplot matrix was constructed to determine correlation coefficients by Pearson's correlation method for all the productivity-enhancing traits and quality traits investigated under study. The multivariate analyses were performed, including basic statistics, correlation analysis, Principal Component Analysis (PCA), and two-way hierarchical clustering using Agglomerative Hierarchical Clustering (AHC) approach to figure out the Euclidean distance matrix for productivity-enhancing traits of cucumber under four treatments. All the analyses were carried out using the statistical software package SAS-JMP Pro 16 (SAS Institute Inc., Cary, NC, USA, 1989–2021).

3. Results

One-way ANOVA for different productivity-enhancing traits, including growth, yield, and quality parameters of cucumber under soilless culture, is shown in Table 3, depicting significant differences among them under different treatments. As shown in the table, the crop's fresh fruit yield per plant was significantly affected ($p < 0.05$) by all treatments. Comparisons for all pairs using Tukey–Kramer (HSD) method were carried out and are presented in Table 4. Application of MF at the ratio of 3 kg m^{-3} (T_2) showed the highest fruit yield per plant, which is (918.3 g) and the lowest (525.8 g) at T_1 (control), and there is a significant difference ($p < 0.05$) between them (Figure 2A). Similarly, as shown in Figure 2B, fruit weight was significantly affected ($p < 0.05$) by all treatments compared to control. The maximum average fruit weight (181.8 g) was recorded at T_3, while the lowest average fruits weight (146.1 g) was at T_1 (control), and they differed significantly. In Figure 2C. Fertilizer application significantly affected ($p < 0.05$) leaf area. The maximum increase in leaf area was 658.9 cm^2 for T_2. The vine length of cucumber was changed during plant growth under fertilizers treatments, as presented in Figure 2D. Significant differences ($p < 0.05$) of the treatments over the control were observed 27 days after transplant, attributed to fertilizer application. The highest increase in vine length was 202.8 cm for T_3. Figure 2E revealed that the vine girth of the cucumber plant was changed during plant growth under fertilizers treatments.

Table 3. Mean Squares from One-way ANOVA for different productivity-enhancing traits of Cucumber under Soilless Culture.

Source of Variation	Treatment	Error	Pr (>F)
Degree of freedom	3	8	-
Fruit Yield Per Plant (g)	94,726.2 **	4802.60	0.0005
Fruit Weight (g)	871.007 **	18.9260	<0.0001
Leaf Area (cm^2)	62,614.9 **	1692.00	<0.0001
Vine Length (cm)	611.590 **	41.4210	0.0013
Vine Girth (mm)	0.66979 **	0.08351	0.0085
Number of Leaves	9.69972 **	0.61650	0.0010
Dry Matter (%)	1.13183 **	0.00631	<0.0001
Total Soluble Solids (%)	0.33155 **	0.01271	0.0002
N-Content (Leaf) (%)	6.99080 **	0.04253	<0.0001
N-Content (Fruit) (%)	1.33254 **	0.01317	<0.0001

** Highly significant ($p < 0.01$). N-Content: Nitrogen contents; Fruit Yield Per Plant, Fruit Weight, Leaf Area, Vine Length, Vine Girth, and Number of Leaves are yield traits; Dry Matter, TSS, N-Content (Leaf), and N-Content (Fruit) are quality traits.

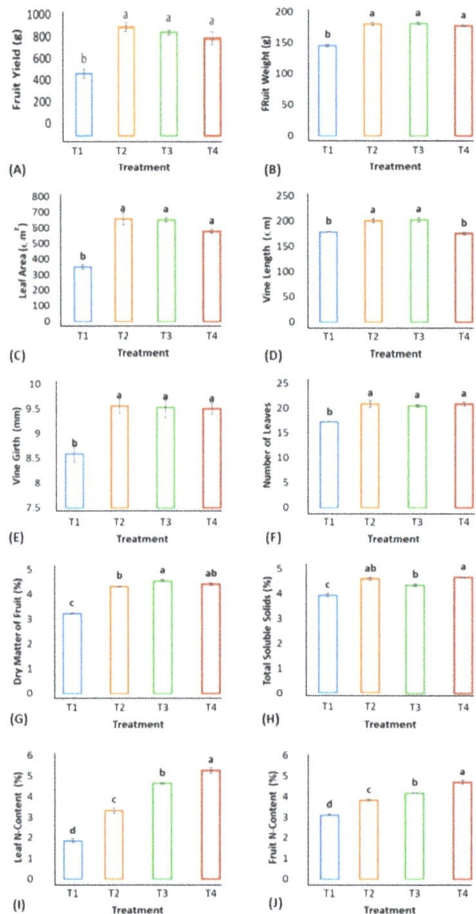

Figure 2. Graphical representation compares productivity-enhancing traits including Fruit Yield (**A**), Fruit Weight (**B**), Leaf Area (**C**), Vine Length (**D**), Vine Girth (**E**), Number of Leaves (**F**), Dry Matter (**G**), Total Soluble Solid (**H**), Leaf N-Content (**I**), and Fruit N-Content (**J**) in cucumber under soilless culture through Tukey–Kramer (HSD) test. Treatments sharing common letters revealed non-significant differences, and different letters depict significant differences ($p < 0.05$).

Table 4. Mean comparison for different productivity-enhancing traits of cucumber for different treatments under soilless culture based on Tukey–Kramer (HSD) method.

Traits	T$_1$ (Control)	T$_2$	T$_3$	T$_4$
Fruit Yield Per Plant (g)	525.8 ± 35.73 b	918.3 ± 37.36 a	875.7 ± 24.3 a	824.7 ± 56.04 a
Fruit Weight (g)	146.13 ± 2.17 b	180.53 ± 3.32 a	181.85 ± 2.74 a	177.71 ± 1.41 a
Leaf Area (cm^2)	350.5 ± 14.95 b	658.8 ± 40.12 a	652.6 ± 14.25 a	579.3 ± 14.81 a
Vine Length (cm)	178.96 ± 1.02 b	201.13 ± 4.89 a	202.8 ± 4.2 a	175.77 ± 3.55 b
Vine Girth (mm)	8.59 ± 0.166 b	9.57 ± 0.158 a	9.54 ± 0.213 a	9.51 ± 0.117 a
Number of Leaves	17.21 ± 0.042 b	20.92 ± 0.686 a	20.58 ± 0.289 a	20.88 ± 0.516 a
Dry Matter (%)	3.23 ± 0.0382 c	4.33 ± 0.0313 b	4.56 ± 0.0441 a	4.43 ± 0.0635 ab
Total Soluble Solids (%)	3.912 ± 0.093 a	4.564 ± 0.065 ab	4.35 ± 0.056 b	4.66 ± 0.03 c
N-Content (Leaf) (%)	1.849 ± 0.098 d	3.344 ± 0.138 c	4.661 ± 0.061 b	5.296 ± 0.156 a
N-Content (Fruit) (%)	3.127 ± 0.033 d	3.84 ± 0.064 c	4.187 ± 0.023 b	4.717 ± 0.109 a

Values sharing the same letters in the same row are not significantly different at ($p < 0.05$). N-Content: Nitrogen contents; Fruit Yield Per Plant, Fruit Weight, Leaf Area, Vine Length, Vine Girth, and Number of Leaves are yield traits; Dry Matter, TSS, N-Content (Leaf), and N-Content (Fruit) are quality traits.

The highest increase in the ratio of the vine girth in comparison with the control was 9.57 cm for treatment (T$_2$). In Figure 2F, significant differences ($p < 0.05$) were observed for the number of leaves applying different fertilizer treatments over control. The maximum number of leaves was observed with T$_2$, i.e., 20.92 leaves. In Figure 2G, the dry matter of the fruit was significantly affected ($p < 0.05$) by all treatments compared with the control. The highest value of the dry matter, which was 4.56%, was obtained at T$_3$, while the lowest value (3.41%) was at T$_1$ (control), and all the treatments differed significantly from control. As shown in Figure 2H, all treatments showed a significant increase ($p < 0.05$) in the TSS of cucumber fruit over the control. The maximum increase in TSS observed was 4.66% with T$_4$. The total nitrogen contents were determined in leaves and fruits, as shown in Figure 2I,J. The results showed that all treatments were significantly different ($p < 0.05$) for the nitrogen contents over the control in both leaves and fruits. The highest value of N-contents in leaves and fruits, i.e., 5.296% and 4.72%, respectively, was achieved by T$_4$. The lowest values obtained for these nitrogen contents in leaves and fruits in control were 1.85% and 3.13%, respectively. The estimations depicted clear and significant differences between all the investigated traits under different treatments.

The correlation matrix of productivity-enhancing traits of cucumber is presented in Figure 3. All the traits depicted a positive trend in correlation with high significance with each other. Maximum values of positive correlation were observed for dry matter with fruit weight (0.97), fruit yield per plant (0.91), and leaf area (0.91). However, vine length revealed a minimum correlation with all the yield and quality traits. The rest of the parameters represented significant positive correlations among themselves.

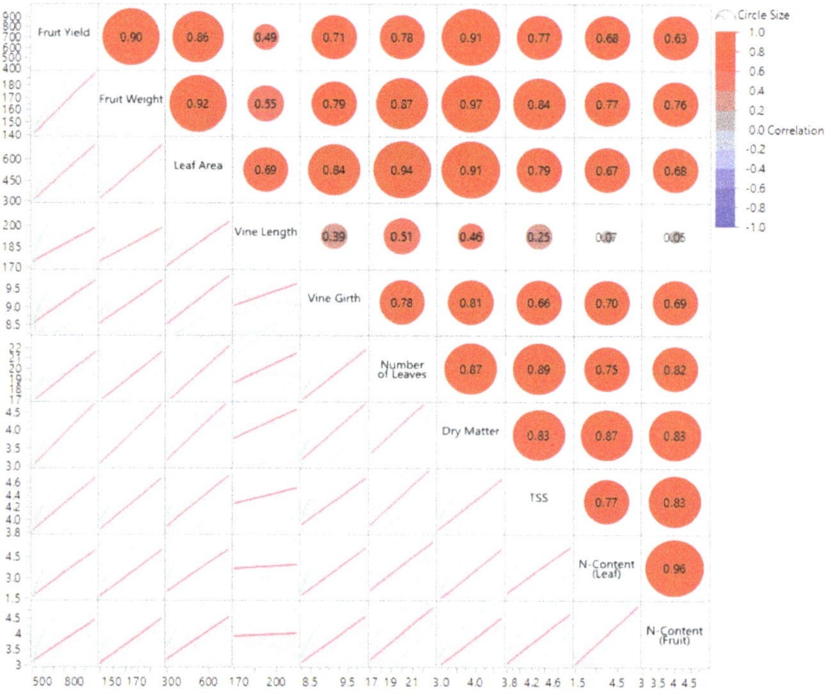

Figure 3. Scatterplot matrix for productivity—enhancing traits of cucumber under 4 treatments. The biplots of each trait with other traits showing significant relationships are presented at the lower side of the diagonal, while on the upper side, the correlation among traits under different treatments is displayed. Different shades of colors show different levels of correlation values and sizes of circles to represent the extent of correlation among these traits under treatments.

Multivariate analysis was performed to further validate the outcome obtained from the univariate method used to distinguish the augmented treatments from conventional ones. The clustering of traits was done based on the discrimination among four treatments depicted in the form of a heat map in Figure 4. The hierarchical clustering was operated using morphological traits data from different treatments via the agglomerative approach. This approach was based on squared Euclidean distance between treatments using Ward's method. This heat plot discriminates the control T_1 from the rest of the augmented treatments T_2, T_3, and T_4, forming two clusters. The two treatments T_2 and T_3 depicted productivity and quality-enhancing outcomes, respectively.

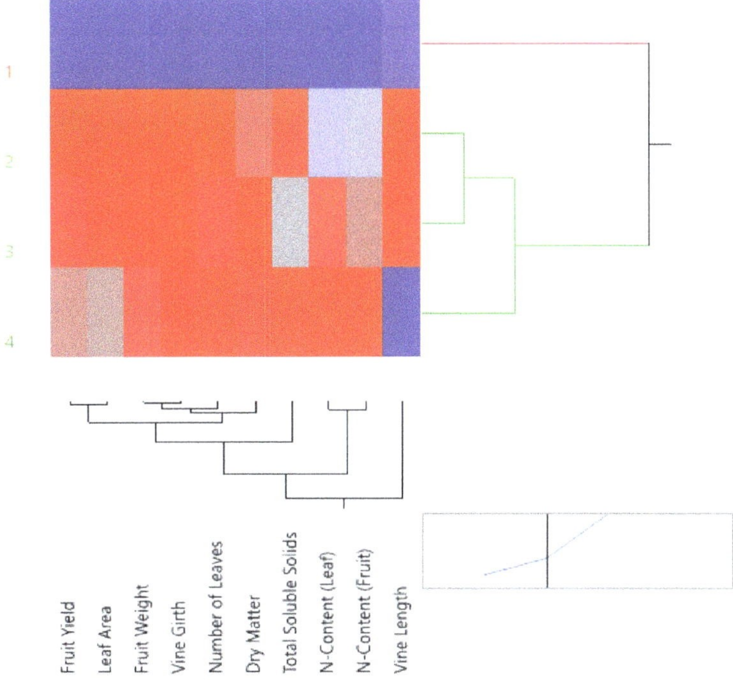

Figure 4. Two-way hierarchical clustering using Agglomerative Hierarchical Clustering (AHC) approach to figure out the Euclidean distance matrix for productivity-enhancing traits of cucumber under 4 treatments. The heat plot was assembled in JMP pro. Based on Ward's minimum variance method, V. 16 (SAS Institute Inc., Cary, NC, USA). Both the vertical and horizontal axes represent clades formation based on division related to traits under different treatments.

According to PCA, 97.5% of the total variation was covered by the first two PCs. The Scree plot demonstrates the first two PCs with maximum coverage of 97.5% of total variation explained (Figure 5a). It was further elaborated by a factor map of squared cosines/coordinates representing long bars for variables' contribution to total variation (Figure 5b). This map chart is also validating the higher contribution of PC1 and PC2. Moreover, the distribution details related to traits and treatments covered by the first two PCs were revealed in the Biplot matrix (Figure 6). It validates clear discrimination of Control T_1 from the rest of the augmented treatments T_2, T_3, and T_4.

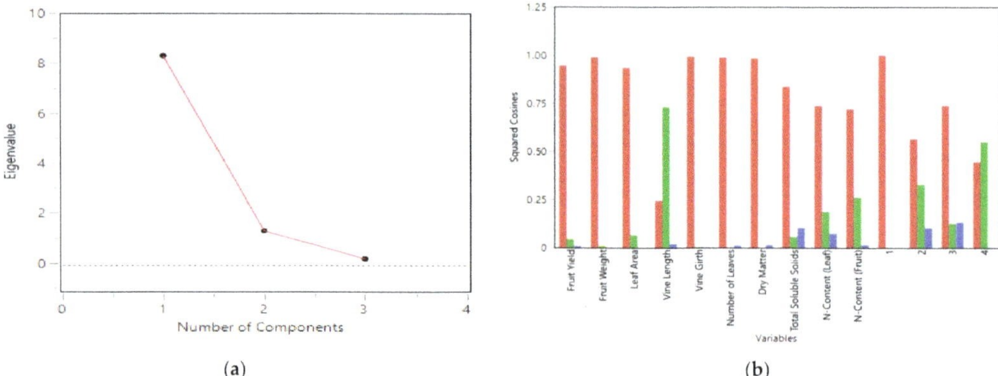

Figure 5. (**a**) Scree plot reveals the number of components covering sufficient variation related to treatments used for investigated traits. (**b**) Squared cosines associated with the principal components for ten investigated traits and four treatments.

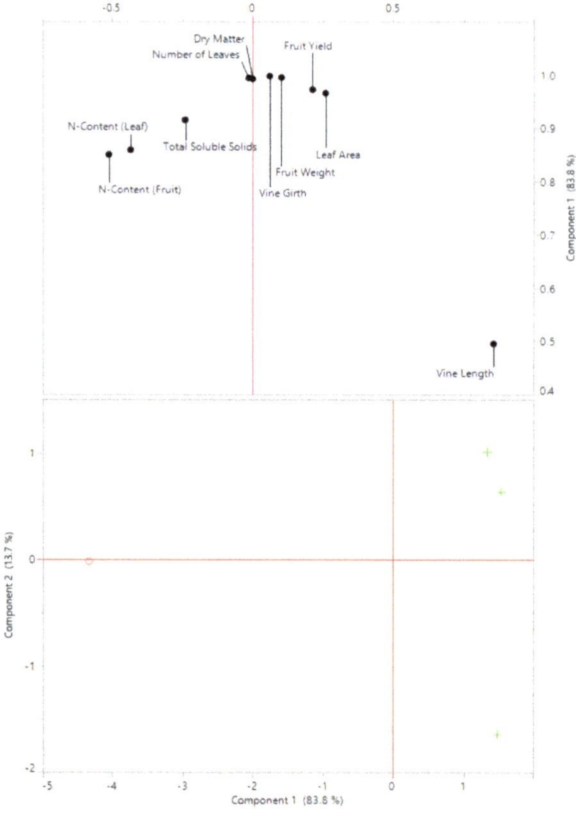

Figure 6. Biplot matrix for principal component analysis revealing the distribution of traits in question under treatments. The upper half of the matrix represents traits, while the lower half depicts treatments placed at different places of the scatterplot.

4. Discussion

Findings of the current study, which was executed to examine the effects of different fertilizer treatments as compared to conventional application of fertilizers under soilless culture, have demonstrated that application of PM combined with MF had a significant effect on the cucumber production by the increase in yield and vegetative growth [50]. It is also evident that we can benefit from poultry manure application, especially improvement in fruit quality resulting from nutrient uptake, increasing organic matter content, and enhancing the activity of microorganisms [51–54]. These nutrients increase vegetative growth and form a robust radical total, thus increasing the number of leaves and vine length. These results agree with previous studies. Increasing amounts of manures in treatments have positively influenced all the traits. Such findings were previously reported by Hamma et al. [55] and Law-Ogbowo [56]. Compared to untreated (control) ones, the highest number of leaves observed for treated cucumber plants, compared to untreated (control) ones, indicated higher nutrient uptake, which ultimately gave rise to a higher leaf area. Higher leaf area was an apparent cause of maximum light interception. Wider leaf area is necessary for maximum light interception for increased photosynthesis and transpiration [57], leading to improved production/yield due to higher translocation of assimilates to the storage organ, i.e., fruit [56].

The application of MF and PM improved the cucumber's growth characteristics, which triggered an increase in the average weight of fruit and net yield. This result agrees with previous works [16,18,21,58–60]. The quality of cucumber fruits was enhanced by mineral contents, dry matter, and total soluble solids. The improvement in the fruit quality of cucumber could mainly be attributed to the combined use of organic and MF. Furthermore, the improved growth of plants under study positively responds to the balanced discharge of additional nutrient elements through moisture made available by applying PM and coco-peat. These findings were also confirmed by previous studies [33,51,55,58].

Among the treatments where poultry manure at 30 kg m^{-3} was combined with MF, the results showed no significant differences. The application of MF was found to increase the growth and yield parameters significantly over the control. However, only minor variations were observed among these treatments. The possible reason could be the slight variation in the dosage of MF, where 3, 5, and 7 kg m^{-3} NPK were applied to T2, T3, and T4, respectively. It is also likely that the crop met the nutrient requirement from the poultry manure combined with the lowest level of NPK. Therefore, more applications of MF might not have significant effects on the measured parameters. Sometimes the overdose of fertilizer negatively affects some parameters [12,61]. For instance, the overdose of nitrogen could prolong the growth stage and delay the fruiting stage. As shown in Figure 2, the total yield among T$_2$, T$_3$, and T$_4$ has slightly decreased as a function of NPK increase. Although the decrease in yield was insignificant, it indicates the negative effect of the overuse of MF on the crop. Our findings are consistent with the previously reported studies [15,56,61–64].

The correlation of parameters investigated with each other represented in Figure 3 depicts a significant positive correlation. All the traits except vine length significantly depend on each other for better performance [65]. All the quality parameters positively correlate with yield parameters and ultimately increase the net yield [66,67]. Maximum correlations were observed for dry matter with fruit yield per plant, fruit weight, and leaf area. This indicates the interdependence of dry fruit matter on traits under study contributing to yield. Our findings follow those of previous studies [42,65,68–71].

Both PCA and AHC multivariate methods are widely adopted today to cluster and represent closeness or distances between variables in the dataset [72]. The visual display of variables data is plotted as factors acquired through the PCA approach [73]. In this study, the first two PCs represented maximum coverage of variation. The AHC method divided the treatments into two distinct clusters, revealing control T1 as discriminant from the rest. The two clusters formation confirms different effects of conventional amount of manure used from augmented amounts used in the study. It was observed that the augmented amounts have a better impact on the productivity and quality of the fruit traits. The T2 improved the productivity traits, while T3 helped to enhance quality traits more. Both univariate and multivariate analysis methods provided the grounds to support the hypothesis that the treatments used on traits were sufficient to cover maximum variation. Similar findings are reported in the study [43].

5. Conclusions

Presently, production of cucumber is dependent mainly on chemical fertilizers which have adverse effects on human health and surrounding environments. This study demonstrated the better and safe use of PM and MF at various combination ratios on cucumber crop. The vegetative growth and yield productivity of the crop were significantly enhanced with organic fertilizer application. The results did not show significant differences between the three experimental mineral fertilizer ratios combined with poultry manure. Hence, the lowest ratio could be recommended as best to achieve better quality and growth. The obtained results can be used as future recommendations in crop cultivation programs focusing on productivity and quality enhancement and the global farming community under soilless cultures of cucumber.

Author Contributions: Conceptualization, W. J, B.N.S.; methodology, H.Y. and B.N.S.; software, B.N.S., Z.S., M.S.I. and S.K.; validation, B.N.S., Z.S. and M.S.I.; formal analysis, B.N.S., M.S.I. and Z.S.; investigation, B.N.S.; data curation, B.N.S.; writing original draft preparation, B.N.S. and Z.S.; writing review and editing, B.N.S., M.S.I., H.W., P.L., T.L. and Q.L.; visualization, B.N.S., M.S.I. and Z.S.; supervision, W.J.; project administration, H.Y.; funding acquisition, W.J.; All authors have read and agreed to the published version of the manuscript.

Funding: This research was financially supported by China's National Key R&D Program 2019YFD1001901, National Key R&D Program of China, 2019YFD1000300, and National Natural Science Foundation of China, 32002115.

Institutional Review Board Statement: Not applicable.

Informed Consent Statement: Not applicable.

Data Availability Statement: Data supporting reported results will be available and provided at reasonable request.

Conflicts of Interest: The authors declare no conflict of interest.

References

1. Patel, C.; Panigrahi, J. Starch glucose coating-induced postharvest shelf-life extension of cucumber. *Food Chem.* **2019**, *288*, 208–214. [CrossRef]
2. Hao, J.; Li, Q.; Yu, H.; Wang, H.; Chai, L.; Miao, T.; Jiang, W. Comparative proteomic analysis of cucumber fruits under nitrogen deficiency at the fruiting stage. *Hortic. Plant J.* **2020**, *7*, 59–72. [CrossRef]
3. Sharma, V.; Sharma, L.; Sandhu, K.S. *Cucumber (Cucumis sativus L.)*; Springer: Singapore, 2020; pp. 333–340.
4. FAOSTAT. Food and Agriculture Organization of the United Nations. *Crop. Prod. Data* **2021**. Available online: http://www.fao.org/faostat/en/#data (accessed on 9 July 2021).
5. Lei, T.L.; Yang, L.J. Overcoming Continuous Cropping Obstacles & mdash: The Difficult Problem. *Sci. Agric. Sin.* **2016**, *49*, 916–918. [CrossRef]
6. Liu, X.; Li, Y.; Ren, X.; Chen, B.; Zhang, Y.; Shen, C.; Wang, F.; Wu, D. Long-Term Greenhouse Cucumber Production Alters Soil Bacterial Community Structure. *J. Soil Sci. Plant. Nutr.* **2020**, *20*, 306–321. [CrossRef]
7. Meena, R.K.; Trivedi, S.; Kumar, M. Protected Cultivation Vegetable: Cucumber. *Biot. Res. Today* **2020**, *2*, 359–361.
8. Singh, M.C.; Singh, J.; Pandey, S.; Mahay, D.; Srivastava, V. Factors Affecting the Performance of Greenhouse Cucumber Cultivation: A Review. *Int. J. Curr. Microbiol. Appl. Sci* **2017**, *6*, 2304–2323. [CrossRef]
9. Ghehsareh, A.M.; Khosravan, S.; Shahabi, A.A. The effect of different nutrient solutions on some growth indices of greenhouse cucumber in soilless culture. *J. Plant. Breed. Crop. Sci.* **2011**, *3*, 321–326.
10. Nikolaou, G.; Neocleous, D.; Kitta, E.; Katsoulas, N. Advances in irrigation/fertigation techniques in greenhouse soilless culture systems (SCS). In *Advances in Horticultural Soilless Culture*; Burleigh Dodds Science Publishing: Cambridge, UK, 2021; pp. 249–275.

11. Yang, Y.; Wang, P.; Zeng, Z. Dynamics of bacterial communities in a 30-year fertilized paddy field under different organic–inorganic fertilization strategies. *Agronomy* **2019**, *9*, 14. [CrossRef]

12. Lu, J.; Yang, R.; Wang, H.; Huang, X. Stress effects of chlorate on longan (*Dimocarpus longan* Lour.) trees: Changes in nitrogen and carbon nutrition. *Hortic. Plant J.* **2017**, *3*, 237–246. [CrossRef]

13. Umekwe, P.; Okpani, F.; Okocha, I. Effects of different rates of NPK 15: 15: 15 and pruning methods on the growth and yield of cucumber (*Cucumis sativus* L.) in Unwana-Afikpo. *IJSR* **2015**, *4*, 36–39.

14. Johnston, A. Food security in the WANA region, the essential need for balanced fertilization. In *International Workshop Proceedings, Ege University, Izmir, Turkey*; International Potash Institute (IPA): Basel, Switzerland, 1997.

15. Singh, M.C.; Kachwaya, D.S.; Kalsi, K. Soilless cucumber cultivation under protective structures in relation to irrigation coupled fertigation management, economic viability and potential benefits-a review. *Int. J. Curr. Microbiol. Appl. Sci.* **2018**, *7*, 2451–2468. [CrossRef]

16. O'Brien, T.A.; Barker, A.V. Growth of peppermint in compost. *J. Herbs Spices Med. Plants* **1996**, *4*, 19–27. [CrossRef]

17. Natsheh, B.; Mousa, S. Effect of organic and inorganic fertilizers application on soil and cucumber (*Cucumis sativa* L.) plant productivity. *Int. J. Agric. For.* **2014**, *4*, 166–170.

18. Agegnehu, G.; Bass, A.M.; Nelson, P.N.; Bird, M.I. Benefits of biochar, compost and biochar–compost for soil quality, maize yield and greenhouse gas emissions in a tropical agricultural soil. *Sci. Total Environ.* **2016**, *543*, 295–306. [CrossRef]

19. Masarirambi, M.T.; Dlamini, P.; Wahome, P.K.; Oseni, T.O. Effects of chicken manure on growth, yield and quality of lettuce (*Lactuca sativa* L.)'Taina'under a lath house in a semi-arid sub-tropical environment. *Agric. Environ. Sci.* **2012**, *12*, 399–406.

20. Ganeshamurthy, A.; Kalaivanan, D.; Selvakumar, G.; Panneerselvam, P. Nutrient management in horticultural crops. *Indian J. Fertil.* **2015**, *11*, 30–42.

21. Singh, M.; Singh, A.; Singh, J.; Singh, K. Economic viability of soilless cucumber cultivation under naturally ventilated greenhouse conditions. *Indian J. Hortic.* **2020**, *77*, 315–321. [CrossRef]

22. Barrett, G.; Alexander, P.; Robinson, J.; Bragg, N. Achieving environmentally sustainable growing media for soilless plant cultivation systems—A review. *Sci. Hortic.* **2016**, *212*, 220–234. [CrossRef]

23. Schmilewski, G. Growing medium constituents used in the EU. In Proceedings of the International Symposium on Growing Media, Nottingham, UK, 2–9 September 2007; pp. 33–46.

24. Huber, J.; Zheng, Y.; Dixon, M. Hydroponic cucumber production using urethane foam as a growth substrate. *Acta Hortic.* **2005**, *697*, 139. [CrossRef]

25. Gul, A.; Engindeniz, S.; Aykut, N. Can closed substrate culture be an alternative for small-scale farmers? *Acta Hortic.* **2007**, *747*, 83. [CrossRef]

26. Janapriya, S.; Palanisamy, D.; Ranghaswami, M. Soilless media and fertigation for naturally ventilated polyhouse production of cucumber (*Cucumis sativus* L.) CV Green Long. *Int. J. Agric. Environ. Biotechnol.* **2010**, *3*, 199–205.

27. Zhang, R.-H.; Zeng-Qiang, D.; Zhi-Guo, L. Use of spent mushroom substrate as growing media for tomato and cucumber seedlings. *Pedosphere* **2012**, *22*, 333–342. [CrossRef]

28. Mazahreh, N.; Nejatian, A.; Mousa, M. Effect of different growing medias on cucumber production and water productivity in soilless culture under UAE conditions. *Merit Res. J. Agric. Sci. Soil Sci.* **2015**, *3*, 131–138.

29. Singh, M.C.; Singh, J.; Singh, K. Optimal operating microclimatic conditions for drip fertigated cucumbers in soilless media under a naturally ventilated greenhouse. *Indian J. Ecol.* **2017**, *44*, 821–826.

30. Gruda, N.S.; Bragg, N. Developments in alternative organic materials for growing media in soilless culture systems. In *Advances in Horticultural Soilless Culture*; Burleigh Dodds Science Publishing: Cambridge, UK, 2021; pp. 73–106.

31. Gül, A.; Kıdoğlu, F.; Anaç, D. Effect of nutrient sources on cucumber production in different substrates. *Sci. Hortic.* **2007**, *113*, 216–220. [CrossRef]

32. RONG, Q.-L.; LI, R.-N.; HUANG, S.-W.; TANG, J.-W.; ZHANG, Y.-C.; WANG, L.-Y. Soil microbial characteristics and yield response to partial substitution of chemical fertilizer with organic amendments in greenhouse vegetable production. *J. Integr. Agric.* **2018**, *17*, 1432–1444. [CrossRef]

33. Xing, Y.; Meng, X. Development and prospect of hydroponics in China. In Proceedings of the International Symposium on Growing Media and Hydroponics, Windsor, ON, Canada, 19–26 May 1997; pp. 753–760.

34. Tejada, M.; Gonzalez, J.; García-Martínez, A.; Parrado, J. Application of a green manure and green manure composted with beet vinasse on soil restoration: Effects on soil properties. *Bioresour. Technol.* **2008**, *99*, 4949–4957. [CrossRef] [PubMed]

35. Opara, E.C.; Zuofa, K.; Isirimah, N.; Douglas, D. Effects of poultry manure supplemented by NPK 15: 15: 15 fertilizer on cucumber (*Cucumis sativus* L.) production in Port Harcourt (Nigeria). *Afr. J. Biotechnol.* **2012**, *11*, 10548–10554. [CrossRef]

36. Bowles, T.M.; Acosta-Martínez, V.; Calderón, F.; Jackson, L.E. Soil enzyme activities, microbial communities, and carbon and nitrogen availability in organic agroecosystems across an intensively-managed agricultural landscape. *Soil Biol. Biochem.* **2014**, *68*, 252–262. [CrossRef]

37. Havlin, J.; Beaton, J.; Tisdale, S.; Nelson, W. Soil acidity and alkalinity. *Soil Fertil. Fertil. Pearson Prentice Hall.* **2005**, *7*, 45–96.

38. Ju, X.-T.; Xing, G.-X.; Chen, X.-P.; Zhang, S.-L.; Zhang, L.-J.; Liu, X.-J.; Cui, Z.-L.; Yin, B.; Christie, P.; Zhu, Z.-L. Reducing environmental risk by improving N management in intensive Chinese agricultural systems. *Proc. Natl. Acad. Sci. USA* **2009**, *106*, 3041–3046. [CrossRef] [PubMed]

39. Shen, W.; Lin, X.; Shi, W.; Min, J.; Gao, N.; Zhang, H.; Yin, R.; He, X. Higher rates of nitrogen fertilization decrease soil enzyme activities, microbial functional diversity and nitrification capacity in a Chinese polytunnel greenhouse vegetable land. *Plant Soil* **2010**, *337*, 137–150. [CrossRef]

40. Zhen, Z.; Liu, H.; Wang, N.; Guo, L.; Meng, J.; Ding, N.; Wu, G.; Jiang, G. Effects of manure compost application on soil microbial community diversity and soil microenvironments in a temperate cropland in China. *PLoS ONE* **2014**, *9*, e108555. [CrossRef]

41. Rashti, M.R.; Wang, W.; Moody, P.; Chen, C.; Ghadiri, H. Fertiliser-induced nitrous oxide emissions from vegetable production in the world and the regulating factors: A review. *Atmos. Environ.* **2015**, *112*, 225–233. [CrossRef]

42. Singh, J.; Singh, M.K.; Kumar, M.; Kumar, V.; Singh, K.P.; Omid, A.Q. Effect of integrated nutrient management on growth, flowering and yield attributes of cucumber (*Cucumis sativus* L.). *Int. J. Chem. Stud.* **2018**, *6*, 567–572.

43. Lotti, C.; Marcotrigiano, A.R.; De Giovanni, C.; Resta, P.; Ricciardi, A.; Zonno, V.; Fanizza, G.; Ricciardi, L. Univariate and multivariate analysis performed on bio-agronomical traits of *Cucumis melo* L. germplasm. *Genet. Resour. Crop. Evol.* **2008**, *55*, 511–522. [CrossRef]

44. Khan, A.; Erum, S.; Riaz, N.; Shinwari, M.I.; Ibrar Shinwari, M. Multivariate Analysis of Potato Genotypes for Genetic Diversity. *bioRxiv* **2020**. [CrossRef]

45. Mangi, N.; Nazir, M.F.; Wang, X.; Iqbal, M.S.; Sarfraz, Z.; Jatoi, G.H.; Mahmood, T.; Ma, Q.; Shuli, F. Dissecting Source-Sink Relationship of Subtending Leaf for Yield and Fiber Quality Attributes in Upland Cotton (*Gossypium hirsutum* L.). *Plants* **2021**, *10*, 1147. [CrossRef]

46. de Boodt, M.; Verdonck, O. The physical properties of the substrates in horticulture. In Proceedings of the III Symposium on Peat in Horticulture, Dublin, Ireland, 28 June–3 July 1971; pp. 37–44.

47. Brown, E. Physical and chemical properties of media component of milled pine bark and sand. *J. Am. Soc. Hort. Sci.* **1975**, *100*, 119–121.

48. Wilkinson, K.M.; Landis, T.D.; Haase, D.L.; Daley, B.F.; Dumroese, R.K. Tropical nursery manual: A guide to starting and operating a nursery for native and traditional plants. *Agric. Handb. Wash. DC US Dep. Agric. For. Serv.* **2014**, *732*, 376.

49. Kim, S.; Abinaya, M.; Park, Y.; Jeong, B. Physiological and biochemical modulations upon root induction in rose cuttings as affected by growing medium. *Hortic. Plant J.* **2018**, *4*, 257–264. [CrossRef]

50. He, M.; Chen, Z.; Sakurai, K.; Iwasaki, K.; Shen, Y.; Zhou, J. Effect of differences in substrate formulations on cucumber growth under soilless bag culture in greenhouse. *Soil Sci. Plant.* **2003**, *49*, 763–767. [CrossRef]

51. Agele, S. Growth and yield of tomato grown on degraded soil amended with organic wastes. In Proceedings of the 35th Conference of the Agricultural Society of Nigeria, Federal University of Agriculture, Abeokuta, Nigeria, 14–17 March 2001; pp. 151–154.

52. Ghosh, P.; Bandyopadhyay, K.; Manna, M.; Mandal, K.; Misra, A.; Hati, K. Comparative effectiveness of cattle manure, poultry manure, phosphocompost and fertilizer-NPK on three cropping systems in vertisols of semi-arid tropics. II. Dry matter yield, nodulation, chlorophyll content and enzyme activity. *Bioresour. Technol.* **2004**, *95*, 85–93. [CrossRef]

53. Ayoola, O.; Adeniyan, O. Influence of poultry manure and NPK fertilizer on yield and yield components of crops under different cropping systems in south west Nigeria. *Afr. J. Biotechnol.* **2006**, *5*, 1386–1392.

54. Adekiya, A.; Agbede, T. Growth and yield of tomato (*Lycopersicon esculentum* Mill) as influenced by poultry manure and NPK fertilizer. *Emir. J. Food Agric.* **2009**, *21*, 10–20.

55. Hamma, I.; Ibrahim, U.; Haruna, M. Effect of poultry manure on the growth and yield of cucumber (*Cucumis sativum* L.) in Samaru, Zaria. *Niger. J. Agric. Food Environ.* **2012**, *8*, 94–98.

56. Law-Ogbomo, K.E.; Osaigbovo, A.U. Growth and yield responses of cucumber (*Cucumis sativum* L.) to different nitrogen levels of goat manure in the humid ultisols environment. *Not. Sci. Biol.* **2018**, *10*, 228–232. [CrossRef]

57. Ma, L.; Gardner, F.; Selamat, A. Estimation of leaf area from leaf and total mass measurements in peanut. *Crop. Sci.* **1992**, *32*, 467–471. [CrossRef]

58. Okoli, P.; Nweke, I. Effect of poultry manure and mineral fertilizer on the growth performance and quality of cucumber fruits. *J. Exp. Biol. Agric. Sci.* **2015**, *3*, 362–367.

59. Ewulo, B.; Ojeniyi, S.; Akanni, D. Effect of poultry manure on selected soil physical and chemical properties, growth, yield and nutrient status of tomato. *Afr. J. Agric. Res.* **2008**, *3*, 612–616.

60. Vo, M.; Wang, C. Effects of manure composts and their combination with inorganic fertilizer on acid soil properties and the growth of muskmelon (*Cucumis melo* L.). *Compost. Sci. Util.* **2015**, *23*, 117–127. [CrossRef]

61. Singh, M.C.; Singh, K.G.; Singh, J.P. Interactive Effects of Fertigation and Varieties on Pant Growth Attributes and Yield of Soilless Cucumbers. *Int. J. Sci. Technol. Soc.* **2020**, *8*, 50.

62. Kim, T.Y.; Lee, S.H.; Ku, H.; Lee, S.Y. Enhancement of Drought Tolerance in Cucumber Plants by Natural Carbon Materials. *Plants* **2019**, *8*, 446. [CrossRef]

63. Fields, J.S.; Gruda, N.S. Developments in inorganic materials, synthetic organic materials and peat in soilless culture systems. In *Advances in Horticultural Soilless Culture*; Burleigh Dodds Science Publishing: Cambridge, UK, 2021; pp. 45–72.

64. Tüzel, Y.; Balliu, A. Advances in liquid-and solid-medium soilless culture systems. In *Advances in Horticultural Soilless Culture*; Burleigh Dodds Science Publishing: Cambridge, UK, 2021; pp. 213–248.

65. Saeed, H.; Waheed, A. A Review on Cucumber. *Int. J. Technol. Res. Sci.* **2017**, *2*, 402–405.

66. Kumar, S.; Kumar, R.; Gupta, R.; Sephia, R. Studies on correlation and path coefficient analysis for yield and its contributing traits in cucumber. *Crop. Improv.* **2011**, *38*, 18–23.
67. Monna, M.N.A.; Robin, A.B.M.A.H.K.; Rabbani, M.G. Genetic variability, correlation and path analysis of Cucumber (*Cucumis sativus* L.). *Bangladesh J. Agril. Sci.* **2006**, *33*, 81–84.
68. Lakshmi, L.; Reddy, S. Studies on correlation and path-coefficient analysis for yield and its contributing characters in Cucumber (*Cucumis sativus* L.). *IJCS* **2018**, *6*, 1649–1653.
69. Ullah, M.; Hasan, M.; Chowdhury, A.; Saki, A.; Rahman, A. Genetic variability and correlation in exotic cucumber (*Cucumis sativus* L.) varieties. *Bangladesh J. Plant. Breed. Genet.* **2012**, *25*, 17–23. [CrossRef]
70. Hasan, R.; Hossain, M.K.; Alam, N.; Bashar, A.; Islam, S.; Tarafder, M.J.A. Genetic divergence in commercial cucumber (*Cucumis sativus* L.) genotypes. *Bangladesh J. Bot.* **2015**, *44*, 201–207. [CrossRef]
71. Arunkumar, K.; Ramanjinappa, V.; Ravishankar, M. Path coefficient analysis in F2 population of cucumber (*Cucumis sativuus* L.). *Plant Arch.* **2011**, *11*, 471–474.
72. Lee, I.; Yang, J. Common clustering algorithms. In *Comprehensive Chemometrics: Chemical and Biochemical Data Analysis*; Elsevier: Oxford, UK, 2009; pp. 577–618.
73. Xu, Q.; Liu, J.-S.; Chen, X.-H.; Li, L.-L.; Go, S.-G.; Chen, Z.-M.; Xiao, J.; Cao, B.-S. Principal Component and Cluster Analysis of Quality Characters of Pickling Cucumber (*Cucumis sativus* L.). *J. Yandzhou Univ.* **2003**, *4*, 78–81.

Article

Impact of the Hydroponic Cropping System on Growth, Yield, and Nutrition of a Greek Sweet Onion (Allium Cepa L.) Landrace

Christos Mouroutoglou [1], Anastasios Kotsiras [1], Georgia Ntatsi [2] and Dimitrios Savvas [2,*

[1] Laboratory of Vegetable Production, Department of Agriculture, University of Peloponnese, 24100 Kalamata, Greece; x.mouroutoglou@uop.gr (C.M.); a.kotsiras@go.uop.gr (A.K.)

[2] Laboratory of Vegetable Production, Department of Crop Science, Agricultural University of Athens, 75 Iera Odos, 11855 Athens, Greece; ntatsi@aua.gr

* Correspondence: dsavvas@aua.gr; Tel.: +30-210-5294510

Abstract: Nerokremmydo of Zakynthos, a Greek landrace of sweet onion producing a large bulb, was experimentally cultivated in a glasshouse using aeroponic, floating, nutrient film technique, and aggregate systems, i.e., AER, FL, NFT, and AG, respectively. The aim of the experiment was to compare the effects of these soilless culture systems (SCSs) on plant characteristics, including fresh and dry weight, bulb geometry, water use efficiency, tissue macronutrient concentrations, and uptake concentrations (UC), i.e., uptake ratios between macronutrients and water, during the main growth, bulbing, and maturation stages, i.e., 31, 62, and 95 days after transplanting. The plants grown in FL and AG yielded 7.87 and 7.57 kg m^{-2}, respectively, followed by those grown in AER (6.22 kg m^{-2}), while those grown in NFT produced the lowest yield (5.20 kg m^{-2}). The volume of nutrient solution (NS) consumed per plant averaged 16.87 L, with NFT plants recording the least consumption. The SCS affected growth rate of new roots and "root mat" density that led to corresponding nutrient uptake differences. In NFT, reduced nutrient uptake was accompanied by reduced water consumption. The SCS and growth stage strongly affected tissue N, P, K, Ca, Mg, and S mineral concentrations and the respective UC. The UC of N and K followed a decreasing trend, while that of Mg decreased only until bulbing, and the UC of the remainder of the macronutrients increased slightly during the cropping period. The UC can be used as a sound basis to establish NS recommendations for cultivation of this sweet onion variety in closed SCSs.

Keywords: soilless culture; nutrient uptake; sweet onion; nerokremmydo; aeroponic; floating; nutrient film technique; bulb

Citation: Mouroutoglou, C.; Kotsiras, A.; Ntatsi, G.; Savvas, D. Impact of the Hydroponic Cropping System on Growth, Yield, and Nutrition of a Greek Sweet Onion (Allium Cepa L.) Landrace. *Horticulturae* **2021**, *7*, 432. https://doi.org/10.3390/horticulturae7110432

Academic Editors: Xun Li, Xiaohui Hu and Shiwei Song

Received: 3 August 2021
Accepted: 14 October 2021
Published: 22 October 2021

Publisher's Note: MDPI stays neutral with regard to jurisdictional claims in published maps and institutional affiliations.

1. Introduction

Onion (Allium Cepa L.) is the second most important vegetable crop after tomato, showing a doubling of production since 2000. The world production (2019) has reached 99.9 Mtons of dry bulbs [1]. Modern paradigms of open-pollinated bulb onion landraces are *Kremmydi Thespion* [2], *Cipolla di Giarratana* [3] and *Jaune des Cévennes* [4] and *Nerokremmydo of Zakynthos* (meaning "an onion that needs plenty of water"), originating from a Greek Ionian sea island. The latter is a Bermuda type, white-colored, mild, and is marketed in the large to colossal categories [5] from mid-July to November in fresh condition; this landrace achieves two to three times the price of standard varieties. It is sown during the second half of November or in December in low-cost tunnel propagation greenhouses, and is thinned, trimmed, and transplanted up to late March–mid April, yielding 3.5–4.0 Kg m^{-2}.

The limiting area (15–20 ha), labor costs for weed and soil-borne diseases management, such as Allium root rot (*Sclerotium cepivorum*) and onion pink root (*Phoma terrestris*) [6,7], are challenges that can be overcome by cultivating in closed soilless culture systems (CSCS). These systems are environmentally friendly, with higher yields and prospect in adverse growing conditions areas [8], or where there are soil-borne pathogens [9]. Furthermore, soilless cultivation can be applied to low to moderate cost greenhouses in Mediterranean countries [10], or even indoor plant production in plant factories with artificial lighting [11].

CSCS comprise many techniques in relation to the root environment, such as water culture systems (WCS) including the floating system (FL), nutrient film technique (NFT), and aeroponics (AER), and aggregate systems (AG) [12,13]. AER is a subgroup of WCS, in which, roots grow inside dark, closed containers, and are sprayed at certain intervals with fine droplets of nutrient solution (NS) to maintain a stable aquatic environment that experiences a minimum of fluctuations in ambient nutrient and oxygen levels [14]. Apart from use in research into spaceflight crop production [15], currently AER is used for the production of potato mini-tubers [16], herbs, medicinal root and aromatic plants, and root cuttings [15,17]. To maximize outcomes, the technique requires precision sensing technology and a strict dosing regimen [18]. Challenges to overcome include provision of a stable power supply, irrigation components, the position of nebulizers, irrigation program, root available space, and root/NS temperature.

Nutrient and water management is simple, making FL one of the most forgiving and stable CSCS, and concomitantly an increasingly popular option for leafy vegetable production [9,19]. In FL, plants are placed in holes on lightweight floating rafts. Roots are totally or partially immersed in the 20–30 cm NS depth, creating a high volume of NS per plant, with a high buffering capacity [19]. The most important benefits of FL include increasing the useful growing space, and the production of clean and healthy plants. It is considered one of the most suitable WCS for crop propagation and the production of short-lived leafy vegetables such as "baby leaf", "microgreen", and herbs. The system seems to be most useful for single harvest crops [20,21]. Attention should be paid to the aeration, depth, and management of the NS [22].

In NFT, plants are cultivated in pipes or gullies and their roots are permanently surrounded by flowing NS, providing a high ratio of NS surface area to solution volume and the absorption of oxygen from ambient air [9,20]. NFT is ideal for short-term crops such as leafy vegetables and herbs, while it is also used for long-term crops, such as cucumbers and tomatoes [23]. Some of the cited advantages [19] include simplified watering, nutrient supply uniformity, rapid turn round of crops, and reduced water and nutrient losses. Cited disadvantages, such as the collapse of crops in the event of power outages, root mass NS impedance in long-term crops, low buffering capacity, and the higher level of management decisions required, should be considered [24].

Aggregate systems are defined as cultivation on porous solid materials other than soil, commonly termed substrates, and are classified into inorganic and organic categories [9]. The composition of solid substrates affects plant physiology, yield and quality [25]. Among the soilless culture systems, the most widely used in South-Eastern European countries for vegetable production, are rockwool, perlite, coir, and pumice. Solid substrates have been used for the production of high-value vegetables, as well as for plant propagation [13]. The biggest challenges in the use of substrates are: ideal selection, i.e., the optimum physicochemical relationship of plant/substrate, purchase costs, available root volume, nutrient management, NS recycling, salinity levels in the root zone, rapid development of nutrient deficiencies, plant/microorganism interactions, reuse, pathogen control, as well as taste and nutritional value [9,26].

A comparison of CSCS, shows fundamental differences in architecture, constituents needed, and sensitivity to failures. When choosing the ideal CSCS, factors such as water and nutrient use efficiency (WUE/NUE), system support, usability, input requirements, production scale, cost and management requirements, and root aeration, should be considered with respect to specific plant species, and their growing and length cycles [19,20,27]. It is also important, when choosing CSCS, that the original systems layout under the same environmental conditions is considered.

An important parameter that differentiates the yield and growth of plants grown in CSCS from that observed in other systems is the morphological and physiological properties of the roots that differ significantly depending on the medium in which they are grown [28]. Compared to other WCS, AER plants are reported to have greater root mass [29–31]. Differences observed between WCS and AG are generally significant, in contrast to those observed between different substrates [28].

Nutrient uptake concentrations (ion-to-water uptake ratios) are of great value for the estimation of plant nutrient needs. They exhibit appreciable stability over time under similar climatic conditions and are plant species and developmental stage specific, provided that the root ionic concentrations are relatively stable over the same time period [32–35]. The information available on the composition of NS for onion production is limited [36,37]. It is important to create suitable NS and cultivation protocols for commercial soilless production of fresh bulb onions under Mediterranean conditions. This should include technical adaptations for bulb production, root aeration, NS composition, and irrigation parameters.

In light of this background, the aim of the present study was to compare the agronomic and nutritional response of the Greek onion landrace *Nerokremmydo of Zakynthos* in four CSCS, aiming to highlight their relative advantages and limitations. Finally, the impact of the CSCS on UCs and nutrient accumulation by onion plants were studied.

2. Materials and Methods

2.1. Systems, Plant Material, Vegetative Growth and Treatments

The experiment was conducted in an unheated glasshouse constructed with a 1% slope and an appropriate industrial floor, located at the University of Peloponnese, Kalamata, Greece (37°03′ N, 22°03′ E, 5 m a.s.l.). Four of the most widely applied closed soilless culture systems (CSCS), i.e., aeroponic, floating, nutrient film technique, and aggregate (AER, FL, NFT, and AG respectively), were used. A randomized complete block design with 3 replicates per system was used (Figure 1), and each replicate occupied an area of 1.60 × 1.80 m (length × width).

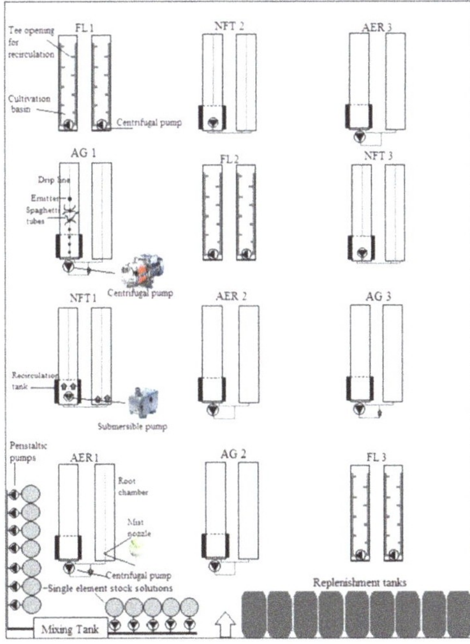

Figure 1. Layout of the 12 experimental units; aeroponic (AER$_i$), floating (FL$_i$), nutrient film technique (NFT$_i$), and aggregate (AG$_i$) I = 1, 2, 3 replicates.

The root chamber of the AER system was made of extruded polystyrene boards while the NFT and AG systems comprised 4 and 2 PE channels, respectively. The three systems were installed on 2 galvanized bases with a 100 L PE tank underneath per replicate used for the recirculation of the 70 L growth nutrient solution (GNS). Each FL replicate comprised of 2 basins made of Ytong® blocks covered with a 0.5 mm thick black soft PE sheet containing 390 L GNS each. White Styrofoam® boards, were used for AER, NFT, and FL, being the coverage and/or the floating supporting media for the onion plants. For the AG system, perlite (PerloflorHydro® with a particle size range of 0.5–2.5 mm) was used. Information on the irrigation hardware and management is shown in Table 1.

Table 1. Hardware and irrigation management.

	AER	**FL**	**NFT**	**AG**
Pump	Pedrollo JCR 15H, Italy	DAB Euroinox 30/30 T, Italy	Eheim 1001.220, Germany	Pedrollo JCR 10H, Italy
Application of GNS	Netafim coolnet Pro 65 microns (4.0 bar)	Perforated pipes at the bottom of the basin	5 mm tubes	$8\,L\,h^{-1}$
Irrigation	Daily 1 min 3 min^{-1}	System worked 2–3 h day^{-1} aiming at O_2 levels between 5–6 ppm	Daily 0.15 $m^3\,h^{-1}$ per channel	4–9 irrigation events day^{-1} based on solar radiation integrator, for a target leaching fraction of 0.4

Automatic replenishment of fresh GNS was done by respective 110 L tanks for AER, NFT, and AG, and directly from the hydroponic head for FL, according to consumption (Figure 1). A threshold value of 3.0 dS m^{-1}, for the recycling GNS was chosen for the replacement of GNS. NS pH was recorded daily and adjusted in the range of 5.5–5.7 by adding nitric acid.

Due to a lack of literature on the composition of nutrient solutions for onion bulb production, adaptations were made based on [36–40]. The composition of NS was the same throughout the experiment for all CSCS until harvest. The pH, EC, macro, and micronutrient concentrations of the tap water and the nutrient solutions used during the experiment are presented at Table 2.

Table 2. pH, electrical conductivity (EC; dS m^{-1}), macronutrient (K, Ca, Mg, N, S, P; mM) and micronutrient (Fe, Mn, Zn, Cu, B, Mo; µM) concentrations in the tap water, in the starter (SNS) and in the growth nutrient solution (GNS).

Parameter	Tap Water	SNS	GNS
pH	7.78	5.6	5.6
EC	0.70	2.00	2.60
$[Ca^{2+}]$	2.30	3.50	4.50
$[K^+]$	0.07	7.00	10.00
$[Mg^{2+}]$	1.28	1.30	1.50
$[Na^+]$	1.09	1.09	1.09
$[NH_4^+]$	0.00	1.10	1.25
$[SO_4^{2-}]$	1.08	1.32	1.50
$[NO_3^-]$	0.01	12.00	17.00
$[H_2PO_4^-]$	0.00	1.80	2.00
$[HCO_3^-]$	4.60	0.79	1.59
$[Cl^-]$	1.55	1.55	1.55
[Fe]	0.00	40.00	40.00
Mn^{2+}	0.00	5.00	5.00
Zn^{2+}	1.07	4.00	4.00
Cu^{2+}	0.00	0.80	0.80
B	0.00	30.00	30.00
Mo	0.00	0.50	0.50

Seeds of Greek onion (Allium Cepa) local landrace *"Nerokremmydo of Zakynthos"*, which can be translated as "water onion of Zakynthos", were procured from local producers of Zakynthos Island. On November 26, seeds were sown following standard cultivation procedures and at 140 DAS, plants were transplanted, at a density of 13.7 plants m^{-2}. To avoid water evaporation, all experimental units were covered with double-sided PE sheets and all plants were forced to bulb above it. Horizontal net was used for plant support. Plants were kept free from pests following standard greenhouse management practices.

Weekly recordings started 10 days after transplanting (DAT) including the number of green leaves, longest leaf length up to tip and its midpoint diameter (LLMD) [37], pseudostem length, [41] neck diameter, bulb diameter, and height. Based on these measurements, plant height and the bulbing ratio, i.e., maximum bulb: neck diameter (BR), were calculated [41]. Bulb shape characteristics were described with the bulb shape index (ratio of diameter to bulb height) [42,43] and sphericity [44]. During growing, bulb shape firstly adopted a prolate ellipsoid (height > diameter) form and then gradually changed to an oblate ellipsoid (height < diameter) form. Considering this, bulb volume, surface, and sphericity were calculated according to Mohsenin [44] and Stroshine [45].

Onion maturity was rated by visual observation, and was considered to be achieved when 70% of the tops had fallen over due to softened necks [41,46,47]. Irrigation was then stopped (July 18—233 DAS/95 days after transplant—DAT) and, after cutting off the roots, the bulbs were left to cure, while the remaining leaves gradually dried [46]. The bulbs were harvested 9 days after termination of irrigation (244 DAS/104 DAT).

2.2. Plant Sampling, Water Use Efficiency, Harvest Indices, Macronutrients and Uptake Concentrations

Twelve plants (leaves, bulbs and roots) were sampled at 0, 31, 62, and 95 DAT, following growth stages. After weighing, samples were dried at 72 °C till constant weight in a Raypa DAF 635 (Spain) oven and ground in a Wiley mill (A 11 basic IKA-Germany, Breisgau) to pass a 40-mesh screen. Water use efficiency (WUE) was expressed as the ratio between plant dry weight and the respective NS volume consumed per growth stage. FP, FB, and DP were the fresh plant, bulb, and dry plant weight to total consumed NS (TCNS) [41,48].

The % plant fresh weight change was calculated based on the plant fresh weight prior to and after curing. Marketable yield (kg m^{-2}) was expressed as the fresh weight of intact bulbs having diameter >70 mm per m^2. Soluble solids were recorded with a hand-held refractometer, (0–90% Brix, 0.5 %, ATC), harvest index (HI) was calculated as the ratio of the dry weight partitioned to the bulb to the total plant dry mass [47]. The % of non-marketable bulbs was calculated as the percentage of the number of bulbs with defects, such as decay, split, mechanical damage, and bulbs with a diameter <70 mm, to the total number of produced bulbs.

Total N was determined by applying the Kjeldahl method [49]. Phosphorus and SO$_4$-S were determined colorimetrically using a Hitachi U2001 spectrophotometer, according to the ammonium molybdate method [50] and turbidimetric method [51], at 460 and 400 nm respectively. Ca, Mg, K, were determined by atomic absorption spectrophotometry (GBC 906A/A) following dry ashing of the plant tissue at 550 °C and extraction with HCl.

The determination of the macronutrient to water uptake ratios, i.e., apparent uptake concentrations (UC), for x = N, P, K, Ca, Mg, and S was based on the amount of x macronutrient recovered from plant tissues and the mean volume of NS per plant that was absorbed by the plants in each circuit of CSCS at the respective growth stage (adapted from Tzerakis et al. [52]), as follows:

$$UC_x = \frac{(C_{xt'} \cdot B_{t'}) - (C_{xt} \cdot B_t)}{V_{up\,(t'-t)}} \qquad (1)$$

where C_x, B, and V_{up} denote the concentrations (mmol g^{-1}) of the x macronutrient (x = N, P, K, Ca, Mg S) in plant dry biomass (g plant^{-1}) and the volume of NS (L plant^{-1}) consumed during the mentioned intervals ($t'-t$ = 31, 62 and 95 DAT).

2.3. Statistics

Statistical analysis was carried out with the Statistica program (StatSoft Inc. Release 12, Tulsa, OK, USA). Means were compared using Duncan's multiple-range test ($p \leq 5\%$). Data are presented in tables including the statistical significance, and in figures drawn using Microsoft® Excel 13. Figures depict means ± SE at the 31st, 59th, and 94th DAT, for growth development, and at the 31st, 62nd, and 95th DAT for fresh weight, % dry matter, mineral concentrations, nutrient solution consumption, water use efficiency, and apparent uptake concentrations of N, P, K, Ca, Mg, and S.

3. Results

3.1. Conditions

The climate conditions and root zone temperature during the experiment are presented in Table 3 and Figure 2, respectively. VPD escalated to 1.61 kPa up to 90 DAT. The mean root zone temperature of the water culture systems climaxed above 32 °C after 76 DAT, while mean maximum root temperature of the same systems exceeded 32 °C, after 20 days, with FL presenting the least fluctuation.

Table 3. Mean hourly values (± standard error) of greenhouse air temperature (Tair, °C), greenhouse air relative humidity (RHair, %), vapor pressure deficit (VPD—kPa) during experimental period (from 15 April 2014 to 29 July 2014) at time intervals (TI) of 15 days from transplant.

TI	Tair	RHair	VPD
1–15	18.70 ± 0.57	68.36 ± 1.36	0.84 ± 0.05
16–30	20.81 ± 0.51	65.13 ± 1.59	1.06 ± 0.09
31–45	22.44 ± 0.63	60.64 ± 1.77	1.31 ± 0.10
46–60	23.59 ± 0.34	63.99 ± 1.77	1.22 ± 0.07
61–75	26.17 ± 0.44	61.90 ± 2.75	1.54 ± 0.16
76–90	26.77 ± 0.22	59.86 ± 1.70	1.61 ± 0.08
91–105	27.18 ± 0.34	65.65 ± 1.57	1.42 ± 0.08

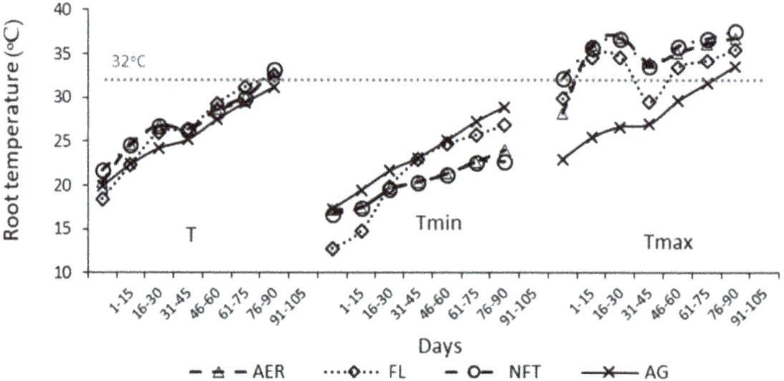

Figure 2. Temperature (°C) in the root zone of the compared CSCS at time intervals (TI) of 15 days from transplant. (I) Mean, (II) Mean minimum, (III) Mean maximum values.

3.2. Vegetative Development

At 31 DAT, AG and FL plants were significantly taller, and exhibited longer leaf midpoint diameter (LLMD) than NFT and AER plants, while the neck diameter was significantly higher in AG and FL than in NFT (Figure 3b,c,e). At the same time, NFT recorded the highest pseudostem lengths but the difference was significant only in comparison with AER and AG. (Figure 3d). At 59 DAT, FL although not differing from AG plants, exceeded NFT plants in terms of pseudostem length and neck diameter (Figure 3d,e). At 94 DAT, AG and FL plants were significantly taller compared to those in the other CSCS, while the neck diameter was significantly larger than in AER and NFT (Figure 3b,e). Leaf number, pseudostem length, and the LLMD of AG and FL plants were higher than in the rest of the compared systems, with NFT plants presenting the leanest leaf (Figure 3a,d,c respectively). As emerged from the weekly recording (Figure 3f), all plants exhibited the maximum values of plant height, LLMD, number of green leaves, and pseudostem diameter at 66 DAT, i.e., 110.75 cm, 18.78 mm, 13.92, and 36.81 mm, respectively.

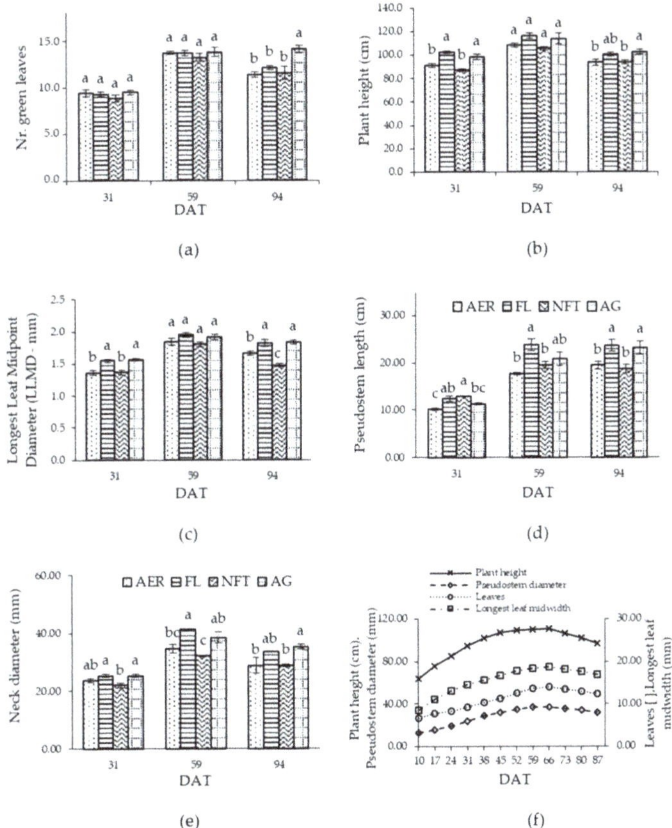

Figure 3. Impact of the CSCS on vegetative development: (**a**) number of green leaves; (**b**) plant height (cm); (**c**) longest leaf midpoint diameter (LLMD); (**d**) pseudostem length (cm); (**e**) neck diameter (mm) at about monthly intervals, i.e., 31, 59, and 94 DAT, (**f**) the average vegetative development of CSCS as indicated by the weekly samplings of plant height (cm), pseudostem width (mm), the number of green leaves, and longest leaf midpoint diameter (mm). Vertical bars indicate ± standard errors of means of ten (10) measurements per replicate. Similar letters indicate non-significant differences at $p \leq 0.05$.

3.3. Bulbing and Plant Features during the Bulbing Phase

According to Figure 4a, 52 DAT, NFT plants showed a significantly higher bulbing ratio and the least shape index in comparison to FL and AG plants. NFT plants entered the bulbing phase ($BR_{NFT} = 2.18 > 2$) at 59 DAT, followed by AER plants ($BR_{AER} = 1.96$). The remainder entered bulbing approximately one week later, i.e., 66 DAT. At that time, bulb diameter, number of green leaves, and shape index were not affected (Figure 4b,c,d). NFT plants showed a higher bulbing ratio than FL and AG plants, while the number of green leaves was significantly lower than that of AG. Additionally, the bulb diameter of FL was greater than in the rest of treatments, while the shape index was not affected (Figure 4a,c,d). The neck diameter of NFT was steadily smaller than that of FL plants throughout the pre- and bulbing period (Figure 4e). For the prevailing conditions, the day length to enter bulbing was more than 14:30 h (Figure 4f). As presented in Figure 5, at the 62nd DAT, plants showed a different root size, and coloration. AER and FL plants developed the largest, and AG plants the smallest, root system. Based on root color (with white being the most functional and brown being the most aged), the root system of AG appeared to be the healthiest, followed by AER and NFT, while the least healthy root system was observed in FL plants. Finally, there was a notable mass of new roots in FL and NFT plants.

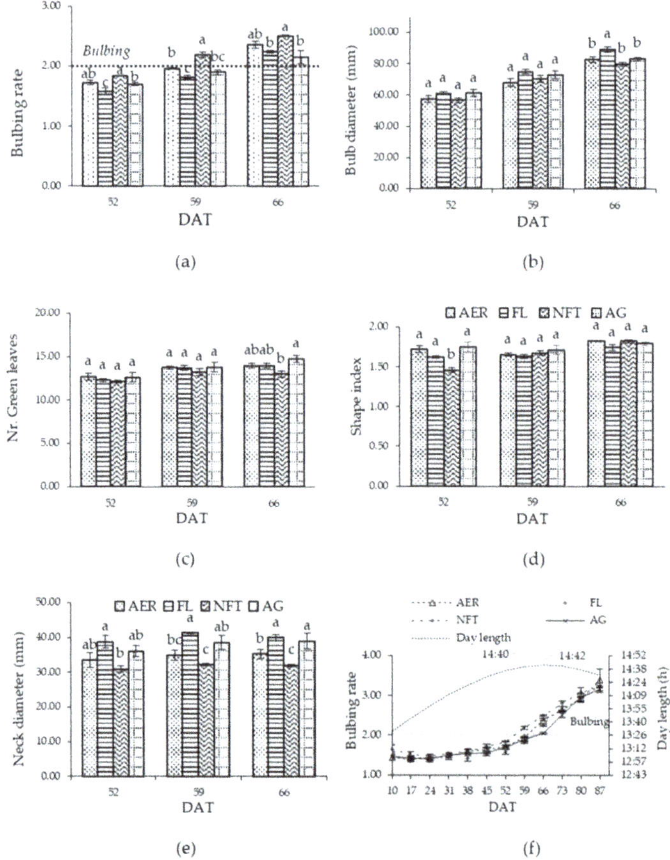

Figure 4. Impact of the CSCS on: (**a**) the bulbing rate described as the quotient of the bulb to minimum sheath diameter ≥ 2; (**b**) the bulb diameter (mm); (**c**) number of green leaves; (**d**) the shape index; (**e**) the neck diameter at the 52nd, 59th, and 66th DAT; (**f**) bulbing rate against time; (days) and day length for the trial location. Vertical bars indicate $\pm$ standard errors of means of ten (10) measurements per replicate. Similar letters indicate non-significant differences at $p \leq 0.05$.

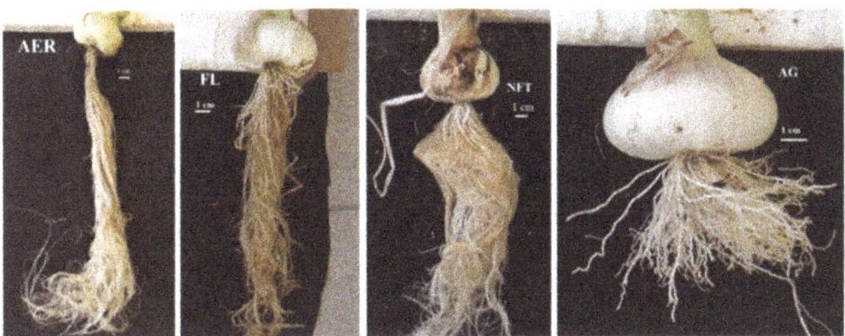

Figure 5. Bulbs and roots of the compared CSCS at 62 DAT (by C.M.).

3.4. Bulb Characteristics

The bulb shape index measured at 31 DAT was significantly higher in AG than in AER and NFT plants, in contrast to sphericity, where AER exceeded FL and AG plants (Figure 6e,f). Thereafter, the bulb shape characteristics were not affected. At the same time, no treatment effect were recorded for the remainder of the bulb characteristics. At 59 DAT, FL bulbs recorded the greatest height without any further effects on the bulbs (Figure 6b). At the end of the experiment (94 DAT), FL and AG plants had formed significantly larger bulbs (diameter, height, surface, and volume) compared to those grown in the AER system, while NFT produced the smallest bulbs (Figure 6a–d).

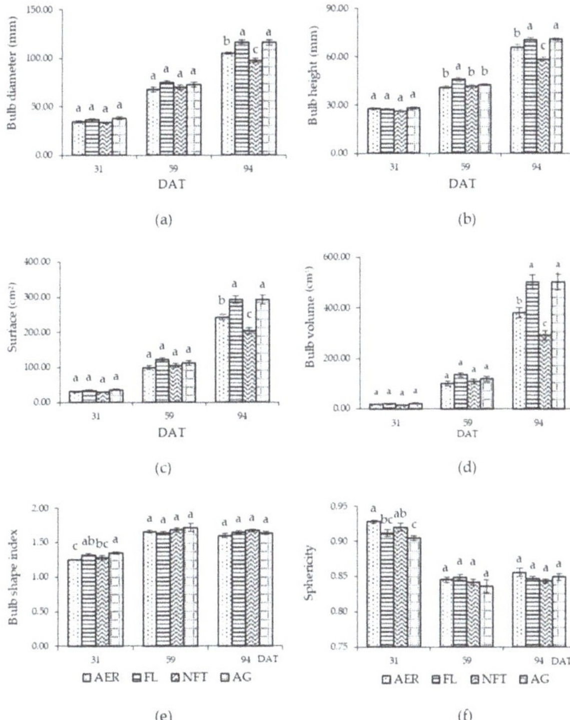

Figure 6. Impact of the CSCS on bulb characteristics: (**a**) the diameter (mm); (**b**) the bulb height (mm); (**c**) the surface of the bulb (cm^2); (**d**) the volume of the bulb (cm^3); (**e**) the bulb shape index and (**f**) the sphericity of the bulb during the 3 samplings (I, II, III), i.e., 31, 59, and 94 DAT. Vertical bars indicate $\pm$ standard errors of means of ten (10) measurements per replicate. Similar letters indicate non-significant differences at $p \leq 0.05$.

3.5. Plant Fresh Weight and % Dry Matter

The fresh weight of FL plants was significantly higher than that measured for the rest of the treatments at 31 DAT. Thereafter, at 62 and 95 DAT, the highest plant fresh weight was recorded in FL and AG plants, while the plants grown in the NFT system recorded the minimum values. Overall, the % dry matter of FL plants was significantly lower than that measured in the other CSCS while, at the 3rd sampling date, NFT outperformed AER and FL plants (Table 4).

Table 4. Impact of CSCS on plant (i) fresh weight (FW—g) and (ii) on % dry matter (% DM) during the 3 samplings (I, II, III), i.e., 31, 62, and 95 days after transplant.

Soilless Culture Systems	FW (g)			DM (%)		
	(I)	**(II)**	**(III)**	**(I)**	**(II)**	**(III)**
AER.	192.76 b	475.25 b	617.91 b	6.04 a	6.92 a	8.22 b
FL.	214.96 a	598.73 a	816.04 a	5.36 b	5.98 b	7.85 c
N.F.T.	177.19 b	343.92 c	523.04 c	6,13 a	6.98 a	8.73 a
AG.	176.19 b	583.75 a	793.99 a	6.17 a	7.06 a	8.44 ab
Statistical Significance						
SYSTEM	**	***	***	**	*	***

Averages (n = 12) where followed by different letters under the same column show statistically significant differences for each factor according to Duncan's multiple range test. *, **, and *** show significant differences at the level of $p \leq 5\%$, 1%, and 0.1% respectively.

3.6. Yield and Quality Features

The % plant fresh weight changed during the curing process and the soluble solids were not affected by the CSCS and averaged 24.5% and 8.22, respectively (Table 5). The marketable yield performance of the four tested CSCS was as follows: FL = AG > AER > NFT. The % dry matter in bulbs was similar for the AG and NFT plants while it was lower in bulbs produced in AER and FL, but the difference was significant only in comparison with AG. The lowest harvest index was recorded in NFT plants but the difference was significant only in comparison with the AER and FL plants. Finally, the highest % of non-marketable bulbs was recorded in the NFT plants, while the AG plants did not produce any non-marketable bulbs.

Table 5. Impact of the CSCS on: (i) the % plant weight change prior and after curing (PWC—%); (ii) the yield (Y—kg m^{-2}); (iii) the bulb % dry matter (% BDM); (iv) the harvest index (HI); (v) the % non-marketable bulbs (%) and (vi) the soluble solids (SS—Brix).

Soilless Culture Systems	PWC (%)	Y (kg m^{-2})	BDM (%)	HI	NMB (%)	SS (Brix)
AER	23.53 a	6.22 b	8.03 b	0.7018 a	5.56 b	8.13 a
FL	25.17 a	7.87 a	7.94 b	0.711 a	5.56 b	8.28 a
NFT	24.49 a	5.20 c	8.16 ab	0.678 b	16.67 a	8.12 a
AG	24.80 a	7.57 a	8.56 a	0.706 ab	0.00 c	8.36 a
Statistical significance						
SYSTEM	N.S.	***	*	*	*	N.S.

Averages (n = 12) where followed by different letters under the same column show statistically significant differences for each factor according to Duncan's multiple range test. N.S., *, and *** show non-significant differences or significant differences at the level of $p \leq 5\%$ and 0.1%, respectively.

3.7. NS Consumption—WUE

The consumption of nutrient solution (Table 6) was significantly less in NFT at 62 DAT, compared to the other three CSCS, while at crop termination no significant difference was found. The WUE of AG plants was significantly greater during the 2nd sampling, while during the 3rd, FL, NFT, and AG did not differ from each other. At harvest, the fresh plant and bulb WUE of AG and the FL plants presented the highest values, while the WUE of dry weight of AG plants was significantly higher than the remainder, followed by FL, and last NFT and AER, without differing from each other.

Table 6. Impact of the CSCS on: (A) the consumption of nutrient solution (CNS—L plant^{-1}) and on the total consumption of nutrient solution (TCNS—L plant^{-1}), B) the water use efficiency (WUE) expressed as the dry weight produced (g) per L of nutrient solution consumed (WUE—g L^{-1}) during the 3 samplings (I, II, III), i.e., 31, 62, and 95 DAT and as the fresh weight of the plant (FP), the bulb (FB) and the dry weight of the plant (DP), (g L^{-1}) at harvest divided by the TCNS.

Soilless Culture Systems	CNS			TCNS	WUE					
	(I)	(II)	(III)	(I)	(I)	(II)	(III)	FP	FB	DP
	L plant^{-1}				g L^{-1}					
AER.	3.57 a	7.23 a	6.93 a	17.73 a	2.96 a	2.95 bc	2.60 b	34.09 b	25.36 b	2.87 c
FL.	3.75 a	7.23 a	7.04 a	18.02 a	2.80 a	3.38 b	4.03 a	44.65 a	31.74 a	3.57 b
N.F.T.	3.13 a	5.58 b	6.38 a	15.09 b	3.15 a	2.36 c	3.39 ab	33.74 b	24.86 b	3.03 c
AG.	3.02 a	7.03 a	6.6 a	16.65 a	3.32 a	4.33 a	3.91 a	46.86 a	32.95 a	4.03 a
Statistical significance SYSTEM	N.S.	***	N.S.	**	N.S.	**	*	***	***	***

Averages (n = 12) where followed by different letters under the same column show statistically significant differences for each factor according to Duncan's multiple range test. N.S., *, **, and *** show non-significant differences or significant differences at $p \leq 5\%$, 1% and 0.1% respectively.

3.8. Macronutrients—Uptake Concentrations

According to Figure 7a, the total-N concentration in the plant tissue recorded a concentration peak at 31 DAT with a decreasing tendency for the rest of the experimental period. Nevertheless, the total-N concentration at the end of the experiment (95 DAT) averaged 39.29 mg g^{-1}, a value larger than the original value of 33.56 mg g^{-1} before the transplant to the CSCS. The CSCS influenced the tissue concentration of total-N at the 2nd and 3rd sampling of the trial with the AG plants presenting the lowest values (40.56 and 35.14, respectively). The UC were affected during the second sampling (62 DAT) with the NFT plants presenting the lowest value (7.56 mmol L^{-1}) (Figure 7b).

The tissue P concentration showed a slight increasing tendency throughout the trial, which was followed by a commensurate increase in the UC of P (Figure 7c,d). P tissue concentrations were not affected by the CSCS during the first 2 samplings (i.e., 31 and 62 DAT). The AG plants showed significantly less P tissue concentration than the AER and NFT plants at the end of the trial (95 DAT). At the same time, AG and FL plants presented the highest UC values (1.01 and 0.98 mmol L^{-1}, respectively).

The CSCS affected the tissue concentration of K. After a sharp increase up to 31 DAT, there was a small decline without falling to the initial levels, with an average of 49.46 mg g-1 at the end of the trial. At 31 DAT, AER plants exhibited significantly larger K concentrations compared to NFT and AG plants (Figure 7e). At 62 DAT, the AG plants exhibited the minimum K tissue concentrations (48.88 mg g^{-1}), while at the end of the experiment, AER plants recorded the highest value compared to all other treatments (53.37 mg g^{-1}). There was a similar decline of the K UC following the decline of the tissue K concentration that averaged 4.26 mmol L^{-1} (Figure 7f). The K UC of the compared CSCS plants showed significant differences during the two last sampling dates. The AG along with the FL plants exceeded AER and NFT plants during the 2nd sampling (i.e., 62 DAT), while the AER plants exhibited the lowest UC at the last sampling date (95 DAT).

The tissue Ca concentrations showed an increasing trend, (Figure 8a) with the AG and FL plants exhibiting a smoother increase and NFT plants recording a decline at the end of the experiment. AER plants recorded the highest value at the 1st and the last sampling dates, without any significant difference compared to NFT plants, which exhibited the highest value during the 2nd sampling, without any significant difference compared to the AER plants. The Ca UCs of AER, FL, and NFT showed a slight increasing pattern during the cropping period, while AG plants seemed to stabilize the Ca UC after 62 DAT (Figure 8b). The Ca UC were influenced by the CSCS only during the 1st period of growing (main growth period); the NFT and AER plants exhibited significantly higher UC values than FL plants.

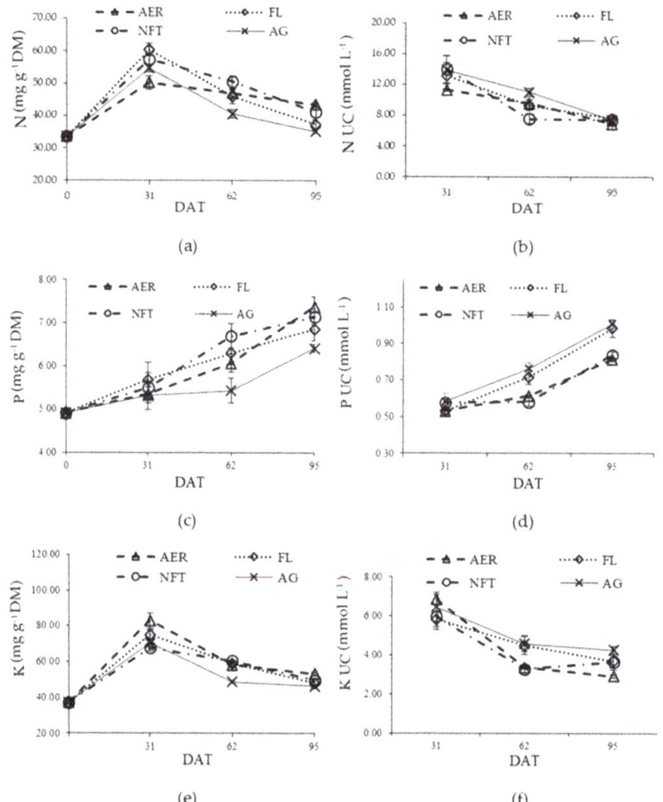

Figure 7. Impact of the different CSCS on the concentrations in the plant dry mass (DM) and on the corresponding apparent uptake concentrations (UC), i.e., mmol of the element uptake per L of water uptake of N, P, and K. Vertical bars indicate ± standard errors of means of twelve (12) samples per replicate.

The tissue Mg concentration presented slight fluctuations during the experiment (Figure 8c). At 31 DAT, there was a small increase in the tissue Mg concentration in AER and NFT plants, while in the rest of the CSCS a decrease was found, with the AG plants recording the lowest value (1.73 mg g^{-1} DW). The concentration of Mg in all plants decreased during the 2nd sampling and increased again at the end of the experiment with the exception of NFT plants, which showed a slight but steady decrease. NFT plants recorded the highest concentrations during the last two samplings. The UC of Mg showed a fluctuation pattern with the highest values recorded during the 1st and the 3rd sampling (0.27 mmol L^{-1} and 0.29 mmol L^{-1}, respectively) (Figure 8d). The lowest values were recorded during the 2nd sampling (i.e., 62 DAT) near the time of bulbing (0.16 mmol L^{-1}).

The tissue S concentration and the UC of S showed a slightly increasing tendency during the experiment (Figure 8e). At 31 DAT, the S tissue concentration of the FL and the AG plants was significantly higher than in the rest of the treatments. There were no further effects recorded during the rest of the experiment. The UC of S was significantly higher in the AG system during the last cropping period (maturation), while it was not influenced by the CSCS at earlier cropping stages (Figure 8f).

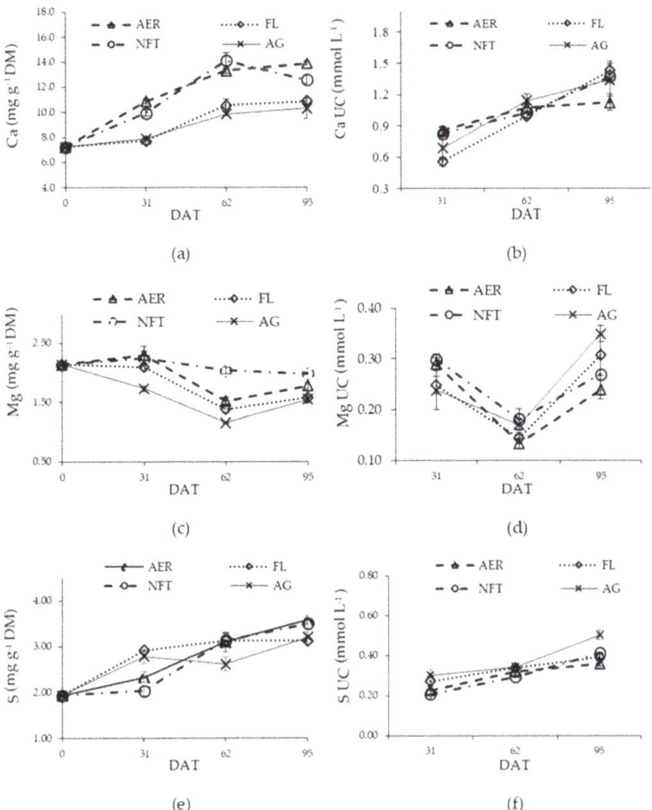

Figure 8. Impact of the different CSCS on the concentrations in the plant dry mass (DM) and on the corresponding apparent uptake concentrations (UC), i.e., mmol of the element uptake per L of water uptake of Ca, Mg, and S. Vertical bars indicate ± standard errors of means of twelve (12) samples per replicate.

4. Discussion

Plant growth of Nerokremmydo of Zakynthos follows the typical growth of an onion plant as described by other researchers [53,54]. Plants of this landrace grown in AG and FL systems appeared to adjust faster to their environment during the first main growth stage (up to 31 DAT). This was reflected by the plant height and LLMD superiority of FL and AG plants (Figure 3b,c). The longer pseudostem of NFT (Figure 3d), resulted in less photosynthetic area, as the larger parts of the leaves are "scaled" next to each other [55]. FL plants developed a rather long pseudostem but this was accompanied by larger leaves. The neck diameter is of primary importance; transverse growth and thickening of the upper part of the sheaths reflect commensurate changes in the transpiration stream. A smaller neck diameter, which was observed in NFT-grown plants, restricts the osmotically active solutes in the outer cells of the sheaths, thus pumping less water and nutrients, which leads to reduced cell swelling [41]. The differences in the shape index and sphericity of early bulbs observed in FL and AG (Figure 6e,f) can be attributed to the higher rate of swelling compared to the rate of elongation of the leaf base, which favored growing in diameter rather than in height. The reasons for the superiority of these two systems in terms of shape index and sphericity are somewhat different. The greater height and shape index in AG plants can be attributed to faster cell division while in FL plants it appears to be a result of water accumulation, as indicated by the larger plant FW and the lower %DM in the FL treatment (Table 4).

Bulbing, i.e., BR > 2 (Figure 4a), was first recorded when the day length exceeded 14.5 h (Figure 4f). According to US standards, [56] the day length needed to bulb classifies onions into short, intermediate, and long-day types when daytime is greater than or equal to 11–12, 13–14, and greater than 14 h, respectively. Brewster [41] in a broader approach, incorporated the 14 h limit within the intermediate category, referring to long-day type when bulbing occurs at >16 h of day light. Taking into account that the ranges serve a rough approach to consider the suitability of a variety for cultivation in a specific climatic zone and season, *Nerokremmydo of Zakynthos* is considered an intermediate bulb onion.

NFT plants started bulbing earlier (Figure 4a), followed by the AER plants (BR = 1.96) which started bulbing one week earlier than in the rest of CSCS (59 DAT); this reflects the definition of BR itself. At 59 DAT, and AER_{BR} and NFT_{BR} approaching or exceeding 2, respectively, there was no significant difference of bulb diameter among the plants of the compared systems. However, the smaller denominator (neck diameter) of NFT and AER (Figure 4e) lead to recording of earlier bulbing. Considering the size of the compared CSCS, and the purpose of the experiment, the choice of monitoring the bulbing rate instead of the "leaf ratio" [41] was the only feasible option. The reasons for the smaller neck diameter should be investigated in the plant root adjustment to the systems architecture and the dominant conditions in the root zone. The NFT system provides little room for root growth, with consequent problems in the uniformity of the NS applied, oxygenation, and temperature [57–59]. It seems that the slight flow obstruction of NS by the growing root may have resulted in a partial impedance of NS supply and a daily partial lack of oxygen [30,60]. Air average temperature after 61 DAT ranged above 25.5 °C (Table 3) and the average duration of such temperatures from bulbing onwards was >12 h day^{-1} (data not shown). Temperatures exceeding 25.5 °C are quite high, leading to onion yield reduction [61]. Additionally, average root ambient temperature followed air temperature, while the mean maximum temperatures exceeded 25.5 °C during the time from 16 to 30 DAT (Figure 2). Based on the maximum and average temperature values, it is concluded that high temperatures did not occur during the whole day (Figure 2), but for systems with low buffer temperature range, such as NFT and secondly AER, these extremes seem to have affected the root growth [62].

According to plant FW, % DM in plant tissues (Table 4), and visual observations (Figure 5), AER favored root growth without a following increase of shoot size and mass; this is confirmed by research results reported by other investigators [63]. The onion has a root system that grows in concentric circles [64], with the older roots confined towards the bottom center of the disk. According to Bosch Serra et. al., [65], the maximum length of the root occurs about 15 days before bulbing, while the appearance of new roots is limited during the growth of the bulb. At 62 DAT, it was visually confirmed that the plants had entered the bulbing stage with roots of different ages and density (Figure 5). AG plants exhibited a typical root system, with few older roots towards the bottom center and some thicker roots which were probably a direct result of the plant response to nutrient supply [66]. In AER and NFT plants, the new roots (white) were already at full length concentrated at the bottom end and entangled with the older roots. Building on the findings of Qiansheng et al. [63] concerning AER drawbacks that are associated with the position of the nebulizers applying the NS, and the time intervals between irrigations, it seems that during root renewal, the older roots of AER and NFT plants may partly impede NS supply with commensurate problems as discussed above. FL plants developed some new white roots, thicker and less branched than the older ones (light brown), a typical characteristic of new onion roots [64]. In FL, the plants adapted better, due to the ample nutrient disposal and the available root space [28]. Any new roots observed after bulbing and until maturation, are probably genotypically and nutritionally controlled; further differences can be explained based on the provision of nutrients and the individual conditions in the compared systems. Since no new leaves were recorded, the role of such new roots was presumably to accumulate assimilates.

The largest increase in the rate of fresh biomass production occurred during the period before bulbing (Table 4). The increase ranged between 48 and 70% for NFT and AG respectively, reflecting the increase in NS consumption (Table 6) and plant biomass accumulation (Table 4). The importance of sufficient water availability before and during bulbing in onion plants has been confirmed experimentally [58,67]. For reasons already discussed, NFT plants had reduced plant biomass and yield performance, suggesting a less efficient transpiration stream, which was confirmed by reduced CNS and commensurate hysteresis in WUE. In contrast, the WUE in AG was highest due to increased plant biomass production. The larger neck in FL and AG plants, in combination with no significant differences in leaf area, presumably point to a thicker leaf morphology at the base of the pseudostem, allowing for better water and nutrient movement. Records of the maximum values of plant height, leaves number, and neck diameter for all the compared CSCS (Figure 3f) coincide with the bulbing date and are in agreement with Brewster [41].

At harvest, based on the number of leaves per plant, the plant height, and the neck diameter, AG plants exhibited a later maturation tendency; nevertheless, 70% of tops fell over due to soft necks. According to Kamenetsky and Rabinowitch [68] the number of leaves has a decisive role in yield success. It is probable that continuing with irrigation in AG would lead to a later harvest but this requires further investigation. In addition to the cumulative effects of water consumption and neck size during the critical period of bulbing, at harvest, AG and FL plants prevailed. It has been reported that plant growth parameters, such as plant height and bulb diameter, correlate positively with bulb yield [47,69], LLMD, pseudostem length, neck diameter (Figure 3c,d,e), bulb diameter, height, surface area, and volume (Figure 6a,b,c,d). Such correlations are consistent with the observed FL and AG superiority to the rest of the compared systems in terms of FW (Table 4) and yield (Table 5).

Moisture concentration in fresh onions is higher than in onions dried for long storage [70]. Additionally, large bulb onions have a higher moisture percentage [71]. Here, the high moisture percentage (>91%) confirm that *Nerokremmydo* should be treated as a fresh mild onion with reduced shelf life [70], implying a relatively short period of availability in local markets. According to our findings (Table 5), bulb %DM averaged 8.17%, which is similar to that of other sweet onions [72,73]. The development of plant % DM (Table 4) shows that its greatest increase occurred after bulbing and up to harvest, in agreement with relevant findings of other researchers [74,75]. The differences in % DM between the CSCS were due to dilution or condensation effects for FL and NFT plants, respectively. Schmautz et. al., [59] have presented similar ranking results (AG > FL %DM) with cherry tomato. At harvest, plant %DM differences followed the sequence NFT $\geq$ AG $\geq$ AER > FL which is not reflective of the bulb measurements; probably due to the inclusion of dry leaves in the calculation of plant %DM. The percentage of DM in bulbs produced in NFT did not differ from that measured in AG plants, probably due to lower TCNS in the former (Table 6). It has been reported that impending water absorption positively affected the dry weight in AER in comparison to AG and soil trial of lemon balm [76]. Temperature differences in the root rhizosphere during bulbing and afterwards, may have also contributed to differences in the DM percentage between bulbs in different treatments.

The yield obtained by cultivating in CSCS surpassed the range of 3.25–3.75 kg m^{-2} reported by local farmers (personal communication) and the national onion yield, i.e., 3.53 kg m^{-2} [77], exceeding them by 38 and 142% for NFT and FL, respectively (Table 5). It is known that yield depends on plant density, variety, and conditions during cultivation. According to Siracusa et al., [3], *Cipolla di Giarratana*, a local onion landrace (Syracuse, Sicily), morphologically similar to *Nerokremmydo of Zakynthos*, yielded 15.1 kg m^{-2}. Considering that the plant density applied in Sicily is ~2.5 times greater than local practice in Zakynthos, and prevailing conditions, the yield difference is reasonable. Trials in Kentucky (U.S.A.) [56] on long day mild onions, *Walla Walla* and *Ailsa Craig*, yielded 3.38 kg m^{-2}. The *Vidalia* onion (GA, USA), a trade name for several short-day mild onion varieties of declared specifications, depending on the year's conditions, reached 7.0 kg m^{-2}. *Jaune des Cévennes* and *Paille des Vertus* (France) yielded 3.3 and 1.3 kg m^{-2}, respectively [73]. FL and AG outperformed the other two CSCS in yield. *Nerokremmydo of Zakynthos* adapted best in two completely different CSCS, supporting the view that system management is of paramount importance for crop success [28]. Water content in FL was the most marked component without compromising the quality as described by non-significant SS (Table 5). The SS averaged 8.22 brix; the results surpassed or lagged behind the mildness of *Cipolla di Giarratana* and *Jaune des Cévennes* (6.73 and 11.9 respectively) [3,73], and exceeded the majority of the white and yellow varieties of mild onion in the U.S.A. [72,78,79].

According to Brewster [41], bulb onions usually have an increased harvest index (HI), with 80% of the shoot being concentrated in the bulb. HI values of *Nerokremmydo* averaged 0.703 (Table 5). It is known that plants with higher water consumption have a lower HI [80]. Similar onion HI values were obtained by Gonçalves et al., [75] and Kassa [81], (0.73 and 0.71, respectively), while in standard cultivars of bulb onions, higher (0.85) [82], or lower values (0.23–0.48) [83] were found. Differences in CSCS may be explained by the combination of different plant FW, size of bulbs, plant size of AG and FL plants, and commensurate bulb dry weight distribution, as a result of better adaptation to the system. According to Schwarz [84], the HI is related to the root:shoot ratio, nutrition, salinity, and climatic conditions, that can induce stress effects, i.e., temperature, light, and radiation.

Bulb onions need curing to restrict humidity losses and to avoid the entrance of pathogens, thereby extending their shelf-life [85,86]. Even in varieties such as *Nerokremmydo of Zakynthos* (Bermuda type) curing is needed [87]. According to our findings (Table 5), % plant moisture reduction (as described by the weight loss of the bulbs) before and after curing, averaged 24.5%, similar to other mild onions [86], for the period traditionally applied for curing. The average % of non-marketable onions (NMB) was 6.96 %, similar to the results of other trials [72,88]. The majority of the NMB of NFT bulbs were under the minimum market limit of 70 mm. Of note is the non-appearance of NMB onions in the AG system.

In the present study, the macronutrients in plant tissues and the uptake concentrations studied were influenced by the CSCS compared and by the plant growth stage. The two, physiologically independent, mechanisms, namely nutrient uptake and water consumption, are determinatively controlled by sunlight, linearly and curvilinearly, respectively [89]. The increasing solar radiation over time, led to the downward trend of N and K UC up to harvest (Figure 7b,f) and Mg UC up to bulbing (Figure 8d) [90]. The downward trend of N concentration in plant tissues is expected as the plant matures [91]. Furthermore, the increased need for nutrients in the early stages of plant growth, which is not accompanied by adequate water absorption due to the size of the root and the leaves, increases uptake concentrations [33,34]. As leaf area increases with time, the demand for nutrients increases. The highest water consumption occurred during the bulbing stage, or shortly thereafter, as has been found by other researchers [38,67]. tissue potassium concentrations exceeded those of all other nutrients within each sampling, throughout the trial. The values and the downward trend of the K concentration agree with the results of Okada et al. [92] in onions grown under controlled conditions (plant factory). According to these authors, the gradual decrease in the tissue concentration of K is due to the increased absorption of water (Table 5), which is required for the swelling of the cells of the thickening sheaths that form the bulb. From the beginning to the end of the trial, P, Ca, and S UC showed a slight upward trend, indicating that the demand for P, Ca, and S during bulb maturation outweighs the corresponding water demand (Figures 7d and 8b,f). Phosphorus exhibited a similar trend in a study concerning the uptake concentrations of sweet pepper [34]. Furthermore, the increasing trend of P absorption in onions has been confirmed by Antoniadis et. al., [93]. Other studies [94,95] have highlighted the increase of enzymes for sucrose metabolism, where P is needed. The initial decrease of the Mg UC before, and its increase after bulbing, may be due to a combination of causes. In addition to the already reported effects of increasing solar radiation over time, and sucrose metabolism, its competition with K (62 DAT) may also have partially affected the reduction in Mg [96,97]. Generally, the C and UC of macronutrients exhibited notable uniformity in changes over time between the compared CSCS.

At bulbing (61 DAT), the difference in nutrient concentration between AG and the water-culture CSCS, was probably due to luxury consumption in the latter. The mean UC of N dropped by 30% of its initial value (31 DAT). The reduced UC of N by NFT plants is explained by the reduced absorption of nutrients and water (Table 5) due to inadequacies in root efficiency, as discussed above. The choice of the NH_4^+: NO_3^- ratio regulates cation absorption, as stated by Abbès et al. [98]. The FL and AG plants showed increased K and P UC values during bulbing. The much larger increase of dry weight, along with the analogous water consumption, support this. Tissue K concentrations ranging around 4.8% measured in AG fall within the sufficiency range suggested by the literature of 3.5–5.5% for sweet onions during the middle of the growing season [72]. Hence any extra uptake of K, while antagonistically interfering with Ca and Mg uptake, could lead to an increase in production costs, without a significant contribution to final production [72],

From maturing and up to harvest, the tested CSCS did not affect the UC of N, Ca, and Mg. At this stage, there is an accumulation of nutrients in the bulb, an evolutionary feature of the bulbs aiming to cope with the adverse conditions of temperature and lack of humidity in summer [41]. The tissue N, P, K, and Ca levels were significantly higher in AER plants compared to those measured in the other three CSCS, due to higher availability of new roots in this system (visual assessment) which continued to appear as the plant adapted to the increased temperatures inside the root growth chamber (Figure 2). The plants in the FL system were also developing new roots, although in terms of tissue nutrient concentrations, they showed an intermediate behavior between AER and NFT, similar to that of AG plants. The P concentration increased to meet the increased metabolic processes of stored bulbs. The higher values of P UC of FL and AG plants, reflect higher growth rates of the bulbs, a process that requires energy consumption, showing similar differences with those recorded in the bulbing phase. The small differences of K UC between the water-culture systems could be explained by the individual needs of K for controlling the open and closure of stomata to balance the varying transpirational stream, especially during the high photosynthetic rates of the day [99]. According to Okada et. al., [92], after commencement of bulbing, a further absorption of K is noticed, in addition to the bulb sinking from leaves. An adequate supply of K has been shown to improve vegetable quality compared to that obtained under K shortage conditions [100,101], especially in onions, and improves shelf life [85]. The UC of S in AG recorded the largest increase between the compared CSCS during the maturation period (Figure 8f). This increase indicates the production of structural proteins, which need S-containing amino acids to be biosynthesized [102].

5. Conclusions

It appears that *Nerokremmydo of Zakynthos*, a Greek intermediate day-length big-bulb variety of mild onion, adapts well in CSCS which eliminate soil problems and nutrient pollution of groundwater, and enhance yield. Water and nutrient availability during bulbing are crucial factors for yield performance. Under the conditions prevailing in the current experiment, FL and AG outperformed the other two systems, followed by AER, while NFT was the least efficient in terms of plant biomass, bulb size, total yield, and WUE. Cultivation in AER and NFT resulted in the highest tissue macronutrient concentrations, probably due to root prevalence, and condensation effects, respectively. UC followed a similar evolution over time in the different CSCS and may be used to establish typical nutrient solution formulations for mild onion cultivation in CSCS. Control of NS temperature and adjustments of CSCS structural components in each system may mitigate the recorded differences.

Author Contributions: Conceptualization, C.M., A.K. and D.S.; Data curation, C.M., A.K. and D.S.; Formal analysis C.M., A.K., G.N. and D.S.; Funding acquisition, A.K., G.N. and D.S.; Investigation, C.M., A.K., G.N. and D.S.; Methodology, C.M., A.K., G.N. and D.S.; Project administration, A.K. and D.S.; Resources, C.M., A.K. and D.S.; Software, C.M., A.K., G.N. and D.S.; Supervision A.K. and D.S.; Validation, C.M., A.K., G.N. and D.S.; Visualization, C.M., A.K., G.N. and D.S.; Writing—original draft, C.M., A.K., G.N. and D.S.; Writing—review & editing, C.M., A.K., G.N. and D.S. All authors have read and agreed to the published version of the manuscript.

Funding: This research received no external funding.

Informed Consent Statement: Not applicable.

Data Availability Statement: The raw data supporting the conclusions of this article will be made available by the authors, without undue reservation.

Conflicts of Interest: The authors declare no conflict of interest.

References

1. FAOSTAT Database. Available online: http://www.fao.org/faostat/en/#data/QC (accessed on 2 March 2021).
2. Dritsas, P. Kremmydi Thespion. Available online: https://www.ecpgr.cgiar.org/in-situ-landraces-best-practice-evidence-based-database/landrace?landraceUid=13272 (accessed on 21 March 2021).
3. Siracusa, L.; Avola, G.; Patanè, C.; Riggi, E.; Ruberto, G. Re-evaluation of traditional mediterranean foods. the local landraces of "cipolla di giarratana" (*Allium Cepa* L.) and long-storage tomato (*Lycopersicon Esculentum* l.): Quality traits and polyphenol content. *J. Sci. Food Agric.* **2013**, *93*, 3512–3519. [CrossRef]
4. Díaz-Pérez, J.C.; Bautista, J.; Gunawan, G.; Bateman, A.; Riner, C.M. Sweet onion (*Allium Cepa* L.) as influenced by organic fertilization rate: 1. plant growth, and leaf and bulb mineral composition. *HortScience* **2018**, *53*, 451–458. [CrossRef]
5. USDA. *United States Standards for Grades of Bermuda-Granex-Grano Type Onions*; USDA: Washington, DC, USA, 2014; p. 11.
6. Cherry, K. Allium Root Rot. Available online: https://projects.ncsu.edu/cals/course/pp728/sclerotium_cepivorum/Sclerotium_cepivorum.html (accessed on 6 March 2021).
7. Frye, J. Pink Root of Onion. Available online: https://projects.ncsu.edu/cals/course/pp728/Phoma/Phoma_terrestris.html (accessed on 6 March 2021).
8. Gruda, N. Do soilless culture systems have an influence on product quality of vegetables? *J. Appl. Bot. Food Qual.* **2009**, *82*, 141–147. [CrossRef]
9. Savvas, D.; Gianquinto, G.; Tuzel, Y.; Gruda, N. Soilless culture. In *Good Agricultural Practices for Greenhouse Vegetable Crops—Principles for Mediterranean Climate Areas*; Baudoin, W., Nono-Womdim, R., Lutaladio, N., Hodder, A., Castilla, N., Leonardi, C., De Pascale, S., Eds.; FAO: Rome, Italy, 2013; pp. 303–354, ISBN 978-92-5-107649-1.
10. Monteiro, J.; Teiel, M.; Baeza, E.; Lopez, J.C.; Karica, M. Greenhouse design and covering materials. In *Good Agricultural Practices for Greenhouse Vegetable Crops—Principles for Mediterranean Climate Areas*; FAO: Rome, Italy, 2013; pp. 35–62, ISBN 978-92-5-107649-1.
11. Son, J.E.; Kim, H.J.; Ahn, T.I. Hydroponic Systems. In *Plant Factory: An Indoor Vertical Farming System for Efficient Quality Food Production*; Kozai, T., Niu, G., Tagaki, M., Eds.; Academic Press: Cambridge, MA, USA, 2016; pp. 213–221, ISBN 978-0-12-801775-3.
12. Lommen, W.J.M. The canon of potato science: 27. Hydroponics. *Potato Res.* **2007**, *50*, 315–318. [CrossRef]
13. Savvas, D.; Gruda, N. Application of soilless culture technologies in the modern greenhouse industry—A review. *Eur. J. Hortic. Sci.* **2018**, *83*, 280–293. [CrossRef]
14. Alshrouf, A. Hydroponics, Aeroponic and Aquaponic as Compared with Conventional Farming. *Am. Sci. Res. J. Eng. Technol. Sci.* **2017**, *27*, 247–255.
15. Lakhiar, I.A.; Gao, J.; Syed, T.N.; Chandio, F.A.; Buttar, N.A. Modern plant cultivation technologies in agriculture under controlled environment: A review on aeroponics. *J. Plant Interact.* **2018**, *13*, 338–352. [CrossRef]
16. Tunio, M.H.; Gao, J.; Shaikh, S.A.; Lakhiar, I.A.; Qureshi, W.A.; Solangi, K.A.; Chandio, F.A. Potato production in aeroponics: An emerging food growing system in sustainable agriculture for food security. *Chil. J. Agric. Res.* **2020**, *80*, 118–132. [CrossRef]
17. Salachas, G.; Savvas, D.; Argyropoulou, K.; Tarantillis, P.A.; Kapotis, G. Yield and nutritional quality of aeroponically cultivated basil as affected by the available root-zone volume. *Emirates J. Food Agric.* **2015**, *27*, 911–918. [CrossRef]
18. Wong, C.E.; Teo, Z.W.N.; Shen, L.; Yu, H. Seeing the lights for leafy greens in indoor vertical farming. *Trends Food Sci. Technol.* **2020**, *106*, 48–63. [CrossRef]
19. Tüzel, Y.; Gül, A.; Tüzel, I.H.; Öztekin, G.B. Different Soilless Culture Systems and Their Management. *J. Agric. Food Environ. Sci.* **2019**, *73*, 7–12.
20. Walters, K.J.; Currey, C.J. Hydroponic greenhouse basil production: Comparing systems and cultivars. *Horttechnology* **2015**, *25*, 645–650. [CrossRef]
21. Kyriacou, M.C.; De Pascale, S.; Kyratzis, A.; Rouphael, Y. Microgreens as a Component of Space Life Support Systems: A Cornucopia of Functional Food. *Front. Plant Sci.* **2017**, *8*, 1–4. [CrossRef]

22. Domingues, D.S.; Takahashi, H.W.; Camara, C.A.P.; Nixdorf, S.L. Automated system developed to control pH and concentration of nutrient solution evaluated in hydroponic lettuce production. *Comput. Electron. Agric.* **2012**, *84*, 53–61. [CrossRef]

23. Hochmuth, R.C.; Cantliffe, D. *Greenhouse Cucumber Production—Florida Greenhouse Vegetable Production Handbook*; University of Florida: Gainesville, FL, USA, 2018; Volume 3.

24. Burrage, S.W. The nutrient film technique (NFT) for crop production in the mediterranean region. *Acta Hortic.* **1999**, *491*, 301–305. [CrossRef]

25. Alsmairat, N.G.; Al-Ajlouni, M.G.; Ayad, J.Y.; Othman, Y.A.; Hilaire, R.S. Composition of soilless substrates affect the physiology and fruit quality of two strawberry (Fragaria × ananassa Duch.) cultivars. *J. Plant Nutr.* **2018**, *41*, 2356–2364. [CrossRef]

26. Bar-Tal, A.; Saha, U.K.; Raviv, M.; Tuller, M. Inorganic and synthetic organic components of soilless culture and potting mixtures. In *Soilless Culture: Theory and Practice Theory and Practice*; Elsevier: Amsterdam, The Netherlands, 2019; pp. 259–301, ISBN 9780444636966.

27. Benton, J.J.J. *Hydroponics A Practical Guide for the Soilless Grower*, 2nd ed.; CRC Press Taylor & Francis Group: Boca Raton, FL, USA, 2004.

28. Rodríguez-Ortega, W.M.; Martínez, V.; Nieves, M.; Simón, I.; Lidón, V.; Fernandez-Zapata, J.C.; Martinez-Nicolas, J.J.; Cámara-Zapata, J.M.; García-Sánchez, F. Agricultural and Physiological Responses of Tomato Plants Grown in Different Soilless Culture Systems with Saline Water under Greenhouse Conditions. *Sci. Rep.* **2019**, *9*, 1–13. [CrossRef]

29. Tabatabaei, S.J. Effects of cultivation systems on the growth, and essential oil content and composition of valerian. *J. Herbs Spices Med. Plants* **2008**, *14*, 54–67. [CrossRef]

30. Blok, C.; Jackson, B.E.; Guo, X.; De Visser, P.H.B.; Marcelis, L.F.M. Maximum plant uptakes for water, nutrients, and oxygen are not always met by irrigation rate and distribution in water-based cultivation systems. *Front. Plant Sci.* **2017**, *8*, 1–15. [CrossRef]

31. Eldridge, B.M.; Manzoni, L.R.; Graham, C.A.; Rodgers, B.; Farmer, J.R.; Dodd, A.N. Getting to the roots of aeroponic indoor farming. *New Phytol.* **2020**, *228*, 1183–1192. [CrossRef]

32. Tzerakis, C.; Savvas, D.; Sigrimis, N. Responses of cucumber grown in recirculating nutrient solution to gradual Mn and Zn accumulation in the root zone owing to excessive supply via the irrigation water. *J. Plant Nutr. Soil Sci.* **2012**, *175*, 125–134. [CrossRef]

33. Neocleous, D.; Savvas, D. Effect of different macronutrient cation ratios on macronutrient and water uptake by melon (*Cucumis melo*) grown in recirculating nutrient solution. *J. Plant Nutr. Soil Sci.* **2015**, *178*, 320–332. [CrossRef]

34. Ropokis, A.; Ntatsi, G.; Kittas, C.; Katsoulas, N.; Savvas, D. Impact of cultivar and grafting on nutrient and water uptake by sweet pepper (*capsicum annuum* l.) grown hydroponically under mediterranean climatic conditions. *Front. Plant Sci.* **2018**, *9*, 1–12. [CrossRef] [PubMed]

35. Ropokis, A.; Ntatsi, G.; Rouphael, Y.; Kotsiras, A.; Kittas, C.; Katsoulas, N.; Savvas, D. Responses of sweet pepper (*Capsicum annum* L.) cultivated in a closed hydroponic system to variable calcium concentrations in the nutrient solution. *J. Sci. Food Agric.* **2021**, *101*, 4342–4349. [CrossRef] [PubMed]

36. Inal, A.; Tarakcioglu, C. Effects of nitrogen forms on growth, nitrate accumulation, membrane permeability, and nitrogen use efficiency of hydroponically grown bunch onion under boron deficiency and toxicity. *J. Plant Nutr.* **2001**, *24*, 1521–1534. [CrossRef]

37. Kane, C.D.; Jasoni, R.L.; Peffley, E.P.; Thompson, L.D.; Green, C.J.; Pare, P.; Tissue, D. Nutrient solution and solution pH influences on onion growth and mineral content. *J. Plant Nutr.* **2006**, *29*, 375–390. [CrossRef]

38. Gamiely, S.; Randle, W.M.; Mills, H.A.; Smittle, D.A.; Banna, G.I. Onion Plant Growth, Bulb Quality, and Water Uptake following Ammonium and Nitrate Nutrition. *HortScience* **1991**, *26*, 1061–1063. [CrossRef]

39. Güneş, A.; Inal, A.; Aktaş, M. Reducing nitrate content of NFT grown winter onion plants (*Allium Cepa* L.) by partial replacement of NO3 with amino acid in nutrient solution. *Sci. Hortic.* **1996**, *65*, 203–208. [CrossRef]

40. De Kreij, C.; Voogt, W.; Baas, R. *Nutrient Solutions and Water Quality for Soilless Cultures*; Applied Plant Research, Division Glasshouse: London, UK, 1999.

41. Brewster, J.L. (Ed.) *Onions and Other Vegetable Alliums, 2nd ed*; Cab International: Wellesbourne, UK, 2008; ISBN 978-1-84593-399-9.

42. Amer Essa, A.H.; Gamea, G.R. Physical and Mechanical Properties of Bulb Onions. *Misr J. Agric. Eng.* **2003**, *20*, 661–676. [CrossRef]

43. Bosekeng, G. Response of Onion (*Allium Cepa* L.) to Swing Date and Plant Population. Ph.D. Thesis, University of the Free State Bloemfontein, Bloemfontein, South Africa, 2012.

44. Mohsenin, N.N. Physical Properties of Plant and Animal Materials: Structure, Physical Characteristics and Mechanical Properties, 2nd ed.Gordon and Breach Science Publishers: London, UK, 1986.

45. Stroshine, R. *Physical Properties of Agricultural Materials and Food Products*; R. Stroshine: West Lafayette, IN, USA, 2004.

46. Nabi, G.; Rab, A.; Sajid, M.; Ullah, F.; Abbas, S.J.; Ali, I. Influence of curing methods and storage conditions on the post-harvest quality of onion bulbs. *Pakistan J. Bot.* **2013**, *45*, 455–460.

47. Abdissa, Y.; Tekalign, T.; Pant, L.M. Growth, bulb yield and quality of onion (*Allium Cepa* L.) as influenced by nitrogen and phosphorus fertilization on vertisol I. growth attributes, biomass production and bulb yield. *African J. Agric. Res.* **2011**, *6*, 3252–3258. [CrossRef]

48. Rouphael, Y.; Colla, G.; Cardarelli, M.; Fanasca, S.; Salerno, A.; Rivera, C.M.; Rea, E.; Karam, F. Water use efficiency of greenhouse summer squash in relation to the method of culture: Soil vs. Soilless. *Acta Hortic.* **2005**, *697*, 81–86. [CrossRef]

49. Horneck, D.; Miller, R. Determination of total nitrogen in plant tissue. In *Methods for Plant Analysis—Tissue Tests—Let Plants Speak*; Kalra, Y.P., Ed.; CRC Press Taylor & Francis Group: Boca Ratoon, FL, USA, 1998; p. 291.

50. Murphy, J.; Riley, J.P. A modified single solution method for the determination of phosphate in natural waters. *Anal. Chim. Acta* **1962**, *27*, 31–36. [CrossRef]

51. Tabatabai, M.A.; Bremner, J.M. A Simple Turbidimetric Method of Determining Total Sulfur in Plant Materials. *Agron. J.* **1970**, *62*, 805–806. [CrossRef]

52. Tzerakis, C.; Savvas, D.; Sigrimis, N.; Mavrogiannopoulos, G. Uptake of Mn and Zn by cucumber grown in closed hydroponic systems as influenced by the Mn and Zn concentrations in the supplied nutrient solution. *HortScience* **2013**, *48*, 373–379. [CrossRef]

53. Addai, I.K.; Anning, D.K. Response of onion (*Allium Cepa* L.) to bulb size at planting and NPK 15:15:15 fertilizer application rate in the Guinea savannah agroecology of Ghana. *J. Agron.* **2015**, *14*, 304–309. [CrossRef]

54. Backes, C.; Bôas, R.L.V.; De Godoy, L.J.G.; Vargas, P.F.; Santos, A.J.M. Determination of growth and nutrient accumulation in bella vista onion. *Rev. Caatinga* **2018**, *31*, 246–254. [CrossRef]

55. Nobel, P.S. Chapter 7—Temperature and Energy Budgets. In *Physicochemical and Environmental Plant Physiology*, 4th ed.; Academic Press: San Diego, CA, USA, 2009; pp. 318–363, ISBN 978-0-12-374143-1.

56. Coolong, T.; Williams, M.A. Overwintering potential of onion in Kentucky. *Horttechnology* **2014**, *24*, 590–596. [CrossRef]

57. Van Os, E.; Gieling, T.H.; Lieth, J.H. Technical equipment in soilless production systems. In *Soilless Culture: Theory and Practice Theory and Practice*; Raviv, M., Lieth, J., Eds.; Elsevier: Amsterdam, The Netherlands, 2008; pp. 157–207.

58. Pérez Ortolá, M.; Knox, J.W. Water Relations and Irrigation Requirements of Onion (*Allium Cepa* L.): A Review of Yield and Quality Impacts. *Exp. Agric.* **2015**, *51*, 210–231. [CrossRef]

59. Schmautz, Z.; Loeu, F.; Liebisch, F.; Graber, A.; Mathis, A.; Bulc, T.G.; Junge, R. Tomato productivity and quality in aquaponics: Comparison of three hydroponic methods. *Water* **2016**, *8*, 533. [CrossRef]

60. Suhl, J.; Oppedijk, B.; Baganz, D.; Kloas, W.; Duijn, B. Van Oxygen consumption in recirculating nutrient film technique in aquaponics. *Sci. Hortic.* **2019**, *255*, 281–291. [CrossRef]

61. Khokhar, K.M. Environmental and genotypic effects on bulb development in onion—A review. *J. Hortic. Sci. Biotechnol.* **2017**, *92*, 448–454. [CrossRef]

62. Papadopoulos, A.P.; Hao, X. Interactions between nutrition and environmental conditions in hydroponics. In *Hydroponic Production of Vegetables and Ornamentals*; Savvas, D., Passam, H.C., Eds.; Embryo Publications: Athens, Greece, 2002; pp. 413–445.

63. Qiansheng, L.; Xiaoqiang, L.; Tang, B.; Mengmeng, G. Growth responses and root characteristics of lettuce grown in Aeroponics, Hydroponics, and Substrate Culture. *Horticulturae* **2018**, *4*, 35. [CrossRef]

64. Rubatzky, V.; Yamaguchi, M. *World Vegetables: Principles, Production, and Nutritive Values*, 2nd ed.; Springer: Dordrecht, The Netherlands, 1997; ISBN 978-1-4615-6015-9.

65. Bosch Serra, A.D.; Torrens, M.B.; Olivé, F.D.; Melines Pagès, M.A. Root growth of three onion cultivars. *Dev. Crop Sci.* **1997**, *25*, 123–133. [CrossRef]

66. Forde, B.; Lorenzo, H. The nutritional control of root development. *Plant Soil* **2001**, *232*, 51–68. [CrossRef]

67. Kadayifci, A.; Tuylu, G.I.; Ucar, Y.; Cakmak, B. Crop water use of onion (*Allium Cepa* L.) in Turkey. *Agric. Water Manag.* **2005**, *72*, 59–68. [CrossRef]

68. Kamenetsky, R.; Rabinowitch, H.D. *Physiology of Domesticated Alliums: Onions, Garlic, Leek, and Minor Crops*, 2nd ed.; Elsevier: Amsterdam, The Netherlands, 2016; Volume 3, ISBN 9780123948083.

69. Nasreen, S.; Haque, M.M.; Hossain, M.A. Nitrogen and Sulphur Fertilization. *Bangladesh J. Agric. Res.* **2007**, *32*, 413–420. [CrossRef]

70. Shultz, S. Commodity of the quarter onions. *J. Agric. Food Inf.* **2010**, *11*, 8–15. [CrossRef]

71. Ko, S.S.; Chang, W.N.; Wang, J.F.; Cherng, S.J.; Shanmugasundaram, S. Storage variability among short-day onion cultivars under high temperature and high relative humidity, and its relationship with disease incidence and bulb characteristics. *J. Am. Soc. Hortic. Sci.* **2002**, *127*, 848–854. [CrossRef]

72. Díaz-Pérez, J.C.; Bautista, J.; Bateman, A.; Gunawati, G.; Riner, C. Sweet onion (*Allium Cepa*) plant growth and bulb yield and quality as affected by potassium and sulfur fertilization rates. *HortScience* **2016**, *51*, 1592–1595. [CrossRef]

73. Pérez, M.R.; Merkt, N.; Zikeli, S.; Zörb, C. Quality aspects in open-pollinated onion varieties from Western Europe. *J. Appl. Bot. Food Qual.* **2018**, *91*, 69–78. [CrossRef]

74. Thangasamy, A. Quantification of Dry-Matter Accumulation and Nutrient Uptake Pattern of Short Day Onion (*Allium Cepa* L.). *Commun. Soil Sci. Plant Anal.* **2016**, *47*, 246–254. [CrossRef]

75. Gonçalves, F.A.R.; De Aquino, P.M.; Duarte, L.O.; De Aquino, R.F.B.A.; Dos Reis, M.R.; De Aquino, L.A. Macronutrient extraction curves of the onion crop. *Semin. Agrar.* **2019**, *40*, 2497–2512. [CrossRef]

76. Son, Y.J.; Park, J.E.; Kim, J.; Yoo, G.; Nho, C.W. The changes in growth parameters, qualities, and chemical constituents of lemon balm (*Melissa officinalis* L.) cultivated in three different hydroponic systems. *Ind. Crops Prod.* **2021**, *163*, 113313. [CrossRef]

77. World's Leading Onion Producing Countries. Available online: https://www.atlasbig.com/en-ca/countries-by-onion-production (accessed on 15 March 2021).

78. Vavrina, C.S.; Smittle, D.A. Evaluating Sweet Onion Cultivars for Sugar Concentrations and Pungency. *HortScience* **1993**, *28*, 804–806. [CrossRef]

79. Russo, V.M. Nutrient content and yield in relation to top breakover in onion developed from greenhouse-grown transplants. *J. Sci. Food Agric.* **2009**, *89*, 815–820. [CrossRef]

80. Cooper, J.; Scherer, H. Nitrogen Fixation. In *Marschner' s Mineral Nutrition of Higher Plants*, 3rd ed.; Marschner, P., Ed.; Academic Press: San Diego, CA, USA, 2012; pp. 389–408.

81. Kassa, A. Evaluation of Yield and Yield Components of Onion (*Allium Cepa* L.) Under Hatseva Condition, Israel AWOKE. *Int. J. Agric. Innov. Res.* **2018**, *7*, 50–58.

82. Kumara, B.R.; Mansur, C.P.; Chander, G.; Wani, S.P.; Alloli, T.B.; Jagadeesh, S.L.; Mesta, R.K.; Satish, D.; Meti, S.; Reddy, S.G. Effect of Potassium Levels, Sources and Time of Application on Storage Life of Onion (*Allium Cepa* L.). *Int. J. Pure Appl. Biosci.* **2018**, *6*, 540–549. [CrossRef]

83. Bettoni, M.M.; Mogor, Á.F.; Pauletti, V.; Goicoechea, N.; Aranjuelo, I.; Garmendia, I. Nutritional quality and yield of onion as affected by different application methods and doses of humic substances. *J. Food Compos. Anal.* **2016**, *51*, 37–44. [CrossRef]

84. Schwarz, M. *Advanced Series in Agricultural Sciences 24 Soilless Culture Management*; McNeal, B.L., Tardieu, F., Van Keulen, H., Van Vleck, D., Eds.; Springer: Berlin/Heidelberg, Germany, 1995; ISBN 978-3-642-79095-9.

85. Petropoulos, S.A.; Ntatsi, G.; Ferreira, I.C.F.R. Long-term storage of onion and the factors that affect its quality: A critical review. *Food Rev. Int.* **2017**, *33*, 62–83. [CrossRef]

86. Maw, B.W.; Mullinix, B.G. Moisture loss of sweet onions during curing. *Postharvest Biol. Technol.* **2005**, *35*, 223–227. [CrossRef]

87. Opara, L. *Onion Post-Harvest Operations*; FAO: Rome, Italy, 2003.

88. El-Desuki, M.; Mahmoud, A.R.; Hafiz, M.M. Response of Onion Plants to Minerals and Bio-fertilizers Application. *Res. J. Agric. Biol. Sci.* **2006**, *2*, 292–298.

89. Sonneveld, C. Composition of nutrient solutions. In *Hydroponic Production of Vegetables and Ornamentals*; Savvas, D., Passam, H., Eds.; Embryo Publications: Athens, Greece, 2002; pp. 179–210.

90. Sonneveld, C.; Voogt, W. *Plant Nutrition of Greenhouse Crops*; Springer: Berlin/Heidelberg, Germany, 2009; Volume 53, ISBN 978-85-781-107-96.

91. Barker, A.V.; Bryson, G.M. Nitrogen. In *Handbook of Plant Nutrition*; Barker, A., Pilbeam, D.J., Eds.; Taylor & Francis Group, LLC: Boca Ratoon, FL, USA, 2007; pp. 22–50.

92. Okada, H.; Abedin, T.; Yamamoto, A.; Hayashi, T.; Hosokawa, M. Production of low-potassium onions based on mineral absorption patterns during growth and development. *Sci. Hortic.* **2020**, *267*, 109252. [CrossRef]

93. Antoniadis, V.; Petropoulos, S.A.; Golia, E.; Koliniati, R. Effect of phosphorus addition on onion plants grown in 13 soils of varying degree of weathering. *J. Plant Nutr.* **2017**, *40*, 2054–2062. [CrossRef]

94. Engels, C.; Kirkby, E.; White, P.J. Mineral Nutrition, Yield and Source–Sink Relationships. In *Marschner's Mineral Nutrition of Higher Plants*; Marschner, P., Ed.; Elsevier: Amsterdam, The Netherlands, 2012; pp. 89–133.

95. Zhang, C.; Zhang, H.; Zhan, Z.; Liu, B.; Chen, Z.; Liang, Y. Transcriptome analysis of sucrose metabolism during bulb swelling and development in onion (Allium Cepa L.). *Front. Plant Sci.* **2016**, *7*, 1–11. [CrossRef] [PubMed]

96. Madakadze, R.M.; Kwaramba, J. Effect of Preharvest Factors on the Quality of Vegetables Produced in the Tropics—Vegetables: Growing Environment and the Quality of Produce. In *Production Practices and Quality Assessment of Food Crops Volume 1 Preharvest Practice*; Dri, R., Jain, S.M., Eds.; Kluwer Academic Publishers: Dordrecht, The Netherlands, 2004; pp. 1–36, ISBN 1-4020-2533-5.

97. Hawkesford, M.; Horst, W.; Kichey, T.; Lambers, H.; Schjoerring, J.; Møller, S.I.; Philip, W. Functions of Macronutrients. In *Marschner' s Mineral Nutrition of Higher Plants*; Marschner, P., Ed.; Elsevier: Amsterdam, The Netherlands, 2012; pp. 135–189.

98. Abbès, C.; Parent, L.E.; Karam, A.; Isfan, D. Effect of NH4+:NO3- ratios on growth and nitrogen uptake by onions. *Plant Soil* **1995**, *171*, 289–296. [CrossRef]

99. Khokhar, K.M. Mineral nutrient management for onion bulb crops—A review. *J. Hortic. Sci. Biotechnol.* **2019**, *94*, 703–717. [CrossRef]

100. Savvas, D.; Ntatsi, G.; Passam, H.C. The European Journal of Plant Science and Biotechnology Plant Nutrition and Physiological Disorders in Greenhouse Grown Tomato, Pepper and Eggplant. *Eur. J. Plant Sci. Biotechnol.* **2008**, *2*, 45–61.

101. Asao, T. (Ed.) *Hydroponics—A Standard Methodology for Plant Biological Researches*; InTech: London, UK, 2012; ISBN 9789535103868.

102. Boyhan, G. Sulfur, its role in onion production and related Alliums. In *Sulfur: A Missing Link between Soils, Crops and Nutrition*; Jez, J., Ed.; American Society of Agronomy; Crop Science Society of America; Soil Science Society of America: Madison, WI, USA, 2008; pp. 183–196.

horticulturae

Article

Efficiency of Reductive Soil Disinfestation Affected by Soil Water Content and Organic Amendment Rate

Rui Zhu [1,2], Xinqi Huang [3,4,5,6], Jinbo Zhang [3,4,5,6], Zucong Cai [3,4,5,6], Xun Li [1,2,6] and Teng Wen [3,4,5,6,*]

1 State Key Laboratory of Soil and Sustainable Agriculture, Institute of Soil Science,
Chinese Academy of Sciences, Nanjing 210008, China; zhurui@issas.ac.cn (R.Z.); xli@issas.ac.cn (X.L.)
2 University of Chinese Academy of Sciences, Beijing 100049, China
3 School of Geography, Nanjing Normal University, Nanjing 210023, China; xqhuang@njnu.edu.cn (X.H.);
zhangjinbo@njnu.edu.cn (J.Z.); zccai@njnu.edu.cn (Z.C.)
4 Jiangsu Engineering Research Center for Soil Utilization & Sustainable Agriculture,
Nanjing Normal University, Nanjing 210023, China
5 Jiangsu Center for Collaborative Innovation in Geographical Information Resource Development
and Application, Nanjing 210023, China
6 Zhongke Clean Soil (Guangzhou) Technology Service Co., Ltd., Guangzhou 510000, China
* Correspondence: wenteng@njnu.edu.cn

Abstract: Reductive Soil Disinfestation (RSD) is a good method which can restore degraded greenhouse soil and effectively inactivate soil-borne pathogens. However, the approach needs to be optimized in order to facilitate its practical application in various regions. In the present work, we investigated the effect of soil water content (60% water holding capacity (WHC), 100% WHC and continuous flooding) and maize straw application rates (0, 5, 10, and 20 g kg^{-1} soil) on the improvement of soil properties and suppression of soil-borne pathogens (*Fusarium oxysporum*, *Pythium* and *Phytophthora*). The results showed that increasing the soil water content and maize straw application rate accelerated the removal of excess sulfate and nitrate in the soil and elevated the soil pH. Elevating the water content and maize straw application rate also produced much more organic acids, which could strongly inhibit soil-borne pathogens. Soil properties were improved significantly after RSD treatment with a maize straw amendment rate of more than 5 g kg^{-1}, regardless of the water content. However, RSD treatments with 60% WHC could not effectively inactivate soil-borne pathogens and even stimulated their growth by increasing the maize application rate. RSD treatments of both 100% WHC and continuous flooding could inactivate soil-borne pathogens and increase the pathogens mortality indicated by cultural cells relatively effectively. The inhibited pathogens were significantly increased with the increasing maize application rate from 5 g kg^{-1} to 10 g kg^{-1}, but were not further increased from 10 g kg^{-1} to 20 g kg^{-1}. A further increased mortality of *F. oxysporum*, indicated by gene copies, was also observed when the soil water content and maize straw application rate were increased. Therefore, RSD treatment with 60% WHC could improve soil properties significantly, whereas irrigation with 100% WHC or continuous flooding was a necessity for effective soil-borne pathogens suppression. Holding 100% WHC and applicating maize straw at 10 g kg^{-1} soil were optimum conditions for RSD field operation to restore degraded greenhouse soil.

Keywords: application rate; nitrate; organic acid; soil-borne pathogens; soil pH; sulfate

Citation: Zhu, R.; Huang, X.; Zhang, J.; Cai, Z.; Li, X.; Wen, T. Efficiency of Reductive Soil Disinfestation Affected by Soil Water Content and Organic Amendment Rate. *Horticulturae* **2021**, *7*, 559. https://doi.org/10.3390/horticulturae7120559

Academic Editor: Giovanni Bubici

Received: 26 October 2021
Accepted: 6 December 2021
Published: 7 December 2021

Publisher's Note: MDPI stays neutral with regard to jurisdictional claims in published maps and institutional affiliations.

1. Introduction

Greenhouse cultivation plays a very important role in vegetable production all over the world. The health status of greenhouse soil directly influences the yield and the quality of the products from greenhouse cultivation [1]. Yet, driven by the economic benefits, farmers usually apply a large amount of chemical fertilizer for intensive vegetable production in greenhouses in China [2,3]. As a result, acidification and salinization widely occur in greenhouse soil, which not only greatly deteriorates the physical and chemical properties of soil, but also induces serious soil-borne disease [3,4]. This problem has become the bottleneck for sustainable development of greenhouse cultivation all over the world, especially in China [5,6].

Reductive soil disinfestation (RSD), also known as biological soil disinfestation (BSD) or anaerobic soil disinfestation (ASD), is an emerging environmentally friendly method and effective alternative to chemical fumigations [7]. This method was first proposed by Shinmura [8] and Blok [9], and was characterized by applying labile carbon to develop microbial-driven anaerobic soil conditions in moist or flooded soil covered with polyethylene mulch [7]. It was proven to control a wide range of soil-borne pathogens and nematodes, as well as improve soil properties and microbial environment [7,10,11].

RSD needs to apply a large amount of organic material to the soil. Various kinds of organic materials have been used for RSD treatment, including rye-grass, straw stalk, wheat bran, syrup, ethanol, animal waste, and so on [12–16]. However, organic material is not always easily accessible; sometimes this entails expensive purchase and delivery costs. In China, crop residues, such as crop straws, which are usually recognized as agricultural production wastes, can be easily obtained in most areas. Their management and recycling were also long-standing problems for sustainable agriculture. Open-field burning is the most common approach to treat crop residues, but this undoubtedly leads to heavy air pollution. For this reason, applying crop straw as an organic amended material not only helps to recycle waste straw, but also reduces the costs of RSD treatment. Previous studies discovered that maize straw performs well as an organic, amended material for RSD, in which application of 5 g maize straw kg^{-1} in saturated soil (100% water holding capacity, WHC) could inactivate more than 90% soil-borne pathogens [17].

RSD also needs large amount of water, but it is not appropriate for flooding in all areas. Soil water holding capacity is affected by soil texture and topography greatly; soil abundant in organic matters can hold more water than sandy soil [18], and it is difficult to implement flooding in hilly areas and uplands. Moreover, water resources are very precious; therefore, it is very necessary to develop a more efficient water-saving method for RSD. Previous studies showed that irrigating soil to 100% WHC and the simultaneous mulching with polyethylene films could create same highly anaerobic environments [18,19] and effectively kill soil-borne pathogens with a maize straw application rate of 5 g kg^{-1} soil, as in flooded soil [11]. However, in view of the cost and practicability, an optimum soil water content and amendment rate have not yet been proposed. Accordingly, the aim of this study was to explore the effects of water content and organic amendment rates of RSD treatment on the suppression of soil-borne pathogens and improvement of soil properties, and then to propose an optimum soil water content and maize straw application rate for RSD field operation.

2. Materials and Methods

2.1. Materials

The soil for the present laboratory experiment was collected from a greenhouse in Baguazhou, Nanjing, China (118.81 E, 32.19 N). The soil was of a clay loamy texture and used as growing substrate of *Artemisia selengensis* for several years. *Fusarium* wilt disease perniciously spread among the field, and the soil was severely degraded in this area. All soil samples collected from this area were sieved through 2 mm sieve before experiment. The maize straw was collected as the organic material from Hongta county, Lilou towns, Bengbu, Anhui, China (117.44 E, 32.87 N). Before mixed into soil, the straw was air-dried, crushed and sieved through 2 mm sieve. The total organic carbon (TOC), total N (TN), easily oxidized carbon (EOC) and C/N of the soil sample and maize straw are shown in Table 1.

Table 1. Total N, total C, easily oxidized carbon (EOC) content and C/N ratio of maize straw and the soil used in this study.

Sample	Total C (g kg^{-1})	Total N (g kg^{-1})	EOC (mg kg^{-1})	C/N
Maize straw	505.1	10.64	1.75	47.47
Soil	23.65	2.68	0.05	8.82

2.2. Experiment Design

Prepared soil samples (250 g soil on air-dry basis), which were thoroughly mixed with maize straw at designed rates, were placed into self-closure bags. The PE bags were 15 cm long, 9 cm wide, and 0.1 mm thick. The soil water content was set at 60% WHC, 100% WHC, and continuous flooding, respectively. The straw amendment rate was set at 0, 5, 10, and 20 g kg^{-1} soil, respectively. Soil samples subjected only to water content modification were referred as CK treatments, while those subjected to both water and maize straw modifications were designed as RSD treatments. Thus, there were 12 treatments which were listed in Table 2. Each treatment set three replicates and all closed bags were incubated randomly at 30 °C. Previous studies showed that organic acids were one of the key factors in the sterilization of RSD [20]. For measuring organic acids in the soil solution during the incubation, a soil solution (about 2 mL) was collected by pre-embedded porous-cup soil solution sampler (DIK-8393; Daiki Rika Kogyo, Saitama, Japan) [21] on days 4, 8, 12, and 15 after incubation. The sampler was completely buried into the mixed soil in each bag for the nondestructive collection of soil solution samples through a syringe with a two-way valve, then by adding water to the objective soil water content [22]. At the end of 15 days of incubation, the physical and chemical properties of soil samples and the populations of *Fusarium oxysporum* spp., *Pythium* spp. and *Phytophthora* spp. in soil samples were determined.

Table 2. Treatments list.

Number	Treatment
1	CK–60% WHC
2	60% WHC + 5 g maize straw kg^{-1} soil
3	60% WHC + 10 g maize straw kg^{-1} soil
4	60% WHC + 20 g maize straw kg^{-1} soil
5	CK–100% WHC
6	100% WHC + 5 g maize straw kg^{-1} soil
7	100% WHC + 10 g maize straw kg^{-1} soil
8	100% WHC + 20 g maize straw kg^{-1} soil
9	CK–continuous flooding
10	Continuous flooding + 5 g maize straw kg^{-1} soil
11	Continuous flooding + 10 g maize straw kg^{-1} soil
12	Continuous flooding + 20 g maize straw kg^{-1} soil

2.3. Soil Analyses

Soil pH was measured in slurry (soil: water = 1:2.5) by a pH electrode (Mettler S220K, Greifensee, Switzerland) [23,24]. Nitrate and ammonium concentrations in soil were extracted by 2 mol L^{-1} KCl solution (soil: water = 1:5) after 250 rpm shaking for 1 h, and then measured by a continuous-flow analyzer (Skalar San++, Breda, Holland) [17]. Soil EC was measured in filtrate (soil: water = 1:5) by a conductivity meter (KangYi Corp., Shanghai, China) after 180 rpm shaking for 20 min [23]. SO$_4$$^{2-}$ concentration in the soil was measured in filtrate (soil: water = 1:5) by ion chromatography (Thermo Dionex ICS 1100, Waltham, MA, USA) [25]. Soil TOC was measured by wet digestion [26]. Soil TN was measured by Macro Kjeldahl method [27]. Soil EOC was measured by the potassium permanganate oxidation method [28]. The maximum water holding capacity of soil was measured as follows: First, the soil was thoroughly mixed with the corresponding amount of maize straw. Second, the soil was thoroughly flooded in a funnel equipped with filter paper for 2 h. Then, free water was leached out for 6–8 h. Until this point, the soil water content was defined as 100% WHC. The amended water holding capacity of soil without maize straw was calculated at the same time. The soil water content was determined by drying the soil at 105 °C to a constant mass.

Organic acids produced during RSD in the soil solution were measured by HPLC (Agilent 1260, Santa Clara, CA, USA) with a modification according to the literature [29]. The column was XDB-C18 (4.6 × 250 mm, Agilen, USA), and the mobile phase was consisted of different gradient dilution of 2.5 mM H_2SO_4 (A) and methanol (B) at a flow rate of 1 mL min^{-1}. The composition of gradients was set as follows: 0–5 min, 95% A plus 5% B; 5–13 min, 95% A plus 5% B; 13–53 min, 85% A plus 15% B; and 53 min stop, 85% A plus 15% B. The wavelength of UV detector was set at 210 nm. The quantification of organic acid was calculated by comparing the time of the peak appearance and the peak area with the standard samples.

2.4. The Population of the Main Soil-Borne Pathogens: Fusarium oxysporum spp., Pythium spp. and Phytophthora spp.

Plate count method was used to count the cultural cells in different soil samples. Ten-times-step dilution was applied when performing the dilution plating procedure. K2 medium [30] was used for the growth of *Fusarium oxysporum* spp. (K_2HPO_4 1 g, $MgSO_4 \cdot 7H_2O$ 0.5 g, KCl 0.5 g, Fe-Na-EDTA 0.01 g, L-Asparagine 2 g, D-galactose 2 g, agar 24 g, distilled water 1000 mL. PCNB 1 g, $Na_2B_4O_7 \cdot 10H_2O$ 0.5 g, Streptomycin 0.12 g, 10% H_3PO_4 for adjusting pH to 3.8–4.0 after sterilization). Maize meal agar was used for the growth of *Pythium* spp. (maize meal agar 20 g, distilled water 1000 mL; Pimarincin 50 mg and Ampicillin 250 mg after sterilization). Oatmeal agar was used for the growth of *Phytophthora* spp. First, by adding 20 g of oatmeal to 900 mL of water, and then boiling the water lightly for 45 min, filtering the solution, adding distilled water and 24 g agar to 1000 mL. Ampicillin (200 mg) and rifampicin (20 mg) was amended after sterilization.

2.5. Extraction of Soil DNA and Quantification of Fusarium oxysporum

Soil DNA was extracted through the Power Soil TM DNA Isolation Kit (MOBIO Laboratories Inc., Carlsbad, CA, USA) according to the kit instructions and [31]. Real-time PCR was performed on the CFX-96 thermocycler (Bio-Rad Laboratories Inc., Hercules, CA, USA). The 20 μL reaction system was designed as follows: 2 μL DNA template, 10 μL SYBR GREEN premix EX Taq (2×, TaKaRa, Dalian, China), 6 μL sterile deionized water, 1 μL primer of F and 1 μL primer of R (ITS1-F (F): 5′-CTTGGTCATTTAGAGGAAGTAA-3′ [32] and AFP308 (R): 5′-CGAATTAACGCGAGTCCCAAC-3′ (10 μmol L^{-1}) [33]). Reaction conditions were designed as follows: Pre-denaturation step at 95 °C for 2 min followed by 40 cycles, including denaturation at 95 °C for 10 s, annealing at 58 °C for 15 s, extended at 75 °C for 20 s. Melt curve profiles were obtained at annealing step in each cycle.

2.6. Statistical Analyses

Differences among treatments were assessed with ANOVA analysis. LSD test and SNK test was applied in SPSS 22.0 (SPSS Inc., Chicago, IL, USA) when ANOVA analysis revealed significant differences at $p < 0.05$. Origin 2016 was used to create diagrams.

3. Results

3.1. Soil pH and EC

The soil pH in all RSD treatments was significantly increased at the end of treatments (Figure 1 and Table 3). The initial soil pH was 4.42. After the RSD treatments, soil pH was slightly increased in CK, and was further increased in RSD treatments. Among treatments with the same maize straw application rates, flooded soil and 100% WHC soil always had higher pH values than 60% WHC soil. Adding more maize straw helped to achieve higher pH values in 100% WHC and flooded treatments. The highest pH (5.34) was observed in the flooded plus 20 g maize straw kg^{-1} soil treatment. In 60% WHC treatments, however, soil pH was not significantly increased with the increasing maize straw application rates from 5 g kg^{-1} to 20 g kg^{-1}. Generally, soil EC had a decreased trend with increasing water content (Figure 1). However, the effects of the water content and straw amendment rate, and their interaction on soil EC, were not significant (Table 3).

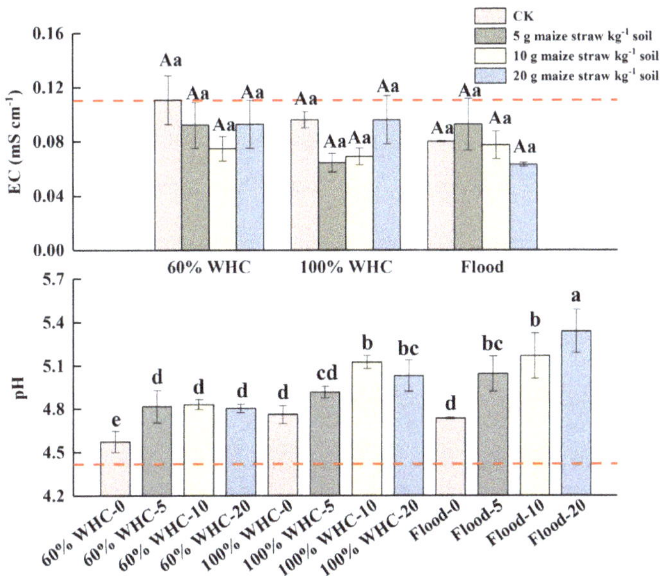

Figure 1. The effects of water content and maize amendment rate on soil EC and pH after RSD treatment. Error bars indicate standard deviations. Different uppercase letters in EC implicate the a significant difference among the same maize straw amended rate under different soil water contents, and lowercase letters within the same water content show significant differences among the various maize straw amended rates (SNK tests, $p < 0.05$). Different lowercase letters show a significant difference among the various treatment of pH (SNK tests, $p < 0.05$). The dashed lines represent the initial value of the original soil.

Table 3. ANOVA analyses of the RSD effects of maize application rates and soil water contents on soil properties and populations of soil-borne pathogens.

Variable	*p*-Value							
	pH	EC	SO_4^{2-}	NH_4^+	NO_3^-	*F. oxysporum*	*Phytophthora*	*Pythium*
Straw rate (S)	<0.001	0.239	<0.001	<0.001	<0.001	<0.001	<0.001	<0.001
Water content (W)	<0.001	0.266	0.365	0.002	0.132	<0.001	<0.001	<0.001
S × W	0.018	0.300	0.217	0.001	0.447	<0.001	<0.001	<0.001

3.2. Soil SO_4^{2-}, NH_4^+ and NO_3^- Contents

Across all RSD treatments, SO_4^{2-} and NO_3^- concentrations in soil were significantly decreased. On the contrary, soil NH_4^+ concentration was increased distinctly in 60% WHC and 100% WHC treatments when compared to CK (Figure 2). The effects of straw amendment on the SO_4^{2-}, NO_3^- and NH_4^+ concentrations were significant, while the significant effect of water content only appeared in soil NH_4^+ (Table 3). In CK treatments, soil SO_4^{2-} and NO_3^- concentrations were effectively reduced ($p < 0.05$) and soil NH_4^+

increased when compared with their initial contents. This trend was much more clear after maize straw was amended. The contents of NH_4^+ were double or triple their initial values, whereas NO_3^- was decreased from 3.36 mg kg^{-1} to almost zero ($p < 0.001$), and SO_4^{2-} was reduced from 302 mg kg^{-1} to 79-128 mg kg^{-1} ($p < 0.001$) in all RSD treatments. Although SO_4^{2-} content shows the lowest in 100% WHC and flood–RSD treatment with the straw amendment rate of 20 g kg^{-1}, no significant differences in NO_3^- and NH_4^+ contents were observed among different straw amendment rates in the same soil water content treatments. On the other hand, despite the fact that NO_3^- and SO_4^{2-} were significantly decreased in RSD treatment, there were no significant differences among different soil water content under the same straw amendment rate. RSD treatment with 60% WHC and 100% WHC showed the same efficiency in lifting soil NH_4^+ content and removing soil NO_3^- content as the continuous flooding treatments.

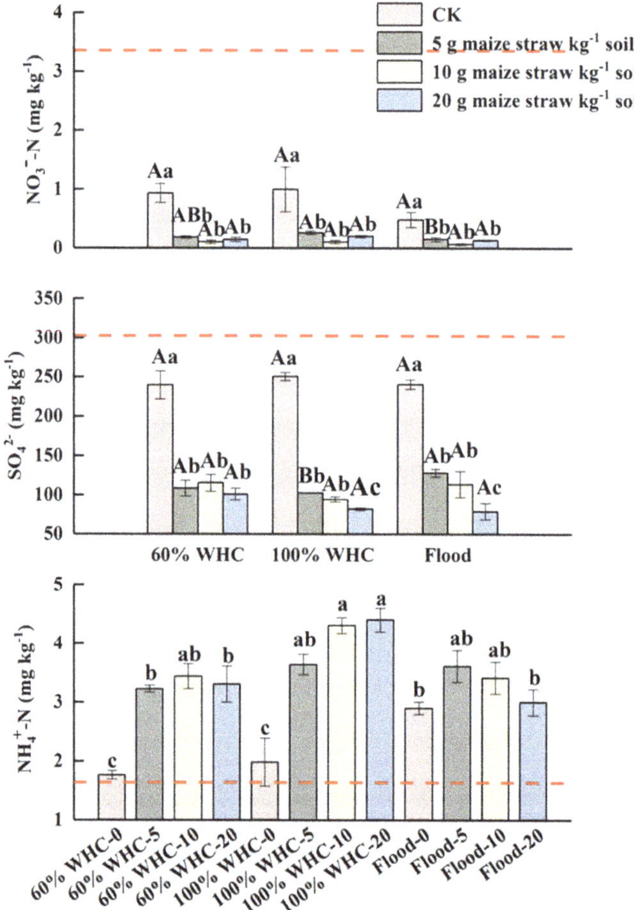

Figure 2. The contents of NO_3^-, SO_4^{2-} and NH_4^+ after RSD treatment. Error bars indicate standard deviations. Different uppercase letters implicate the significant difference among the same maize straw amended rate under the different soil water contents of NO_3^- and SO_4^{2-}, and lowercase letters within the same water content show a significant difference among the various maize straw amended rates (SNK tests, $p < 0.05$). Different lowercase letters show a significant difference among the various treatments of NH_4^+ (SNK tests, $p < 0.05$). The dashed lines represent the initial value of the original soil.

3.3. Dynamics of Organic Acids during the RSD Treatment

A great quantity of organic acids, which were toxic to soil-borne pathogens [34], were produced in all RSD treatments, but none of them were detected in CKs (Figure 3). Acetic acid exhibited the highest concentration among the four detectable acids, followed by propionic acid and butyric acid, while isovaleric acid was the most minor acid and only showed low concentrations in the flooded RSD treatment amended with a high amount of maize straw. The elevating application rate of maize straw helped to produce more organic acids. The concentrations of acetic acid in RSD treatments with maize amendment rate of 20 g kg^{-1} were fivefold to tenfold compared to those with 5 g kg^{-1}. Moreover, organic acids concentrations maintained high levels throughout the treating period in RSD applied with high amended rates of maize straw, but declined to almost zero after 15 days' incubation in those with low amended rates of maize straw.

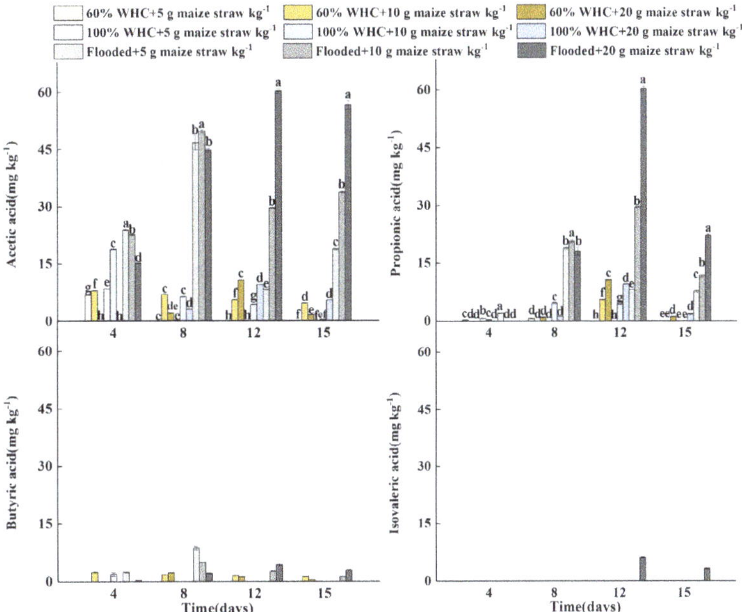

Figure 3. Dynamics of organic acids during the RSD treatments. Error bars indicate standard deviations. Different lowercase letters show significant differences among the various maize straw amended rates and soil water contents. ($p < 0.05$). CKs were not displayed in the figure because no organic acid was detected throughout the incubation period.

3.4. Community of Soil-Borne Pathogens

The populations of soil-borne pathogens (*F. oxysporum* spp., *Phytophthora* spp. and *Phyrium* spp.) were counted through selective cultural mediums. Compared with CK, in which the populations of pathogens were only slightly reduced or even increased from the initial values, RSD treatments by applying maize straw into 100% WHC or flooded soil dramatically killed pathogens ($p < 0.001$, Figure 4). Both water content and straw amendment rate and their interaction significantly affected the populations of studied soil-borne pathogens (Table 3). However, the water content in the soil has a smaller effect on the content of pathogenic microorganisms than the addition of straw to 100% WHC and flooded soil. The highest mortalities of all three pathogens were found in treatments applying with maize straw at 20 g kg^{-1}, which inactivated more than 96% of pathogens (*F. oxysporum* spp., *Phytophthora* spp. and *Phyrium* spp.) in both flooded and 100% WHC soil. Applying maize straw at 5 g kg^{-1} into flooded soils also effectively reduced up to 90% of *F. oxysporum* spp. and *Pythium*, but no statistically significant differences were observed between this treatment and CK in the 100% WHC condition. Increasing the straw applying rate to 10 and 20 g kg^{-1} in both flooded and 100% WHC soil significantly decreased the populations of studied soil-borne pathogens by more than 95%, and the differences between the straw applying rate of 10 and 20 g kg^{-1} were not significant. In 60% WHC soil; however, the highest pathogens mortality was less than 60% even when the highest rate of maize straw (20 g kg^{-1}) was applied. Instead, it seemed that the more maize straw that was used, the less the pathogens were suppressed. In the treatments of 60% WHC + 20 g maize straw kg^{-1} soil, the populations of three pathogens increased twofold, fivefold and onefold, respectively, compared with the CK ($p < 0.05$).

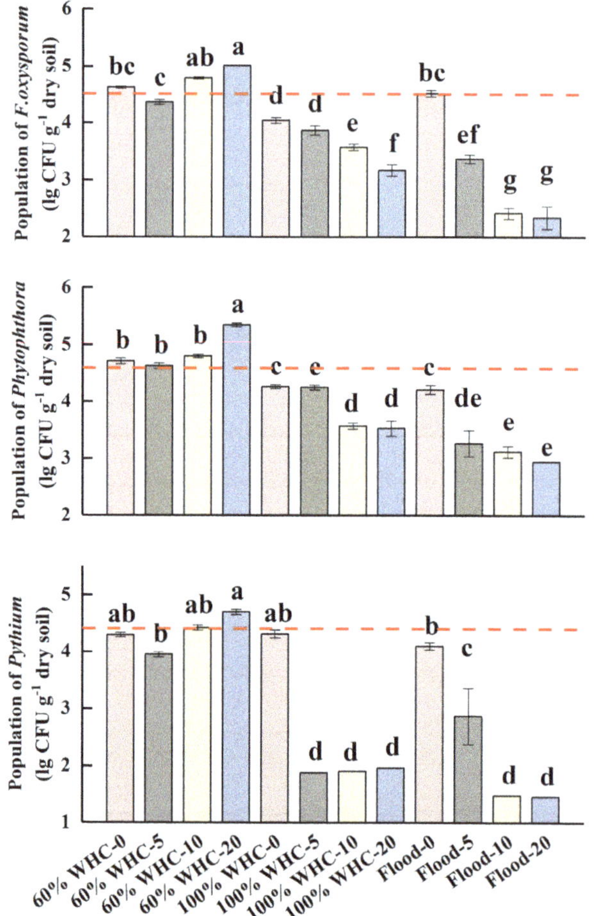

Figure 4. The populations of *F. oxysporum*, *Phytophthora* and *Pythium* in soils after RSD treatment. Error bars indicate standard deviation. Different lowercase letters show significant differences among the various treatment (SNK tests, $p < 0.05$). The dashed lines represent the initial value of the original soil.

Real-time PCR results of *F. oxysporum* were consistent with the plate-count results mentioned above (Figure 5). Applying maize straw at a rate of 5 g kg^{-1} into 100% WHC or flooded soil could reduce the populations of the pathogen beyond 95%, and over 98% of pathogens were killed by enhancing the maize straw application rate to 10 g kg^{-1} or 20 g kg^{-1}. The lowest density of *F. oxysporum* was found in the treatment of flooding coupled with a 20 g maize straw kg^{-1} application rate (5.55 log copies g^{-1} dry soil) while they remained at high levels throughout the experiment (about 7.55~7.82 log copies g^{-1} dry soil) in CKs and 60% WHC RSD treatment, as well as in variants with 100% WHC and flooded soil.

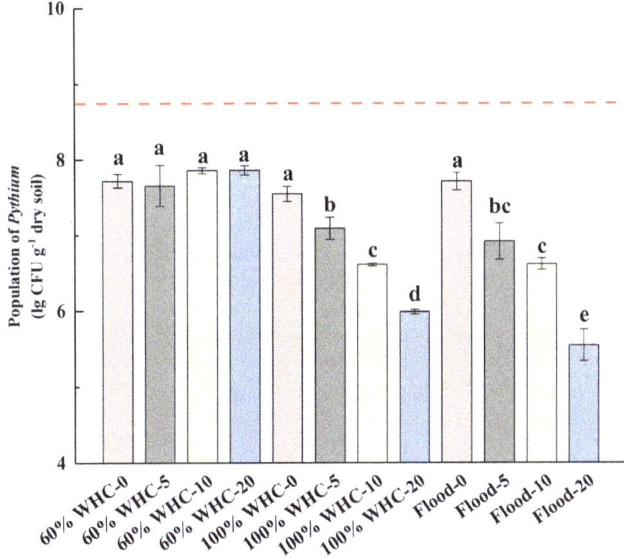

Figure 5. The population of *F. oxysporum* in soils after RSD treatments through real-time PCR. Error bars indicate standard deviations. Different lowercase letters show significant differences among the various treatments (SNK tests, $p < 0.05$). The dashed lines represent the initial value of the original soil.

4. Discussion

Reductive Soil Disinfestation is a highly efficient and environmentally friendly method of remediating the degraded greenhouse soil over a short period [12]. So far, it has been applied in the field by amending organic material into flooded or water-saturated soil (irrigating soil to 100% WHC), and then mulching plastic film to achieve the desired effect [9,19,35]. Maize straw, alfalfa, wheat bran and other easily decomposable agricultural waste could effectively suppress soil-borne pathogens in RSD treatments [23,36,37]. In this study, aiming to promote RSD application in fields where water and straw are difficult and costly to access, we optimized the RSD-treating conditions from different soil water contents (60% WHC, 100% WHC and continuous flooding, respectively) and maize straw application rates (0, 5, 10, and 20 g kg^{-1}, respectively).

Our results showed that all the RSD treatments with different soil water contents and maize straw application rates could effectively remove excessive soil SO_4^{2-} and NO_3^-, but increase NH_4^+ in degraded soil (Table 3). A relatively low original soil EC could be the reason that RSD treatments did not reduce soil EC significantly. The results showed significant interactions between water content and the maize straw application rate on soil pH and NH_4^+ (Table 3). Applying a low amount of maize straw (5 g kg^{-1}) and irrigating to 60% WHC significantly reduced soil SO_4^{2-} and NO_3^- contents and increased NH_4^+ content, but no further changes in these variables were observed when the maize straw application rate was increased from 5 g kg^{-1} to 20 g kg^{-1} (Figures 1 and 2). As we know, RSD creates a strong anaerobic environment in soil where nitrate can transform into N_2O or N_2 mainly through denitrification, and sulfate can be reduced to H_2S or assimilated to organic sulfur by soil microbes [10]. Previous research suggested that inputting a large amount of organic matter into anaerobic soil by flooding or irrigating to 100% WHC could dramatically decrease soil Eh over several days [17]. Here, although 60% WHC could not fulfill the requirement of completely anaerobic soil, we still found that soil SO_4^{2-} and NO_3^- experienced a sharp decline after amending maize straw. Moreover, our previous work revealed that increasing the application rate of maize straw might accelerate the speed of Eh decline, but the final Eh were not significantly different [17]. In other words, the low amendment rate of maize straw in 60% WHC could induce a similar strong reductive environment to reduce nitrate and sulfate contents in the end.

RSD treatment can also remediate severe soil acidification [7]. In this study, soil pH after RSD treatments was markedly increased and displayed a rising trend with the increasing soil water content (Figure 1). Furthermore, there was a significant interaction between the maize application rate and water content regarding soil pH ($p < 0.05$, Table 2). In 60% WHC condition, there were no significant differences of soil pH among the RSD treatments with maize straw application rates of 5, 10 and 20 g kg^{-1}, whereas in flooded RSD treatments, soil pH was increased with the increasing maize straw application rate from 5 g kg^{-1} to 20 g kg^{-1} (Figure 1). Organic amendment decomposition yielded energy and soluble organic carbon, which could promote soil microbe proliferation and accelerate the consumption of soil oxygen simultaneously. Then, the anaerobic soil environment was established. Anaerobic microbes transported electrons to oxidized substances (SO_4^{2-}, NO_3^-, Fe^{2+}, Mn^{2+}) in soil through anaerobic respiration. The reduction process of these oxidized substances was concomitant with the consumption of H^+, which eventually increased soil pH [38]. The results suggested that if soil pH needed to be elevated further, a higher soil water content and larger organic material application rate were required for RSD treatment. However, our previous research compared the 100% WHC and flooded RSD treatments with equal organic material input, and clearly pointed out that their different effects on soil pH were entirely diminished after 25 days of treatment [17]. The implication was that the RSD treatment with 60% WHC could achieve the same high pH levels as 100% WHC or flooded treatment when the incubation time was extended.

RSD needs to apply a large amount of organic materials to soil, where microorganisms are the major agents used to degrade fresh organic matter [39]. Previous studies showed that a huge amount of organic acid was produced by anaerobes in an anaerobic soil environment, especially acetic acid [40], as well as the antagonistic activity of soil microorganisms in the mechanisms of RSD [20]. Our experiment showed the same trend that RSD treatments produced a large amount of organic acid while the proliferation of soil-borne pathogens was suppressed simultaneously after 15 days of incubation. Treatment with organic matter produced acetic acid, propionic acid, butyric acid and isovaleric acid throughout the culture period. Acetic acid was the major organic acid produced by RSD treatment, and was increased with the increasing amended amount of maize straw (from 5 g kg^{-1} soil to 20 g kg^{-1} soil). Continuous flooding treatments with maize straw produced much higher acetic acid and propionic acid in soil solutions than those in 60% WHC and 100% WHC conditions, especially in 20 g maize straw kg^{-1} treatment. Previous studies showed that organic acids in soil produced during RSD incubation could effectively suppress soil-borne pathogens [34,41]. This may be the reason for the better sterilizing effect of flooding coupled with 20 g maize straw kg^{-1} soil treatment.

However, unlike RSD treatments of continuous flooding or 100% WHC, in which more than 90% soil-borne pathogens were suppressed, only 30~58% pathogens were inactivated in RSD treatments with 60% WHC, even lower than that of CK (flooding alone). Applying more maize straw did not increase the mortality of pathogens, but promoted their propagation instead when the water content was set at 60% WHC (Figure 4). This result is not consistent with existing studies on RSD. A previous study showed that the incorporation of organic matter in soil may be utilized by strains of the saprophytic fungus, which led to a growth in soil pathogen population density [42].

All three fungal species tested in this research were mostly saprophytic. This must be why these populations increased at a 60% WHC condition at high organic amendment rates. Generally, RSD sterilize soil-borne pathogens via three methods: (1) creating a strong reductive and completely anaerobic environment in soil to prevent the aerobe reproduction [7]; (2) producing toxic and volatile fatty acids to inactivate soil-borne pathogens during the process of organic materials anaerobic fermentation [43]; (3) inducing compositional shifts of microbes by diverse *Clostridium* sp. becoming the dominant species and improving soil biodiversity [44]; (4) inducing proliferation and dynamic compositional changes in the Firmicutes community, including methyl sulfide compounds, through a metabolic method [45]. The three ways work together to achieve the suppression of soil-borne pathogens in soil [9]. At 60% WHC amended with maize straw, the anaerobic environment was less sufficient than that in 100% WHC or flooded soil. What is more, organic acids contents were kept at very low levels during the whole RSD treatment period and were not dependent on maize straw application rates from 5 g kg^{-1} to 20 g kg^{-1} (Figure 3). This could be one of the most important reasons why RSD treatment with soil water content of 60% WHC could not effectively inactivate soil-borne pathogens. Conversely, when soil water content was set at 60% WHC for RSD treatment, the populations of soil-borne pathogens were even increased when the maize straw application reached 20 g kg^{-1} (Figure 4). This result implied that the trapped soil air at a soil water content of 60% WHC could temporarily support soil-borne pathogens, as aerobes [44], to grow and reproduce, provided that they were rich in carbon and nitrogen [34].

5. Conclusions

RSD is an economic and effective method to restore degraded greenhouse soil. Although applying maize straw into 60% WHC soil is adequate to alleviate soil acidification and salinization, we found that it could not effectively inactivate soil-borne pathogens. Developing a complete anaerobic soil condition by irrigating soil to at least 100% WHC is necessary in order to suppress soil-borne pathogens; therefore, 100% WHC must be held and 10 g kg^{-1} of maize straw must be applied (18 Mg per ha per 15 cm depth in the field).

Soil was recommended for the RSD field operation to both improve soil properties and suppress soil-borne pathogens.

Author Contributions: Conceptualization, Z.C. and J.Z.; methodology, Z.C., T.W. and R.Z.; software, R.Z.; validation, T.W.; formal analysis, R.Z.; resources, Z.C, X.H. and T.W.; data curation, T.W., X.H. and X.L.; writing—original draft preparation, R.Z.; writing—review and editing, T.W. and R.Z.; visualiza-tion, R.Z.; supervision, T.W. and X.H.; project administration, Z.C. and X.H.; funding acquisition, Z.C. and X.H. All authors have read and agreed to the published version of the manuscript.

Funding: This work was financially supported by the Key-Area Research and Development Program of Guangdong Province (2020B0202010006), the National Natural Science Foundation of China (42090065), and the Priority Academic Program Development (PAPD) of Jiangsu Higher Education Institutions.

Institutional Review Board Statement: Not applicable.

Informed Consent Statement: Not applicable.

Conflicts of Interest: The authors declare no conflict of interest.

References

1. Lal, R. Soil management in the developing countries. *Soil Sci.* **2000**, *165*, 57–72. [CrossRef]
2. Ju, X.; Kou, C.; Zhang, F.; Christie, P. Nitrogen balance and groundwater nitrate contamination: Comparison among three intensive cropping systems on the North China Plain. *Environ. Pollut.* **2006**, *143*, 117–125. [CrossRef] [PubMed]

3. Shi, W.M.; Yao, J.; Yan, F. Vegetable cultivation under greenhouse conditions leads to rapid accumulation of nutrients, acidification and salinity of soils and groundwater contamination in South-Eastern China. *Nutr. Cycl. Agroecosyst.* **2009**, *83*, 73–84. [CrossRef]
4. Zhou, D.P.; Chu, C.B.; Liu, F.F.; Fan, J.Q.; Jiang, Z.F.; Wu, S.H. Effect of asparagu's cultivation years on physio-chemical properties, microbial community and enzyme activities in greenhouse soil. *Plant Nutr. Fert. Sci.* **2012**, *2*, 459–466.
5. Li, J.; Pu, L.; Han, M.; Zhu, M.; Zhang, R.; Xiang, Y. Soil salinization research in China: Advances and prospects. *J. Geogr. Sci.* **2014**, *24*, 943–960. [CrossRef]
6. Singh, A. Soil salinization management for sustainable development: A review. *J. Environ. Manag.* **2021**, *277*, 111383. [CrossRef] [PubMed]
7. Momma, N.; Kobara, Y.; Uematsu, S.; Kita, N.; Shinmura, A. Development of biological soil disinfestations in Japan. *Appl. Microbiol. Biotechnol.* **2013**, *97*, 3801–3809. [CrossRef]
8. Shinmura, A.; Sakamoto, N.; Abe, H. Control of Fusarium root rot of Welsh onion by soil reduction. *Jpn. J. Phytopathol* **1999**, *65*, 352–353.
9. Blok, W.J.; Lamers, J.G.; Termorshuizen, A.J.; Bollen, G.J. Control of soilborne plant pathogens by incorporating fresh organic amendments followed by tarping. *Phytopathology* **2000**, *90*, 253–259. [CrossRef] [PubMed]
10. Zhu, T.; Dang, Q.; Zhang, J.; Müller, C.; Cai, Z. Reductive soil disinfestation (RSD) alters gross N transformation rates and reduces NO and N_2O emissions in degraded vegetable soils. *Plant Soil* **2014**, *382*, 269–280. [CrossRef]
11. Wen, T.; Huang, X.; Zhang, J.; Zhu, T.; Meng, L.; Cai, Z. Effects of water regime, crop residues, and application rates on control of Fusarium oxysporum f. sp. cubense. *J. Environ. Sci.* **2015**, *31*, 30–37. [CrossRef] [PubMed]
12. Goud, J.K.C.; Termorshuizen, A.J.; Blok, W.J.; Van Bruggen, A.H. Long-term effect of biological soil disinfestation on Verticillium wilt. *Plant Dis.* **2004**, *88*, 688–694. [CrossRef] [PubMed]
13. Butler, D.M.; Rosskopf, E.N.; Kokalis-Burelle, N.; Albano, J.P.; Muramoto, J.; Shennan, C. Exploring warm-season cover crops as carbon sources for anaerobic soil disinfestation (ASD). *Plant Soil* **2012**, *355*, 149–165. [CrossRef]
14. Kobara, Y.; Uematsu, S.; Tanaka-Miwa, C.; Sato, R.; Sato, M. Possibility of the new soil fumigation technique with ethanol solution. In Proceedings of the 2007 Annual Research Conference on Methyl Bromide Alternatives and Emissions Reduction, San Diego, CA, USA, 29 October–1 November 2007; p. 74.
15. Momma, N.; Usami, T.; Amemiya, Y.; Shishido, M. Factors involved in the suppression of *Fusarium oxysporum* f. sp. *lycopersici* by soil reduction. *Soil Microorg. (Jpn.)* **2005**, *59*, 27–33.
16. Shinmura, A. Principle and effect of soil sterilization method by reducing redox potential of soil. *PSJ Soilborne Dis. Workshop Rep.* **2004**, *22*, 2–12.
17. Wen, T.; Huang, X.; Zhang, J.; Cai, Z. Effects of biological soil disinfestation and water regime on suppressing Artemisia selengensis root rot pathogens. *J. Soils Sediments* **2016**, *16*, 215–225. [CrossRef]
18. Snyder, G.H. Agricultural flooding of organic soils. In *Bulletin/University of Florida. Agricultural Experiment Station (USA)*; University of Florida, Agricultural Experiment Stations: Gainesville, FL, USA, 1987.
19. Momma, N.; Momma, M.; Kobara, Y. Biological soil disinfestation using ethanol: Effect on *Fusarium oxysporum f.* sp. *lycopersici and soil microorganisms. J. Gen. Plant Pathol.* **2010**, *76*, 336–344. [CrossRef]
20. Kubo, C. Analysis of factors involved in sterilization effect by soil reduction. *Jpn. J. Phytopathol.* **2005**, *71*, 281–282. [CrossRef]
21. Matsuoka, K.; Moritsuka, N.; Kusaba, S.; Hiraoka, K. Concentrations of natural stable Cs in organs of blueberry bushes grown in three types of soils treated with acidification or fertilization. *Hortic. J.* **2019**, *88*, 31–40. [CrossRef]
22. Matsuoka, K.; Moritsuka, N.; Kusaba, S.; Hiraoka, K. Effects of soil type and soil treatment on solubilization of 13 elements in the root zone and their absorption by blueberry bushes. *Hortic. J.* **2018**, *87*, 155–165. [CrossRef]
23. Meng, T.; Yang, Y.; Cai, Z.; Ma, Y. The control of Fusarium oxysporum in soil treated with organic material under anaerobic condition is affected by liming and sulfate content. *Biol. Fertil. Soils* **2018**, *54*, 295–307. [CrossRef]
24. Lu, R.K. *Soil and Agro-Chemistry Analysis Methods*; China Agricultural Science and Technology Press: Beijing, China, 2000.
25. Meng, T.; Zhu, T.; Zhang, J.; Cai, Z. Effect of liming on sulfate transformation and sulfur gas emissions in degraded vegetable soil treated by reductive soil disinfestation. *J. Environ. Sci.* **2015**, *36*, 112–120. [CrossRef]
26. Schumacher, B.A. Methods for the determination of total organic carbon (TOC) in soils and sediments. *US Environ Prot Agency.* **2002**, *32*, 25.
27. Bremner, J. Determination of nitrogen in soil by the Kjeldahl method. *J. Agric. Sci.* **1960**, *55*, 11–33. [CrossRef]
28. Blair, G.J.; Lefroy, R.D.B.; Lise, L. Soil carbon fractions based on their degree of oxidation, and the development of a carbon management index for agriculture systems. *Aust. J. Agric. Res.* **1995**, *46*, 1459–1466. [CrossRef]
29. Ling, N.; Huang, Q.; Guo, S.; Shen, Q. Paenibacillus polymyxa SQR-21 systemically affects root exudates of watermelon to decrease the conidial germination of *Fusarium oxysporum f.* sp. *niveum. Plant Soil* **2011**, *341*, 485–493. [CrossRef]
30. Sun, E.; Su, H.; Ko, W. Identification of *Fusarium oxysporum f.* sp. *cubense* race 4 from soil or host tissue by cultural characters [Bananas, wilt]. *Phytopathology (USA)* **1978**, *68*, 1672–1673.
31. Dineen, S.M.; Aranda, R.; Anders, D.L.; Robertson, J.M. An evaluation of commercial DNA extraction kits for the isolation of bacterial spore DNA from soil. *J. Appl. Microbiol.* **2010**, *109*, 1886–1896. [CrossRef] [PubMed]
32. Gardes, M.; Bruns, T.D. ITS primers with enhanced specificity for basidiomycetes-application to the identification of mycorrhizae and rusts. *Mol. Ecol.* **1993**, *2*, 113–118. [CrossRef]

33. Lievens, B.; Brouwer, M.; Vanachter, A.C.; Lévesque, C.A.; Cammue, B.; Thomma, B.P. Quantitative assessment of phytopathogenic fungi in various substrates using a DNA macroarray. *Environ. Microbiol.* **2005**, *7*, 1698–1710. [CrossRef] [PubMed]
34. Huang, X.; Wen, T.; Zhang, J.; Meng, L.; Zhu, T.; Cai, Z. Toxic organic acids produced in biological soil disinfestation mainly caused the suppression of *Fusarium oxysporum f.* sp. cubense. *BioControl* **2015**, *60*, 113–124. [CrossRef]
35. Muramoto, J.; Shennan, C.; Fitzgerald, A.; Koike, S.; Bolda, M.; Daugovish, O.; Rosskopf, E.; Kokalis-Burelle, N.; Butler, D. Effect of anaerobic soil disinfestation on weed seed germination. In Proceedings of the Annual International Research conference on Methyl Bromide Alternatives and Emissions Reductions, Orlando, FL, USA, 11–14 November 2008; Available online: https://mbao.org/prev_year (accessed on 25 October 2021).
36. Huang, X.; Liu, L.; Wen, T.; Zhu, R.; Zhang, J.; Cai, Z. Illumina MiSeq investigations on the changes of microbial community in the Fusarium oxysporum f. sp. cubense infected soil during and after reductive soil disinfestation. *Microbiol. Res.* **2015**, *181*, 33–42. [CrossRef] [PubMed]
37. Momma, N. Biological soil disinfestation (BSD) of soilborne pathogens and its possible mechanisms. *JARQ* **2008**, *42*, 7–12. [CrossRef]
38. Ponnamperuma, F.N. The chemistry of submerged soils. In *The Chemistry of Submerged Soils*; Academic Press: Cambridge, MA, USA, 1972; Volume 24, pp. 29–96.
39. Chandrasekaran, S.; Yoshida, T. Effect of organic acid transformations in submerged soils on growth of the rice plant. *Soil Sci. Plant Nutr.* **1973**, *19*, 39–45. [CrossRef]
40. Na, P.; Kairong, W.; Buresh, R. Effect of rice straw incorporation on concentration of organic acids and available phosphorus in soil under different water regimes. *Acta Pedol. Sin.* **2006**, *43*, 347.
41. Simmons, C.W.; Higgins, B.; Staley, S.; Joh, L.D.; Simmons, B.A.; Singer, S.W.; Stapleton, J.J.; VanderGheynst, J.S. The role of organic matter amendment level on soil heating, organic acid accumulation, and development of bacterial communities in solarized soil. *Appl. Soil Ecol.* **2016**, *106*, 37–46. [CrossRef]
42. Henry, P.M.; Haugland, M.; Lopez, L.; Munji, M.; Watson, D.C.; Gordon, T.R. The potential for Fusarium oxysporum f. sp. fragariae, cause of fusarium wilt of strawberry, to colonize organic matter in soil and persist through anaerobic soil disinfestation. *Plant Pathol.* **2020**, *69*, 1218–1226. [CrossRef]
43. Momma, N.; Yamamoto, K.; Simandi, P.; Shishido, M. Role of organic acids in the mechanisms of biological soil disinfestation (BSD). *J. Gen. Plant Pathol* **2006**, *72*, 247–252. [CrossRef]
44. Mowlick, S.; Hirota, K.; Takehara, T.; Kaku, N.; Ueki, K.; Ueki, A. Development of anaerobic bacterial community consisted of diverse clostridial species during biological soil disinfestation amended with plant biomass. *Soil Sci. Plant Nutr.* **2012**, *58*, 273–287. [CrossRef]
45. Hewavitharana, S.S.; Klarer, E.; Reed, A.J.; Leisso, R.; Poirier, B.; Honaas, L.; Rudell, D.R.; Mazzola, M. Temporal dynamics of the soil metabolome and microbiome during simulated anaerobic soil disinfestation. *Front. Microbiol.* **2019**, *10*, 2365. [CrossRef]

horticulturae

Article

High NH_4^+/NO_3^- Ratio Inhibits the Growth and Nitrogen Uptake of Chinese Kale at the Late Growth Stage by Ammonia Toxicity

Yudan Wang, Xiaoyun Zhang, Houcheng Liu, Guangwen Sun, Shiwei Song * and Riyuan Chen *

College of Horticulture, South China Agricultural University, Guangzhou 510642, China; ydwang@stu.scau.edu.cn (Y.W.); safiyazhang@163.com (X.Z.); liuhch@scau.edu.cn (H.L.); sungw1968@scau.edu.cn (G.S.)

* Correspondence: swsong@scau.edu.cn (S.S.); rychen@scau.edu.cn (R.C.); Tel.: +86-20-85280228 (S.S.); +86-20-85280228 (R.C.)

Abstract: The aim of this study was to determine the effects of various NH_4^+/NO_3^- ratios in a nutrient solution on the growth and nitrogen uptake of Chinese kale under hydroponic conditions. The four NH_4^+/NO_3^- ratios in the nutrient solution were CK (0/100), T1 (10/90), T2 (25/75), and T3 (50/50). An appropriate NH_4^+/NO_3^- ratio (10/90, 25/75) promoted the growth of Chinese kale. T2 produced the highest fresh and dry weight among treatments, and all indices of seedling root growth were the highest under T2. A high NH_4^+/NO_3^- ratio (50/50) promoted the growth of Chinese kale seedlings at the early stage but inhibited growth at the late growth stage. At harvest, the nutrient solution showed acidity. The pH value was the lowest in T3, whereas NH_4^+ and NH_4^+/NO_3^- ratios were the highest, which caused ammonium toxicity. Total N accumulation and N use efficiency were the highest in T2, and total N accumulation was the lowest in T3. Principal component analysis showed that T2 considerably promoted growth and N absorption of Chinese kale, whereas T3 had a remarkable effect on the pH value. These findings suggest that an appropriate increase in NH_4^+ promotes the growth and nutrient uptake of Chinese kale by maintaining the pH value and NH_4^+/NO_3^- ratios of the nutrient solution, whereas excessive addition of NH_4^+ may induce rhizosphere acidification and ammonia toxicity, inhibiting plant growth.

Keywords: growth rate; root morphology; ammonium to nitrate ratio; nutrient solution composition; nitrogen efficiency

Citation: Wang, Y.; Zhang, X.; Liu, H.; Sun, G.; Song, S.; Chen, R. High NH_4^+/NO_3^- Ratio Inhibits the Growth and Nitrogen Uptake of Chinese Kale at the Late Growth Stage by Ammonia Toxicity. *Horticulturae* **2022**, *8*, 8. https://doi.org/10.3390/horticulturae8010008

Academic Editor: Pietro Santamaria

Received: 24 November 2021
Accepted: 20 December 2021
Published: 22 December 2021

1. Introduction

Nitrogen is an essential nutrient in plant growth. It is the main component of many important organic compounds and participates in many physiological and biochemical processes in plants [1]. The forms of nitrogen that plants can absorb and utilize include ammonium (NH_4^+), nitrate (NO_3^-), nitrite (NO_2^-), soluble protein, and free amino acids; higher plants mainly absorb NH_4^+ and NO_3^- [2]. However, the processes of absorption, storage, transportation, and assimilation of the two types of N in plants are very different. The different forms of N affect the growth and development of plants and ultimately affect their yield and quality.

Most crops prefer to absorb NH_4^+ rather than NO_3^- under the condition of hydroponics because a single NO_3^- nutrient consumes a large amount of energy during the reduction process. The absorption of NO_3^- increases the pH value of the solution, which leads to an insufficient supply of iron and other trace elements and decreases the content of chlorophyll, thus affecting the yield and quality [3]. The energy cost of NH_4^+ absorption and assimilation is lower than that of NO_3^- [4]. However, single NH_4^+ nutrition causes many problems, such as ammonium toxicity, blocked leaf expansion, reduced organic acid synthesis, and decreased osmotic regulation [5].

An increasing number of studies have shown that adding a mixture of NH_4^+ and NO_3^- in appropriate proportions is more advantageous for crop growth and development and that NH_4^+ and NO_3^- can interact when both N forms are provided together [4,6]. Compared with the addition of NH_4^+ or NO_3^- alone, appropriate NH_4^+/NO_3^- ratio nutrition can significantly promote plant growth, increase plant biomass, soluble sugar, soluble protein, and vitamin C content, and reduce nitrate content in strawberry [7], mini-Chinese cabbage [8], flowering Chinese cabbage [9], and Chinese kale [10]. In addition, it is well-known that the absorption of NH_4^+ can induce net release of H^+ and acidify the rhizosphere, whereas the absorption of NO_3^- can increase H^+ uptake through the H^+ cotransport system in PM and alkalize the rhizosphere [11,12]. A plant's uptake of different forms of N and its transport and assimilation mechanisms depend on its NH_4^+/NO_3^- ratios. Our recent research further showed that an appropriate NH_4^+/NO_3^- ratio triggers plant growth and nutrient uptake of flowering Chinese cabbage by optimizing pH value in nutrient solution [13].

Chinese kale (*Brassica alboglabra* L. H. Bailey) is an important vegetable in south China. The flower stalk and leaves of Chinese kale are rich in anticarcinogenic compounds and antioxidants, including glucosinolates, carotenoids, vitamin C, and total phenolics [14,15]. Chinese kale best absorbs NO_3^- under hydroponic conditions and easily accumulates nitrate. In our previous study, an appropriate increase of NH_4^+ in nutrient solution enhancement improved the yield and quality of Chinese kale and significantly reduced the nitrate content in its product organs [16]; however, the physiological mechanism of this regulation is still unclear.

Based on previous research, this study further investigates the effects of different NH_4^+/NO_3^- ratios on plant growth, seedling root morphology, nutrient solution composition, and plant nutrient absorption to reveal the physiological mechanism by which different NH_4^+/NO_3^- ratios regulate the growth and N uptake of Chinese kale.

2. Materials and Methods

2.1. Plant Material

The experiment was conducted in a greenhouse at the College of Horticulture, South China Agricultural University from March 2013 to January 2014. The average temperature from colonization to the end of the experiment was 24–30/20–24 °C (day/night). Chinese kale seeds 'lvbao' were provided by Guangzhou Academy of Agriculture Science. Chinese kale seeds were sown in a perlite medium. Seedlings with one developed true leaf and one core were watered with a 1/4 dose of Hoagland nutrient solution every 4 days. After 1 month, three consistent seedlings with three developed leaves and one core were selected and transplanted into hydroponic containers with 5.5 L of nutrient solution. There were 10 replications in each treatment arranged in a randomized complete block design.

2.2. Treatments

In this experiment, four different NH_4^+/NO_3^- ratios were set based on 1/2 dose of Hoagland sloe NO_3^- nutrient solution formula (Table 1): CK, 0/100; T1, 10/90; T2, 25/75; and T3, 50/50. General formula for mineral elements: B, 0.5 mg L^{-1}; Mn, 0.5 mg L^{-1}; Zn, 0.05 mg L^{-1}; Cu, 0.02 mg L^{-1}; Mo, 0.01 mg L^{-1}. Fe was supplied by EDTANaFe at a concentration of 50 mg L^{-1}. Additionally, 0.2 g L^{-1} ampicillin (excellent grade pure) was added to inhibit microbial activity. Chinese kale seedlings were transplanted and watered with a 1/4 dose of Hoagland nutrient solution (adjusted according to Table 1), adding 3/4 dose of mother liquor after 12 days, and adding pure water to the original volume of 5.5 L. Pure water was added to the original volume every 3 days. Electrical conductivity and pH were measured during the experiment. In nutrient solutions with different NH_4^+/NO_3^- ratios, NH_4^+ was supplied by NH_4Cl. KCl or $CaCl_2$ was added to maintain a constant concentration of K^+ and Ca^{2+} among the treatments. All nutrient solutions were aerated for 15 min per hour using a controlled pump.

Table 1. Nutrient solution with different NH_4^+/NO_3^- ratios (Unit: mM).

Treatments	NH_4^+/NO_3^-	KNO_3	$Ca(NO_3)_2 \cdot 4H_2O$	KH_2PO_4	$MgSO_4 \cdot 7H_2O$	$(NH_4)_2SO_4$	K_2SO_4	$CaCl_2$
CK	0/100	2.5	2.5	0.5	1.4	-	-	-
T1	10/90	1.75	2.5	0.5	1.4	0.375	0.375	-
T2	25/75	0.625	2.5	0.5	1.4	0.9375	0.9375	-
T3	50/50	-	1.875	0.5	1.4	1.875	1.25	0.625

The seeds of Chinese kale were sown in a medium containing 0.5% agar. After 5 days, seedlings with a radicle length of 1.5 cm were transplanted into hydroponic containers with 1.1 L of nutrient solution for different NH_4^+/NO_3^- ratio treatments.

2.3. Parameter Measurements

Chinese kale seedlings were harvested when they reached marketable maturity, plant height, and stem diameter, and their fresh and dry weight (determined after 1 h at 120 °C and 48 h at 75 °C in a drying oven) were measured (3 biological replicates per treatment, 12 plants per replicate). The fresh weight of the product organ (flower stalk above the 4th node) is called the economic yield. The fresh and dry weight were measured at 9, 15, 21, and 27 days after treatment to calculate the growth rate. The growth rate was measured by dividing the difference in fresh weight before and after sampling by the number of days.

The seeds of Chinese kale were accelerated to bud on a medium containing 0.5% agar. After 5 days, seedlings with a radicle length of 1.5 cm were selected and treated with different NH_4^+/NO_3^- ratios. After 2 weeks, the root morphological indices were analyzed (3 biological replicates per treatment, 12 plants per replicate). The root samples were stained with 0.16% neutral red solution, scanned using an applied digital scanner (LA2400), and quantitatively analyzed using the WinRHIZO Pro LA2400 software (Regent Instruments, Quebec City, QC, Canada) for total root length, root surface area, root volume, and average root diameter.

Total N content was determined according to the method suggested by Avery and Rhodes [17], and total N content was multiplied by the dry weight of the whole plant to calculate N accumulation. N loss (NL), N loss rate (NLR), mean residence time of N (MRT), N productivity (NP), and N use efficiency (NUE) were calculated as described by Eckstein and Karlsson [18] using the following formulae: $NL = [(NS_{applied} - NS_{remain}) - (N_{harvest} - N_{transplant}) \times n]/n$; $NLR = NL/(NS_{applied} - NS_{remain})$; $MRT = (N_{harvest} - N_{transplant})/[(lnN_{harvest} - lnN_{transplant}) \times (NL/t)]$; $NP = [(W_{harvest} - W_{transplant})/(T_{harvest} - T_{transplant})] \times [(lnN_{harvest} - lnN_{transplant})/(N_{harvest} - N_{transplant})]$; $NUE = NP \times MRT$. In the above formulae, W represents the dry weight of the plant, T represents the sampling time, N represents the amount of N absorbed by seedlings, NS represents the amount of N in nutrient solution, n represents the number of seedlings in each hydroponic bucket, and t represents the number of days in the whole growth period. Total P and K were determined at 660 nm using a spectrophotometer and atomic absorption spectrophotometer [19].

2.4. Data Analysis

The data were analyzed by one-way analysis of variance (ANOVA) using SPSS 19.0. The differences between treatments were compared using the least significant difference (LSD) with a significance level of $p < 0.05$. The tables and figures were created using Excel 2013 and SigmaPlot 11.0, respectively. Multivariate principal component analysis (PCA) was performed using OriginPro 9.0.

3. Results

3.1. Growth and Biomass

The growth of Chinese kale was significantly affected by the different NH_4^+/NO_3^- ratios (Figure 1 and Figure S1). Compared with CK, plant height increased by 18.89% and

24.70% in T1 and T2, respectively (Figure 1A). Similarly, the stem diameter was the highest in T2, increasing by 13.89%, whereas there was no significant difference between CK and T3 (Figure 1B). The economic yield was the highest in T2 and lowest in T3 at the harvest stage of Chinese kale (Figure 1C). To further study the dynamic changes in Chinese kale growth under different NH_4^+/NO_3^- ratios, fresh and dry weights were measured at 9, 15, 21, and 27 days after treatment. T1, T2, and T3 significantly increased the biomass of Chinese kale during the growth stage (0–21 d) compared to the control. However, in contrast to T1 and T2, the fresh weight and dry weight of T3 at harvest time decreased by 17.57% and 12.60%, respectively, compared with CK (Figure 1D,E). This result is consistent with economic yield. At the early stage of Chinese kale growth (0–15 d), the root/shoot ratio was the highest in CK, followed by T1, and the difference between T2 and T3 was not significant, indicating that the treatment of increasing NH_4^+ in the nutrient solution would reduce the dry matter distribution of the roots. At the harvest stage of Chinese kale (27d), the root/shoot ratio of T3 was 1.13, 1.24, and 1.12 times that of CK, T1, and T2, respectively, indicating that the effect of high NH_4^+/NO_3^- ratio treatment on the shoot was greater than that on the root at the late growth stage of Chinese kale (Figure 1F).

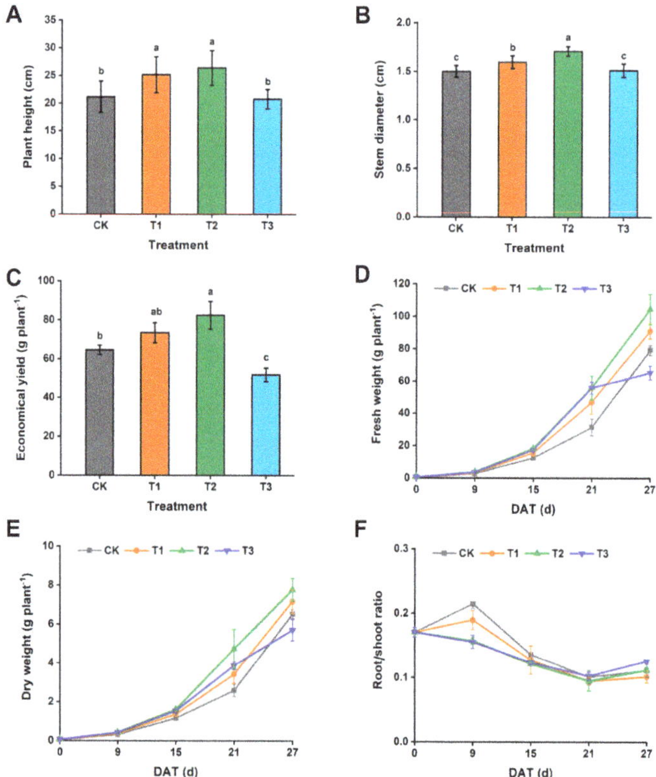

Figure 1. Effect of different NH_4^+/NO_3^- ratios on the growth and biomass of Chinese kale. (**A**) plant height at the harvest stage. (**B**) stem diameter at the harvest stage. (**C**) economical plant yield at the harvest stage. (**D**) dynamic changes in fresh weight. (**E**) dynamic changes in dry weight. (**F**) dynamic changes in the ratio of root to shoot fresh weight. CK = 0/100, T1 = 10/90, T2 = 25/75, T3 = 50/50. DAT, days after treatment. The data represent mean $\pm$ SE (n = 3). Different letters in Figure 1C–F indicate significant differences at $p < 0.05$.

Furthermore, we analyzed the effects of different NH_4^+/NO_3^- ratios on the growth rate of Chinese kale seedlings and found that the growth of Chinese kale was slow in the early growth stage and fast in the late growth stage (Table 2). During the period of 0–9 d after treatment, seedling growth rates were highest in T2 and T3, with no significant difference between them, followed by T1, and were the lowest in the control. During the period of 9–15 d after treatment, the change trend of the four treatments was consistent with that of 0–9 days. During the period of 15–21 d after treating, the growth rates of seedlings reached the maximum in T2 and T3, which were 519.6 mg d^{-1} $plant^{-1}$ and 388.4 mg d^{-1} $plant^{-1}$, respectively. During the period of 21–27 d after treatment, the seedling growth rates reached the maximum in CK and T1, which were 656.1 mg d^{-1} $plant^{-1}$ and mg d^{-1} $plant^{-1}$, respectively. In conclusion, at 0–21 d, the growth rate of seedlings was the highest in T2, followed by T3. However, the difference between them was not significant: at 0–27 d, the growth rate of seedlings was the lowest in T3, which was only 87.45%, 79.47%, and 73.08% that of the control, T1, and T2, respectively, indicating that the growth was significantly inhibited.

Table 2. Effects of different NH_4^+/NO_3^- ratios on the growth rate of Chinese kale (Unit: mg d^{-1} $plant^{-1}$).

Treatments	NH_4^+/NO_3^-	0~9	9~15	15~21	21~27	0~21	0~27
CK	0/100	27.7 ± 2.8 b	141.8 ± 4.9 c	237.5 ± 47.5 c	656.1 ± 17.0 a	120.2 ± 15.3 c	239.0 ± 8.1 bc
T1	10/90	33.7 ± 2.0 ab	167.1 ± 17.5 b	341.3 ± 65.9 b	625.1 ± 27.0 a	159.7 ± 22.2 bc	263.0 ± 22.3 ab
T2	25/75	40.5 ± 3.9a	197.2 ± 4.2 a	519.6 ± 156.0 a	509.0 ± 76.4 b	222.2 ± 47.7 a	286.0 ± 21.4 a
T3	50/50	39.9 ± 5.3 a	186.8 ± 9.3 a	388.4 ± 17.5 ab	304.9 ± 57.7 c	181.4 ± 9.7 ab	209.0 ± 20.4 c

Data are presented as mean ± SE (n = 3). Different letters indicate significant differences at $p < 0.05$.

To study the effects of different NH_4^+/NO_3^- ratios on the root growth of Chinese kale seedlings, we measured the main root length, root total length, root surface area, root volume, and average root diameter (Table 3, Figure S2). The main root length of Chinese kale seedlings in T1 and T2 was slightly higher than that in CK, and that of T3 was slightly lower than that of CK. Compared with CK, the total root lengths of Chinese kale seedlings in T1 and T2 were 1.38 and 1.91 times than those of CK, respectively. Root volumes of Chinese kale seedlings in T1 and T2 were 1.74 and 2.87 times that of CK, respectively. The root surface area and average root diameter of Chinese kale seedlings were maximum in T2. We concluded that the optimal root growth and root indices of Chinese kale seedlings occurred under T2 (NH_4^+/NO_3^- = 25/75).. However, a high NH_4^+/NO_3^- ratio (50/50) may inhibit the main root elongation of Chinese kale seedlings to a certain extent. In addition, root activity showed no significant difference between the control, T1, and T2; however, it decreased by 20.85% in T3 compared with the control. (Figure S3).

Table 3. Effects of different NH_4^+/NO_3^- ratios on root morphological indices of Chinese kale seedlings.

Treatments	NH_4^+/NO_3^-	Main Root Length (cm $plant^{-1}$)	Root Total Length (cm $plant^{-1}$)	Root Surface Area (cm^2 $plant^{-1}$)	Root Volume (cm^3 $plant^{-1}$)	Average Root Diameter (mm)
CK	0/100	6.91 ± 0.77 ab	65.8 ± 11.5 c	13.7 ± 3.2 c	0.23 ± 0.07 cd	0.66 ± 0.04 b
T1	10/90	7.23 ± 0.38 ab	90.8 ± 2.7 b	20.9 ± 5.1 b	0.40 ± 0.19 b	0.73 ± 0.16 ab
T2	25/75	7.72 ± 0.84 a	126.0 ± 8.9 a	31.9 ± 4.2 a	0.66 ± 0.16 a	0.81 ± 0.09 a
T3	50/50	6.63 ± 0.49 b	61.4 ± 12.4 c	14.5 ± 3.7 c	0.28 ± 0.09 bc	0.75 ± 0.07 ab

Data are presented as mean ± SE (n = 3). Different letters indicate significant differences at $p < 0.05$.

3.2. Dynamic Changes in Nutrient Solution Composition

To elucidate the mechanism by which different NH_4^+/NO_3^- ratios affect the growth of Chinese kale, we conducted dynamic monitoring of the physicochemical properties of the nutrient solution and N content (Figure 2). The initial electrical conductivity (EC) value of the nutrient solution increased as the NH_4^+ proportion increased. The EC values were 629, 673, 732, and 879 μs cm^{-1} in the control, T1, T2, and T3, respectively (Figure 2A). With the growth of Chinese kale, both the EC value and ion concentration of the nutrient solution gradually decreased. During the entire growth period of the plant, the decrease in EC value in T2 was the largest, and the decrease in T1 and T3 was also greater than that of the control, but the decrease in EC value in T3 was the lowest during the harvest period. This indicates that an appropriate NH_4^+/NO_3^- ratio could promote the ion absorption of Chinese kale plants, and the NH_4^+/NO_3^- ratio (50/50) could promote ion absorption of plants at the early and middle stages but significantly inhibit ion absorption of plants at the late growth stage.

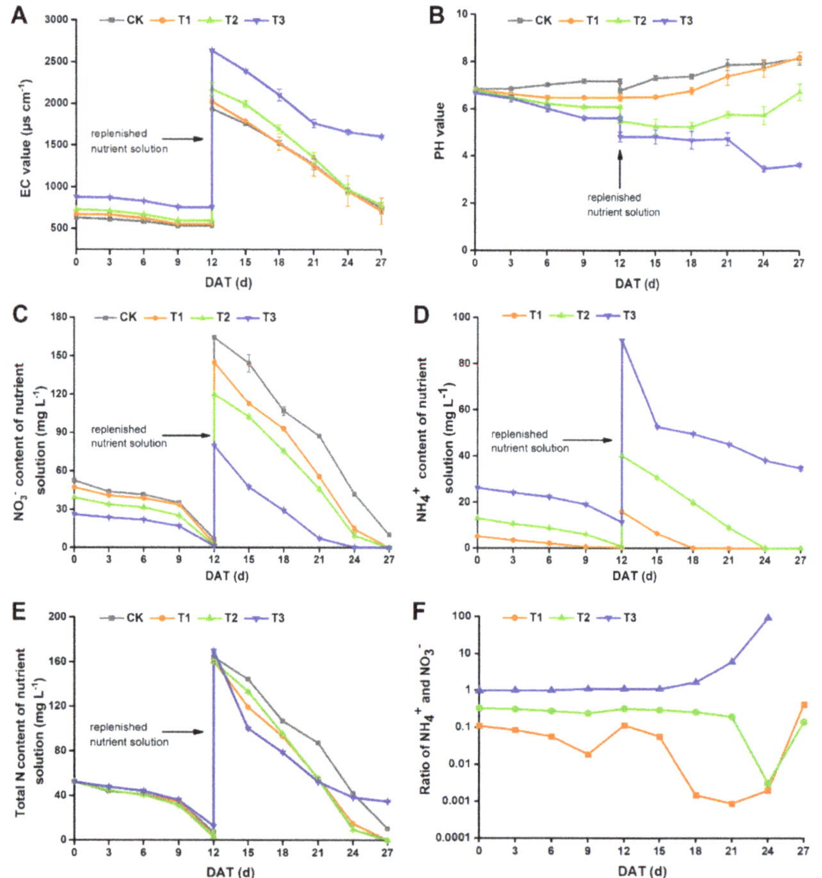

Figure 2. Changes in electrical conductivity value (**A**), pH value (**B**), NO_3^- content (**C**), NH_4^+ content (**D**), total N content (**E**), and NH_4^+/NO_3^- ratios (**F**) in nutrient solution under different NH_4^+/NO_3^- ratios. CK = 0/100, T1 = 10/90, T2 = 25/75, T3 = 50/50. DAT, days after treatment. The data represent mean ± SE (*n* = 3).

The initial pH value of the nutrient solution decreased as the NH_4^+ proportion increased; however, the difference was not significant, both were at approximately 6.75 (Figure 2B). During the entire growth period, the pH value in the CK was mostly weakly alkaline. Before and after the replenishment of the nutrient solution, the average pH of the nutrient solution showed a gradual increase and reached 8.15 at harvest time. The nutrient solution in T1 was weakly acidic at the early stages, and the pH value was stable at approximately 6.5. At 21 d, the pH value was greater than 7.0, the nutrient solution became weakly alkaline, and the pH of the nutrient solution further increased to 8.20 at harvest time. The nutrient solution in T2 was always weakly acidic, and the pH value decreased gradually at the early stage and reached the lowest value at 5.24 at 18 d. The pH value gradually increased and reached 6.73 at harvest time. The nutrient solution in T3 was acidic during the entire growth period, the pH of the nutrient solution decreased gradually, and the pH of the nutrient solution was 3.63 at the harvest stage. In combination with plant biomass, Chinese kale was found to be insensitive to environmental pH and could grow normally in a nutrient solution pH range of 4.73–8.15. An appropriate acidic environment was more conducive to the growth of Chinese kale; however, an over-acidic environment (pH < 4) could significantly inhibit the growth of Chinese kale and reduce the growth rate of plants.

During the plant growth period, the NO_3^- content in the four treatments decreased gradually as NH_4^+ proportion increased (Figure 2C). After 12 d of treatment, the NO_3^- content in the control, T1, T2, and T3 decreased by 45.74, 43.99, 37.38, and 25.12 mg L^{-1}, respectively, compared with that before treatment. After the replenishment of nutrient solution, the NO_3^- content in the control, T1, T2, and T3 was 164.26, 145.01, 120.12 and 79.88 mg L^{-1}, respectively. At harvest, the NO_3^- content in the control, T1, T2, and T3 decreased by 154.00, 145.00, 120.05, and 79.77 mg L^{-1}, respectively, compared with that after the replenishment of nutrient solution. The proportions of NO_3^- absorbed by the four treatments (CK, T1, T2, and T3) to the total NO_3^- were 95.11%, 99.99%, 99.96%, and 99.90%, respectively, indicating that the three treatments (T1, T2, and T3) all promoted NO_3^- uptake by Chinese kale plants, and there was no significant difference between the different treatments. Similarly, with an increase in NH_4^+ proportion, the NH_4^+ content in the three treatments also decreased gradually (Figure 2D). After 12 d of treatment, the NH_4^+ content in T1, T2, and T3 decreased by 4.94, 12.26, and 14.74 mg L^{-1}, respectively, compared with that before treatment. After nutrient solutions were added, the NH_4^+ content in T1, T2, and T3 were 16.06, 40.24, and 90.26 mg L^{-1}, respectively. After 3 d, they decreased by 9.53, 9.52, and 37.55 mg L^{-1}, respectively. Thereafter, they slowly decreased to 0.03, 0.01, and 34.77 mg L^{-1}, respectively, at harvest. The proportions of NH_4^+ absorbed by the three treatments to the total NH_4^+ were 99.86%, 99.98%, and 66.89%, respectively, indicating that low and medium NH_4^+/NO_3^- ratios could significantly promote the absorption of NH_4^+ by Chinese kale compared with the high NH_4^+/NO_3^- ratio.

With the growth of the plants, the total N content in the four treatments decreased gradually with an increase in NH_4^+ proportion (Figure 2E). At harvest, the total N content in the control, T1, T2, and T3 was decreased by 154.00, 161.03, 160.28, and 135.26 mg L^{-1}, respectively, compared with that after the replenishment of nutrient solution. The greatest reduction in total N content was observed in T1 and T2, and the final content at harvest was 0.04 and 0.08 mg L^-, respectively. This indicates that a low NH_4^+/NO_3^- ratio (10/90) significantly promoted the absorption of total N by Chinese kale, whereas a high NH_4^+/NO_3^- ratio inhibited the absorption of total N. In addition, the three treatments (T1, T2, and T3) promoted the absorption of *p* by Chinese kale, whereas the high NH_4^+/NO_3^- ratio significantly inhibited the absorption of K by Chinese kale at the late growth stage (Figure S4).

We further analyzed the NH_4^+/NO_3^- ratio in the four treatments (Figure 2F). Before and after the replenishment of the nutrient solution, the NH_4^+/NO_3^- ratio in T1 showed a decreasing and then an increasing trend. It decreased to a minimum of 0.001 at 21 d and increased to a maximum of 0.445 at the harvest stage. The NH_4^+/NO_3^- ratio in T2 remained between 0.243 and 0.435, which decreased significantly to 0.003 at 24 d of treatment and then increased to 0.214 at the harvest stage. The NH_4^+/NO_3^- ratio in T3 gradually increased during plant growth, and it increased rapidly after the replenishment of nutrient solution and reached a maximum of 95.144 at the harvest stage.

3.3. N Content and N Use Efficiency

The total N content of Chinese kale first increased and then decreased during the growth period (Figure 3A). The N content of Chinese kale increased rapidly from 0 d to 9 d of growth, and T2 and T3 were significantly higher than those of the control and T1. The N content of Chinese kale increased slowly from 9 d to 15 d of growth, and T3 was higher than that of the other three treatments. Except for the control, the N content of Chinese kale plants decreased significantly from 15 to 21 d of growth. The N content in T3 increased by 5.40%, 5.40%, and 6.93% compared with CK, T1, and T2, respectively, at 27 d. In addition to total N content, seedling dry weight also had a significant effect on total N accumulation. Total N accumulation in Chinese kale increased continuously during the growth period (Figure 3B). In the middle and late periods of plant growth, the total N accumulation was the highest in T2, and the total N accumulation of control plants was the lowest in the period from 0 d to 25 d of growth, whereas the total N accumulation of T3 plants was the lowest at 27 d, which was reduced by 7.88%, 16.13%, and 21.62% compared with CK, T1, and T2, respectively.

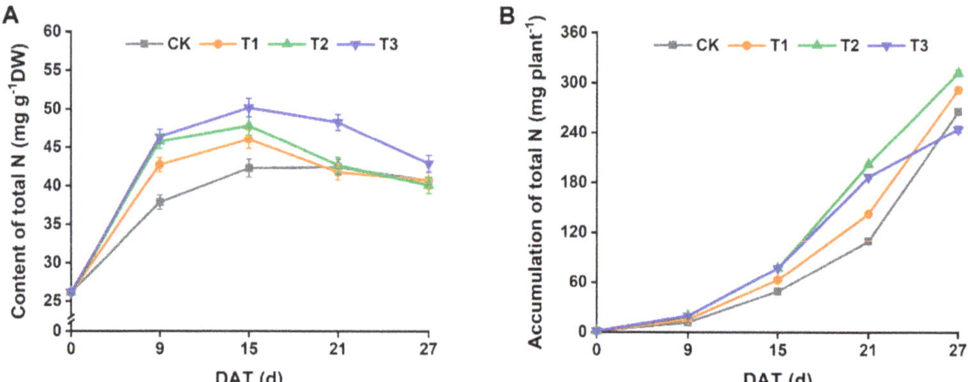

Figure 3. The content (**A**) and accumulation (**B**) of total N affected by different NH_4^+/NO_3^- ratios during the growth period of Chinese kale. CK = 0/100, T1 = 10/90, T2 = 25/75, T3 = 50/50. DW, dry weight; DAT, days after treatment. The data represent mean $\pm$ SE ($n = 3$).

To further study the N utilization of Chinese kale plants in response to different NH_4^+/NO_3^- ratios, we analyzed the N loss, N productivity, N residence time, and N use efficiency (Table 4). We found that there was a certain amount of N loss in Chinese kale production under hydroponic conditions. Compared with the control, N loss in T1, T2, and T3 was reduced by 7.37%, 27.40%, and 23.71%, respectively, and the N loss rate was reduced by 11.83%, 30.90%, and 12.97%, respectively, with significant differences between treatments. N retention time was prolonged in all three treatments; however, it was most significant in T2. N productivity was the highest in T2, followed by T1, and lowest in T3; however, there was no significant difference between the different treatments. Owing to the difference in N loss rate, N productivity, and N retention time, N use efficiency was significantly different between the four treatments. N use efficiency was the highest in T2, followed by T1; however, there was no significant difference between T1 and T3. The results showed that an appropriate NH_4^+/NO_3^- ratio (25/75) significantly reduced N loss, prolonged N retention time, and improved N productivity and N use efficiency compared with the control.

Table 4. Rate of N loss, N residence time, N productivity, and N use efficiency in nutrient solutions with different NH_4^+/NO_3^- ratios during the growth period of Chinese kale.

Treatments	NH_4^+/NO_3^-	N Loss (mg Plant^{-1})	Rate of N Loss (%)	N Residence Time (d)	N Productivity (mg mg^{-1} d^{-1})	N Use Efficiency (mg mg^{-1})
CK	0/100	102.1 ± 5.1 a	27.9 ± 1.2 a	13.9 ± 0.1 c	4.6 ± 0.6 a	63.3 ± 1.7 c
T1	10/90	94.6 ± 1.5 b	24.6 ± 0.4 b	16.1 ± 0.4 b	4.7 ± 0.2 a	75.1 ± 5.4 b
T2	25/75	74.1 ± 3.2 c	19.3 ± 0.8 c	21.8 ± 0.3 a	4.8 ± 0.7 a	104.1 ± 4.1 a
T3	50/50	77.9 ± 3.3 c	24.3 ± 0.9 b	17.0 ± 0.4 b	4.3 ± 0.6 a	72.3 ± 4.2 b

Data are presented as mean ± SE (n = 3). Different letters indicate significant differences at $p < 0.05$. Significant differences among the treatments were determined using SPSS 17.0.

3.4. Principal Component Analysis

Principal component analysis (PCA) was used to visualize the effects of different NH_4^+/NO_3^- ratios on the growth and N uptake of Chinese kale (Figure 4). The computed model captured 84.5% of the total observed variance with the first two principal components (PCs). Four treatments (CK, T1, T2, and T3) were distributed in distinct quadrants in the PCA scatter plot. All indexes of plant growth and N uptake were the best in the T2 treatment, and biomass and root morphology were strongly correlated with N use efficiency. In addition, the NH_4^+/NO_3^- ratio in the nutrient solution and root/shoot ratio were the highest in T3 and were significantly negatively correlated with the pH value.

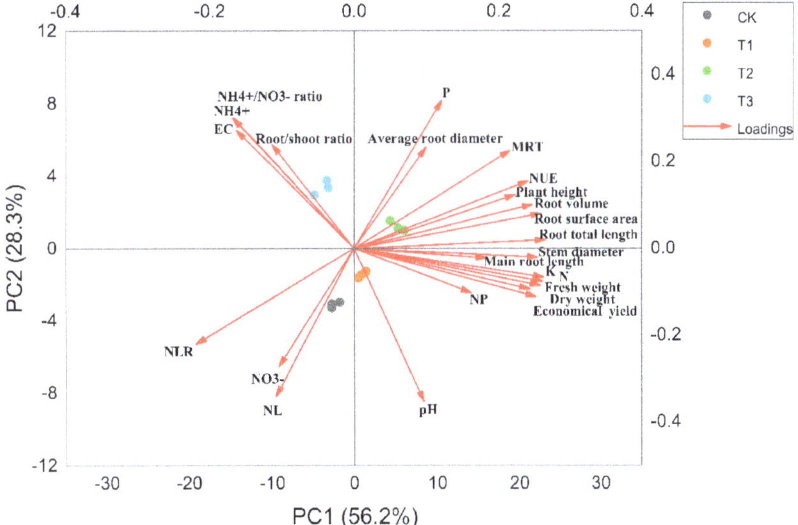

Figure 4. Principal component analysis showing differences and correlations in the investigated parameters of Chinese kale under different NH_4^+/NO_3^- ratio.

4. Discussion

The main forms of inorganic N, NH_4^+ and NO_3^- can be absorbed and utilized by plants [20,21]. To date, there have been many reports on the effects of different N forms and ratios on crop biomass [22]. Our previous study showed that the biomass of Chinese kale was largest when the nutrient solution increased NH_4^+ by 25% to 30%; when the nutrient solution increased NH_4^+ by 45% to 50%, the biomass of kale was significantly lower than that of the control treatment with only NO_3^- [9]. In this study, the high NH_4^+/NO_3^- ratio (50/50) significantly increased the biomass of Chinese kale and promoted plant growth at the early stage; however, it showed a strong inhibitory effect at the late growth stage

(21–27 d). During this period, the plant growth rate was 304.9 mg d^{-1} plant^{-1}, which was only 46.5% of plant growth in the control with only NO$_3^-$ (Figure 1, Table 2). This showed that the effects of different NH$_4^+$/NO$_3^-$ ratios on plant growth could also be closely related to the duration of treatment and growth stage of plants, and that a high concentration of NH$_4^+$ applied for a short time would not cause ammonium toxicity [23,24]. In this study, 2.25, 5.60, and 11.23 mmol L-1 Cl$^-$ was added to the nutrient solution of T1, T2, and T3, respectively, being less than the salt-osmotic stress threshold level of approximately 40 mmol L^{-1} NaCl for most plants [25]. Therefore, the amount of Cl$^-$ in the nutrient solution used in this study was insufficient for salt stress.

Plants mainly absorb nutrients through their roots. Of all mineral nutrients, nitrogen has the greatest influence on root morphology, growth, and distribution in the medium [26]. The absorption of NO$_3^-$ by plants results in rhizosphere alkalization, while the absorption of NH$_4^+$ results in rhizosphere acidification [27]. In the present study, the nutrient solution was weakly acidic under a medium NH$_4^+$/NO$_3^-$ ratio (25/75), and the root morphology index was the highest (Table 3). Root morphology and biomass were significantly positively correlated with pH under a medium NH$_4^+$/NO$_3^-$ ratio (Figure 4). Appropriate ratios of NH$_4^+$/NO$_3^-$ promote root growth and increase root dry weight [28,29]. Pepper treated with a 25:75 ratio increased root length, surface area, and root volume and tips [30]. Therefore, an appropriate NH$_4^+$/NO$_3^-$ ratio (25/75) may promote root growth by maintaining the pH value, thereby increasing biomass. The pH of the rhizosphere also affects the absorption of NH$_4^+$ and NO$_3^-$ by plants [31]. With the growth and development of plants, the pH value of the nutrient solution gradually decreased under a medium NH$_4^+$/NO$_3^-$ ratio (25/75). The NO$_3^-$ was quickly and completely absorbed, and only NH$_4^+$ remained in the nutrient solution at the late growth stage of Chinese kale (Figure 2). When the proportion of NH$_4^+$ in the nutrient solution was more than 75%, cabbage growth was reduced by 87% due to the accumulation of a large amount of free ammonia in the leaves [32]. The pH value and NH$_4^+$ proportion showed a significant negative correlation under a high NH$_4^+$/NO$_3^-$ ratio (50/50) (Figure 4). However, the pH value was acidic, which was not conducive to the root development of Chinese kale seedlings. In addition, the accumulation of large amounts of NH$_4^+$ in Chinese kale leaves resulted in ammonium toxicity, which inhibits plant growth.

The change in pH value in the rhizosphere results in different absorption of NH$_4^+$ and NO$_3^-$, which affects the content and accumulation of N in plants [9]. In this study, the total N content of the three treatments (T1, T2, and T3) was significantly increased at 0–15 d after treatment compared to that of the control treatment with NO$_3^-$ alone (Figure 3A). This is consistent with previous studies on Chinese flowering cabbage [13]. Total N accumulation increased continuously throughout the growth period of the Chinese kale. At the late plant growth stage, the total N accumulation of the high NH$_4^+$/NO$_3^-$ ratio (50/50) was the lowest (Figure 3B). This is similar to the results of previous studies [33]. There was a negative correlation between N accumulation and biomass under high NH$_4^+$/NO$_3^-$ ratios (Figure 4). NH$_4^+$ at higher concentrations caused toxicity in bamboo, as it inhibits root growth and N accumulation [34]. Therefore, a high NH$_4^+$/NO$_3^-$ ratio reduced the uptake and accumulation of N in plants at the late growth stage of Chinese kale, thus inhibiting plant growth. Different NH$_4^+$/NO$_3^-$ ratios also significantly affect the absorption of other nutrients [35]. In this study, a high concentration of NH$_4^+$ promoted the absorption of P but inhibited the absorption of K by Chinese kale at the late stage (Figure S4). There is a strong correlation between pH change in a solution and net cation or anion uptake in geranium and petunia [36]. The elevation of the NH$_4^+$ ratio in the fertilizer solution decreased the soil solution pH at 35 days after sowing, resulting in an increase in tissue P and a decrease in K content in French marigold "Orange Boy" [37]. This may be caused by the interaction between ions and the pH change in the rhizosphere.

This study adopts the definition of nutrient use efficiency by Berendse and Aerts [38], and divides nitrogen use efficiency into N productivity and average retention time of N. N productivity reflects a plant's rapid growth strategy. The average retention time of nitrogen reflects a plant's nutrient retention strategy [39]. In this study, N loss in the nutrient solution increased by NH_4^+ treatment was significantly lower than that of the control with NO_3^- alone (Table 4). In our recent research, the contents of NH_4^+, NO_3^-, and total N in the nutrient solution without plant culture were almost constant during the growth period [13]. Therefore, in a closed hydroponic system, the loss of N may be related to the N metabolism of plants or caused by changes in the nutrient solution composition during plant cultivation. Combined with the experimental conditions, this part of the N loss in this experiment should be attributed to gaseous loss. The volatile forms of gaseous nitrides include NH_3, N_2O, and NO. Previous studies have found that, compared with NH_4^+, when NO_3^- is used as the nitrogen source, the volatilization of N_2O in wheat and of NO in rice is enhanced [40–42]. However, the main loss forms of nitrides during kale production are still unclear, and further experiments are needed. In this study, the medium NH_4^+/NO_3^- ratio (25/75) had the highest N use efficiency, unlike *Brassica napus*, which had the highest N use efficiency with a high NH_4^+/NO_3^- ratio (75/25) [43]. We observed that the NH_4^+/NO_3^- ratio in T2 was maintained at approximately 0.15 (Figure 2D), which may have contributed to balanced N absorption. The NH_4^+/NO_3^- ratio in T3 increased rapidly at the late stage of plant growth (Figure 2D), which may have caused insufficient assimilation of NH_4^+ after its absorption or protein degradation owing to NH_4^+ toxicity, thereby reducing N productivity. Thus, applying appropriate ratios of NH_4^+/NO_3^- is an important way to improve plant N use efficiency [44].

5. Conclusions

An appropriate NH_4^+/NO_3^- ratio (25/75) increased biomass and promoted the growth of Chinese kale, whereas high NH_4^+/NO_3^- (50/50) promoted the growth of Chinese kale seedlings at the early growth stage but inhibited growth at the late growth stage. With the prolongation of treatment time, the pH value of T3 nutrient solution decreased continuously, whereas the NH_4^+/NO_3^- ratio was the highest in T3 at the harvest stage. In addition, T2 had the highest total N accumulation and N use efficiency, whereas total N accumulation was lowest in T3. Thus, an appropriate NH_4^+/NO_3^- ratio (25/75) promotes the growth and N uptake of Chinese kale by maintaining the pH value of the nutrient solution, whereas excessive addition of NH_4^+ may induce rhizosphere acidification and ammonia toxicity, thereby inhibiting plant growth. This study provides a theoretical basis for the effects of different NH_4^+/NO_3^- ratios on plant growth and N absorption and utilization.

Supplementary Materials: The following supporting information can be downloaded at: https://www.mdpi.com/article/10.3390/horticulturae8010008/s1, Figure S1: Plant morphology of Chinese kale treated with different NH_4^+/NO_3^- ratios; Figure S2: The root morphology of Chinese kale seedlings treated with different NH_4^+/NO_3^- ratios; Figure S3: The root activity of Chinese kale under different NH_4^+/NO_3^- ratios; Figure S4: Effect of different NH_4^+/NO_3^- ratios on the content and accumulation of total K (A,B) and P (C,D) in the growth period of Chinese kale.

Author Contributions: R.C. and S.S. conceived and designed the study X.Z. carried out the experiments. Y.W. analyzed the data and wrote the manuscript. H.L. and G.S. reviewed and edited the manuscript. All authors have read and agreed to the published version of the manuscript.

Funding: This work was supported by the Key-Area Research and Development Program of Guangdong Province (2020B0202010006), the Guangdong Provincial Special Fund for Modern Agriculture Industry Technology Innovation Teams (2021KJ131), and the China Agriculture Research System of MOF and MARA.

Institutional Review Board Statement: Not applicable.

Informed Consent Statement: Not applicable.

Data Availability Statement: Not applicable.

Conflicts of Interest: The authors declare no conflict of interest.

References

1. Li, Y.F.; Chang, D.; Zhang, X.; Shi, H.Z.; Yang, H.J. RNA-Seq, physiological, and biochemical analysis of burley tobacco response to nitrogen deficiency. *Sci. Rep.* **2021**, *11*, 16802. [CrossRef]
2. Zhu, Y.Y.; Li, J.; Zeng, H.Q.; Liu, G.; Di, T.J.; Shen, Q.R.; Xu, G.H. Involvement of plasma membrane H^+-ATPase in adaption of rice to ammonium nutrient. *Rice Sci.* **2011**, *18*, 335–342. [CrossRef]
3. Arnold, A.; Sajitz-Hermstein, M.; Nikoloski, Z. Effects of varying nitrogen sources on amino acid synthesis costs in Arabidopsis thaliana under different light and carbon-source conditions. *PLoS ONE* **2015**, *10*, e0116536. [CrossRef]
4. Hachiya, T.; Sakakibara, H. Interactions between nitrate and ammonium in their uptake, allocation, assimilation, and in plants. *J. Exp. Bot.* **2017**, *68*, 2501–2512. [CrossRef]
5. Esteban, R.; Ariz, I.; Cruz, C.; Moran, J.F. Mechanisms of ammonium toxicity and the quest for tolerance. *Plant Sci.* **2017**, *248*, 92–101. [CrossRef] [PubMed]
6. Bittsánszky, A.; Pilinszky, K.; Gyulai, G.; Komives, T. Overcoming ammonium toxicity. *Plant Sci.* **2015**, *231*, 184–190. [CrossRef] [PubMed]
7. Tabatabaei, S.J.; Yusefi, M.; Hajiloo, J. Effects of shading and NO_3^-: $NH4^+$ ratio on the yield, quality and N metabolism in strawberry. *Sci. Hortic.* **2008**, *116*, 264–272. [CrossRef]
8. Hu, L.; Yu, J.; Liao, W.; Zhang, G.; Xie, J.; Lv, J.; Xiao, X.; Yang, B.; Zhou, R.; Bu, R. Moderate ammonium: Nitrate alleviates low light intensity stress in mini Chinese cabbage seedling by regulating root architecture and photosynthesis. *Sci. Hortic.* **2015**, *186*, 143–153. [CrossRef]
9. Song, S.W.; Yi, L.Y.; Zhu, Y.N.; Liu, H.C.; Sun, G.W.; Chen, R.Y. Effects of ammonium and nitrate ratios on plant growth, 405 nitrate concentration and nutrient uptake in flowering Chinese cabbage. *Bangladesh J. Bot.* **2017**, *46*, 1259–1267.
10. Zhu, Y.N.; Li, G.; Liu, H.C.; Sun, G.W.; Chen, R.Y.; Song, S.W. Effects of partial replacement of nitrate with different nitrogen forms on the yield, quality, and nitrate content of Chinese kale. *Commun. Soil Sci. Plant* **2018**, *49*, 1384–1393. [CrossRef]
11. Zeng, H.Q.; Liu, G.; Kinoshita, T.; Zhang, R.P.; Zhu, Y.Y.; Shen, Q.R.; Xu, G.H. Stimulation of phosphorus uptake by ammonium nutrition involves plasma membrane H^+-ATPase in rice roots. *Plant Soil.* **2012**, *357*, 205–214. [CrossRef]
12. Tian, W.H.; Ye, J.Y.; Cui, M.Q.; Chang, J.B.; Liu, Y.; Li, G.X.; Wu, Y.R.; Xu, J.M.; Harberd, N.P.; Mao, C.Z. A transcription factor STOP1-centered pathway coordinates ammonium and phosphate acquisition in Arabidopsis. *Mol. Plant* **2021**, *14*, 1554–1568. [CrossRef] [PubMed]
13. Zhu, Y.N.; Qi, B.F.; Hao, Y.W.; Liu, H.C.; Sun, G.W.; Chen, R.Y.; Song, S.W. Appropriate $NH4^+/NO_3^-$ Ratio Triggers Plant Growth and Nutrient Uptake of Flowering Chinese Cabbage by Optimizing the pH Value of Nutrient Solution. *Front. Plant Sci.* **2021**, *12*, 656144. [CrossRef]
14. Jiang, D.; Lei, J.J.; Cao, B.H.; Wu, S.Y.; Chen, G.J.; Chen, C.M. Molecular Cloning and Characterization of Three Glucosinolate Transporter (GTR) Genes from Chinese Kale. *Genes* **2019**, *10*, 202. [CrossRef]
15. Lei, J.J.; Chen, G.J.; Chen, C.M.; Cao, B.H. Germplasm Diversity of Chinese Kale in China. *Hortic. Plant J.* **2017**, *3*, 101–104. [CrossRef]
16. Song, S.W.; Liao, G.X.; Liu, H.C.; Sun, G.W.; Chen, R.Y. Effect of ammonium and nitrate ratios on growth and yield of Chinese kale. *Appl. Mech. Mater.* **2012**, *142*, 32–36. [CrossRef]
17. Lu, R.K.; Shi, Z.Y.; Shi, J.P. Status Evaluation and Dynamic Change of farmland nutrient balance in 6 provinces in South China. *Chin. J. Agric. Sci.* **2000**, *2*, 63–67.
18. Eckstein, R.L.; Karlsson, P.S. The effect of reproduction on nitrogen use-efficiency of three species of the carnivorous genus Pinguicula. *J. Ecol.* **2001**, *89*, 798–806. [CrossRef]
19. Dong, T.F. Soil nutrients and their ecological stoichiometry of Pinus yunnanensis forest along an elevation gradient. *Chin. J. Ecol.* **2021**, *40*, 672–679.
20. Konnerup, D.; Brix, H. Nitrogen nutrition of Canna indica: Effects of ammonium versus nitrate on growth, biomass allocation, photosynthesis, nitrate reductase activity and N uptake rates. *Aquat. Bot.* **2010**, *92*, 142–148. [CrossRef]
21. Ding, L.; Lu, Z.F.; Gao, L.M.; Guo, S.W.; Shen, Q.R. Is nitrogen a key determinant of water transport and photosynthesis in higher plants upon drought stress? *Front. Plant Sci.* **2018**, *9*, 1143. [CrossRef]
22. Santos, J.H.D.S.; Bona, F.D.D.; Monteiro, F.A. Growth and productive responses of tropical grass Panicum maximum to nitrate and ammonium supply. *Rev. Bras. Zootecn.* **2013**, *42*, 622–628. [CrossRef]
23. Ashraf, M.; Naz, U.; Abid, M.; Shahzad, S.M.; Aziz, A.; Akhtar, N.; Naeem, A.; Mühlinget, K.H. Salinity resistance as a function of $NH_4^+:NO_3^-$ ratio and its impact on yield and quality of tomato (*Solanum lycopersicum* L.). *J. Plant Nutr. Soil Sci.* **2021**, *184*, 246–254. [CrossRef]
24. Sung, J.K.; Lee, N.R.; Choi, J.M. Impact of Pre-planting $NH_4^+:NO_3^-$ Ratios in Inert Media on the Growth of Chinese Cabbage Plug Seedlings. *Korean J. Hortic. Sci.* **2016**, *34*, 736–745.
25. Munns, R.; Tester, M. Mechanisms of salinity tolerance. *Annu. Rev. Plant Bio.* **2008**, *59*, 651–681. [CrossRef] [PubMed]
26. Marschner, H.; Kirkby, E.A.; Cakmak, I. Effect of mineral nutritional status on shoot-root partitioning of photoassimilates and cycling of mineral nutrients. *J. Exp. Bot.* **1996**, *47*, 1255–1263. [CrossRef] [PubMed]
27. Sarasketa, A.; González-Moro, M.B.; González-Murua, C.; Marino, D. Nitrogen source and external medium pH interaction differentially affects root and shoot metabolism in *Arabidopsis*. *Front. Plant Sci.* **2016**, *7*, 29. [CrossRef]

28. Chen, W.; Luo, J.K.; Jiang, H.M.; Shen, Q.R. Effects of different nitrogen forms on biomass and nitrate content of different mini-Chinese cabbage varieties. *Acta Pedol. Sinica.* **2004**, *41*, 420–425.
29. Wang, B.; Shen, Q.R.; Lai, T.; Shen, Q.R. Study on the effect of nutrient solution with different ammonium nitrate ratio on the growth and development of lettuce. *Acta Pedol. Sinica.* **2007**, *44*, 561–565.
30. Zhang, J.; Lv, J.; Dawuda, M.M.; Xie, J.M.; Yu, J.; Li, J.; Zhang, X.; Tang, C.; Wang, C.; Gan, Y. Appropriate Ammonium-Nitrate Ratio Improves Nutrient Accumulation and Fruit Quality in Pepper (*Capsicum annuum* L.). *Agronomy* **2019**, *9*, 683. [CrossRef]
31. Baracaldo, A.D.P.; Daz, M.C.; Flrez, V.J.; Gonzlez, C.A. Efecto de la disminución de N total y aumento de NH_4^+ en la fórmula de fertirriego en el cultivo de clavel. *Rev. Colomb. Cienc. Pec.* **2018**, *12*, 658–667. [CrossRef]
32. Zhang, F.C.; Kang, S.Z.; Li, F.S.; Zhang, J.H. Growth and Major Nutrient Concentrations in Brassica campestris Supplied with Different NH_4^+:NO_3^- Ratios. *J. Integr. Plant Biol.* **2007**, *49*, 455–462. [CrossRef]
33. Yin, H.N.; Li, B.; Wang, X.F.; Xi, Z.M. Effect of Ammonium and Nitrate Supplies on Nitrogen and Sucrose Metabolism of Cabernet Sauvignon (*Vitis vinifera cv.*). *J. Sci. Food Agric.* **2020**, *100*, 10574. [CrossRef]
34. Zou, N.; Shi, W.M.; Hou, L.H.; Kronzucker, H.J.; Huang, L.; Gu, H.M.; Yang, Q.P.; Deng, G.; Yang, G. Superior growth, N uptake and NH_4^+ tolerance in the giant bamboo Phyllostachys edulis over the broad-leaved tree Castanopsis fargesii at elevated NH_4^+ may underlie community succession and favor the expansion of bamboo. *Tree Physiol.* **2020**, *40*, 1606–1622. [CrossRef] [PubMed]
35. Feng, H.; Fan, X.; Miller, A.J.; Xu, G. Plant nitrogen uptake and assimilation: Regulation of cellular pH homeostasis. *J. Exp. Bot.* **2020**, *71*, 4380–4392. [CrossRef]
36. Dickson, R.W.; Fisher, P.R.; Argo, W.R.; Jacquesc, D.J.; Sartaina, J.B.; Trenholma, L.E.; Yeager, T.H. Solution Ammonium: Nitrate ratio and cation/anion uptake affect acidity or basicity with floriculture species in hydroponics. *Sci. Hortic.* **2016**, *200*, 36–44. [CrossRef]
37. Jong, M.C. Influence of Pre-Plant Micronutrient Sources and Post-Plant NH_4^+:NO_3^- Ratios in Fertilizer Solution on Growth and Nutrient Uptake of Marigold in Plug Culture. *Hortic. Environ. Biote.* **2007**, *48*, 314–319.
38. Berendse, F.; Aerts, R. Nitrogen-use-efficiency: A biologically meaningful definition? *Funct. Ecol.* **1987**, *1*, 293–296.
39. Aerts, R. Interspecific competition in natural plant communities: Mechanisms, trade-offs, and plant-soil feedbacks. *J. Exp. Bot.* **1999**, *50*, 29–37. [CrossRef]
40. Chen, N.C.; Inanaga, S. Nitrogen Losses in Relation to Rice Varieties, Growth Stages, and Nitrogen Forms Determined with the 15N Technique. In *Controlling Nitrogen Flows and Losses*; Wageningen Academic Publishers: Wageningen, The Netherlands, 2004; pp. 496–497.
41. Smart, D.R.; Bloom, A.J. Wheat leaves emit nitrous oxide during nitrate assimilation. *Proc. Natl. Acad. Sci. USA* **2001**, *98*, 7875–7878. [CrossRef]
42. Xu, S.G.; Chen, N.C.; Zhou, J.M.; Wu, Q.T.; Bi, D.; Lu, W.S. Regulation mechanism of light and nitrogen application on nitrogen oxide (NO and NO2) exchange between leaves of rice at tillering stage. *Chin. J. Rice Sci.* **2009**, *23*, 297–303.
43. Li, S.; Zhang, H.; Wang, S.L.; Shi, L.; Xu, F.S.; Wang, C.; Cai, H.M.; Ding, G.D. The rapeseed genotypes with contrasting NUE response discrepantly to varied provision of ammonium and nitrate by regulating photosynthesis, root morphology, nutritional status, and oxidative stress response. *Plant Physiol. Biochem.* **2021**, *166*, 348–360. [CrossRef] [PubMed]
44. Rouphael, Y.; Kyriacou, M.C.; Petropoulos, S.A.; Pascale, S.D.; Colla, G. Improving vegetable quality in controlled environments. *Sci. Hortic.* **2018**, *234*, 275–289. [CrossRef]

 *horticulturae*

Article

Effects of NIR Reflective Film as a High Tunnel-Covering Material on Fruit Cracking and Biomass Production of Tomatoes

Hiroko Yamaura [1], Shinichi Furuyama [2], Nobuo Takano [1], Yuka Nakano [1], Keiichi Kanno [1], Takashi Ando [3], Ichiro Amasaki [3], Yukie Watanabe [3], Yasunaga Iwasaki [1,4] and Masahide Isozaki [1,*]

[1] Institute of Vegetable and Floriculture Science, National Agricultural and Food Research Organization (NARO), Tsukuba 305-8519, Japan; yamaurah090@affrc.go.jp (H.Y.); ampf.ishigaki5050@gmail.com (N.T.); yuka88@affrc.go.jp (Y.N.); k.kanno@affrc.go.jp (K.K.); iwasakiy@meiji.ac.jp (Y.I.)

[2] Kamikawa Agricultural Experiment Station, Hokkaido Research Organization (HRO), Pippu 078-0397, Japan; furuyama-shinichi@hro.or.jp

[3] Fujifilm Corporation, Tokyo 107-0052, Japan; takashi.ando@fujifilm.com (T.A.); ichiro.amasaki@fujifilm.com (I.A.); yukie.a.watanabe@fujifilm.com (Y.W.)

[4] Department of Agriculture, Meiji University, Kawasaki 214-8571, Japan

* Correspondence: isozakim915@affrc.go.jp

Citation: Yamaura, H.; Furuyama, S.; Takano, N.; Nakano, Y.; Kanno, K.; Ando, T.; Amasaki, I.; Watanabe, Y.; Iwasaki, Y.; Isozaki, M. Effects of NIR Reflective Film as a High Tunnel-Covering Material on Fruit Cracking and Biomass Production of Tomatoes. *Horticulturae* **2022**, *8*, 51. https://doi.org/10.3390/horticulturae8010051

Academic Editors: Xiaohui Hu, Shiwei Song and Xun Li

Received: 1 December 2021
Accepted: 2 January 2022
Published: 6 January 2022

Publisher's Note: MDPI stays neutral with regard to jurisdictional claims in published maps and institutional affiliations.

Abstract: Tomatoes require higher irradiance, although the incidence of physiological disorders in fruit increases at high temperatures. Near-infrared (800–2500 nm) (NIR) reflective materials are effective tools to suppress rising air temperatures in greenhouses. We examined the physiological and morphological changes in tomato growth and fruit quality when grown in a high tunnel covered with NIR reflective film (NR) and in another covered with polyolefin film (PO; control). There was no relationship between the fruit cracking rate and mean daytime temperature under NR. The fruit temperature at the same truss was lower and the increase in air temperature was slow under NR. Fruit dry matter (DM) content under NR was also significantly decreased. These findings suggest that the reduction in fruit cracking under NR results from a decrease in fruit DM content as a consequence of lower fruit temperature and a decrease in total DM (TDM). Total fruit yield did not differ, whereas TDM was significantly decreased under NR. This was considered to result from a lower transmitted photosynthetic photon flux density (400–700 nm) (PPFD) and LAI, and lower photosynthetic capacity in single leaves because of a decrease in both total nitrogen and chlorophyll content. We conclude that NR film reduces fruit cracking in exchange for a slight reduction in TDM.

Keywords: net assimilation rate (NAR); leaf nitrogen; fruit temperature; dry matter; chlorophyll

1. Introduction

Tomatoes are one of the most commercially valuable vegetables in the world. In recent years, the demand for tomatoes has increased in Asia and their production and cultivated area have been rising [1]. Tomatoes require high irradiance [2]; however, the air temperature in greenhouses tends to increase to levels inappropriate for tomato growth, when the amount of solar radiation entering the greenhouse increases in high-temperature regions and seasonally. As a result, the incidence of physiological disorders in fruit, such as fruit cracking, yellow-shoulder fruits, and increased blossom-end rot, may cause a decrease in marketable fruit yield [3]. Therefore, it is necessary to develop material and management strategies for the greenhouse to improve fruit quality in high-temperature regions and seasons.

Near-infrared (800–2500 nm) (NIR) reflective materials can be an effective tool for growing tomatoes under such conditions. A simulation study of NIR-filtering materials, used to cover a greenhouse in the Netherlands, indicated that tomato production in the summer could increase by 8.6% because the air temperature inside the greenhouse would be lower by 2.0 °C at noon and crop transpiration would also decrease [4]. In fact, air temperatures in a greenhouse with a NIR-filtering material could be reduced by up to 4.0 °C in Thailand [5] and 3.0 °C in Southern Spain [6] without reducing yields during the summer season. Stanghellini et al. [7] reported that the use of NIR reflective film reduced crop transpiration, increased the water use efficiency of rose crops, and maintained rose yields. Accordingly, it has been suggested that NIR reflective films can suppress increases in air temperature in greenhouses and reduce crop transpiration, which results in maintained crop yields, even at high temperatures.

Moreover, NIR reflective films are effective at improving fruit quality at high temperatures [8–10]. In fact, Nakayama et al. [10] reported that NIR reflective film reduced the rate of fruit cracking by 70% compared with controls in June, which is the month presenting the highest solar radiation. Fruit cracking is related to high air temperature, high solar radiation, the difference between daytime and nighttime temperatures, and fruit temperature [11–14]. Therefore, NIR reflective film reduces fruit cracking by suppressing the increase in air temperature in the greenhouse. Similarly, the NIR reflective film may also suppress the increase in fruit temperature. Blanchard and Runkle [15] reported that the temperatures of leaves and shoot-tips in cucumbers decreased under NIR-reflecting materials. Lobos et al. [16] also reported that grape bunches covered with shading nets with low NIR transmittance reduced fruit temperature and improved fruit quality. The occurrence of physiological disorders in fruit increase with higher fruit temperature in tomatoes [14]. Thus, lowering the fruit temperature can prevent fruit cracking. However, there have been no reports on the measurement of the fruit temperature of crops grown in greenhouses covered with NIR reflective film.

Similarly, there have been few case studies giving a detailed analysis of the effects of NIR-reflecting materials on whole-plant growth from morphological and physiological perspectives. Plants display an adaptive response to environmental changes in light and temperature, not only in individual leaves but also at the whole-plant level. For example, physiological responses at the individual leaf level include a decrease in chlorophyll (chl) content per fresh weight under high-temperature conditions [17] and an increase in the proteins involved in light-harvesting [18]. The chl a/b ratio tends to increase but the chl content decreases under low light [19]. In addition to these physiological responses, tomatoes show morphological responses at the individual level. For example, the specific leaf area increases with rising daytime temperature [20] but decreases with high light intensity [21]. Similarly, the net assimilation rate (NAR) increases with increasing temperature [22], and the leaf area ratio increases in low light [23]. As described above, morphological changes at the individual whole-plant level are important for adapting to environmental changes, in addition to physiological changes at the individual leaf level.

NIR reflective films reduce NIR transmittance, the thermal factor that increases air temperature in the greenhouse. However, they may reduce the amount of light available for photosynthesis of the wavelengths from 400 to 700 nm [24]. It is important to analyze the effects of NIR reflective materials on dry matter (DM) production in tomatoes and the morphological and physiological aspects at the whole-plant level, when considering the possibility of expanding the seasons using NIR reflective film in tropical and sub-tropical regions, including Asia.

Therefore, the purpose of this study was to identify the causes of reduced fruit cracking during the summer and autumn and the morphological and physiological changes in plant growth on total yield and fruit quality under NIR reflective film. We examined the microclimate during the daytime and fruit temperature in a high tunnel covered with NIR reflective film and analyzed the relationship with fruit cracking. Then, we analyzed a growth strategy for plants under NIR reflective film. Later in this paper, we will discuss improvements in fruit quality with respect to DM production and the high tunnel microclimate.

2. Materials and Methods

2.1. Plant Materials and Growing Condition PAR

The experiments were conducted from 6 August to 6 November 2020 in two high tunnels (area, 112 m^2; wire height, 2.5 m) located at the National Agricultural and Food Research Organization, Japan (36.0° N, 140.1° E). The roof cover of one high tunnel was NIR reflective film (NR) (thickness, 0.10 mm; MF-450; Fujifilm Co. Ltd., Tokyo, Japan) and the other was agricultural polyolefin film (PO) as a control (thickness, 0.15 mm; Diastar; Mitsubishi Chemical Agri Dream Co., Ltd., Tokyo, Japan). NR transmits 0.71 in the photosynthetically active radiation (PAR) range (400–700 nm), 0.39 in the NIR range, and 0.31 in the ultraviolet (UV) range (300–400 nm). PO transmits 0.91 in the PAR range, 0.91 in the NIR range, and 0.68 in the UV range (Figure 1). The side covers of both high tunnels were made of PO (Tekinashi5, C.I. Kasei Co., Ltd., Tokyo, Japan). In each high tunnel, four cultivation beds were arranged in a north–south direction, with a distance between beds of 1.6 m.

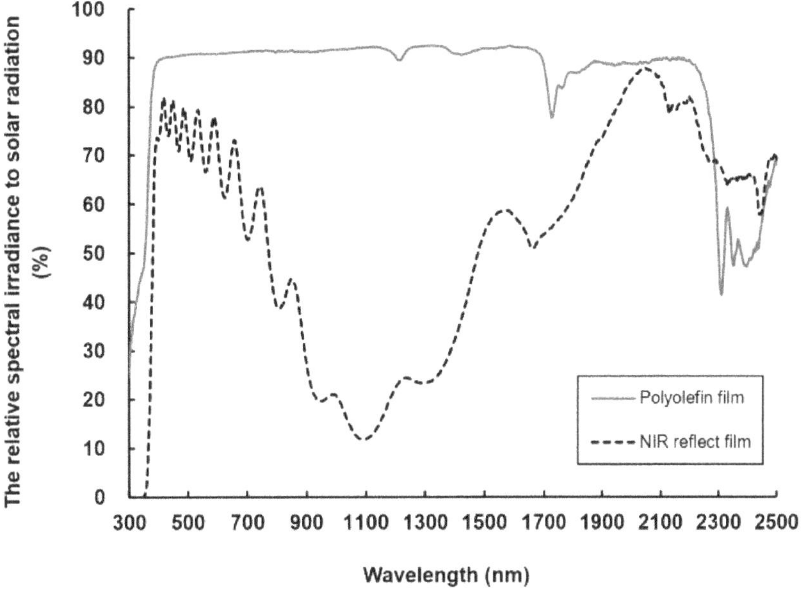

Figure 1. The relative spectral irradiance of polyolefin film (PO) and NIR reflective film (NR) to solar radiation in the present study.

Tomato seeds ("Momotaro Hope", Takii Co., Ltd., Kyoto, Japan) were sown into 72-cell plug trays with a commercial substrate (Tanebaido No.1, Sumitomo Forestry Landscaping Co., Ltd., Tokyo, Japan) on 6 July 2020 and germinated in the darkness for 3 days. Seedlings were grown in a temperature-controlled chamber equipped with fluorescent tubes (NAE Terrace, Mitsubishi Chemical Agri Dream Co., Ltd., Tokyo, Japan) for 1 week. The chamber was operated at 350 μmol m^{-2} s^{-1} photosynthetic photon flux density (400–700 nm) (PPFD), with a day/night temperature of 25 °C/18 °C (daytime 16 h), and 1000 ppm CO$_2$ concentration. The seedlings were irrigated every day with a commercial nutrient solution (High Tempo; Mitsubishi Chemical Agri Dream Co., Ltd.; 9.7 mM NO$_3{}^-$, 5.9 mM K$^+$, 2.6 mM Ca^{2+}, 0.9 mM Mg^{2+}, 2.3 mM H$_2$PO$_4$, 0.9 mM NH$_4{}^+$, 0.7 mM SO$_4{}^{2-}$) at an electrical conductivity (EC) of 1.5 dS m^{-1}. The seedlings were then transplanted to rockwool cubes and arranged at a density of 27.8 plants m^{-2} in nursery high tunnels. They were supplied with a commercial nutrient solution (OAT House-SA; OAT Agrio Co., Ltd., Tokyo, Japan;

17.6 me·L^{-1} NO$_3$, 4.4 me·L^{-1} P, 10.2 me·L^{-1} K, 8.2 me·L^{-1} Ca, and 3.0 me·L^{-1} Mg at EC of 2.6 dS m^{-1}), adjusted to an EC of 1.0 dS m^{-1}.

On 6 August 2020, the seedlings were transplanted onto rockwool slabs in a cultivation bed at 0.125 m intervals (mean plant density, 5.0 plants m^{-2}) under NR or PO. A 300-liter plastic tank for irrigation was prepared in each high tunnel. The concentration of OAT House-SA was homogenized in the tank, and it was gradually increased to an EC of 1.8 dS m^{-1} and adjusted to around pH 6.0 using a fertilizer management system (PCE-11, CEM Corporation, Tokyo, Japan). The concentration of OAT House-SA was gradually increased to an EC of 1.8 dS m^{-1}. The daily draining volume was approximately 20–30% of the total volume of the nutrient solution supplied, and the drainage was discarded. Fruit set was promoted by spraying the flowers with 2-methyl-4-chlorophenoxyacetic acid (Tomato Tone; ISK Bio-sciences K.K., Tokyo, Japan) twice a week. The number of fruits per truss was adjusted to four by pruning and the auxiliary buds were also pruned. At 43 days after transplanting (DAT), the plants were pinched out to generate two leaves above the fourth truss.

In the PO-covered high tunnel, a movable shading screen (Venus Rassell #50, Koizumi Jute Mills Ltd., Kobe, Japan) was closed when the inside air temperature was above 35.0 °C, then opened when the air temperature was below 33.0 °C. This screen transmits 50% of the solar radiation. For both high tunnels, the night temperature was maintained by cooling from the day of transplanting to 37 DAT (18–20 °C) or by heating from 56 DAT onward (15–22 °C). The side vents in both high tunnels were set to start to open at an inside air temperature of 20 °C and to be fully open at 28 °C. A fog cooling system was operated when the external solar radiation was greater than 0.4–0.5 kW m^{-2} and the inside air temperature was above 28 °C. From 44 DAT, CO$_2$ fertilization was applied at 2.0 g m^{-2} h^{-1} at an external solar radiation level of $\geq$0.1 kW. These operations were controlled by a ubiquitous environmental control system device (UECS-Pi, WaBit Co., Ltd., Fukuoka, Japan). This device also recorded air temperature, relative humidity, external solar radiation, and CO$_2$ concentration every minute. The transmitted PPFD was calculated from the operation rate of the movable shading screen. First, a pyranometer (MS-602, EKO instrument Japan Inc., Tokyo, Japan) was set outside the high tunnels to confirm that there was no difference between the amount of solar radiation recorded per hour and the amount of solar radiation data per hour recorded by the Areological Observatory at Tateno (36.1° N, 140.1° E) from the Japan Meteorological Agency (JMA) [25]. Secondly, the quantum sensors (LI-190, Li-Cor Inc., Lincoln, NE, USA) were also set inside each high tunnel to investigate the correlation between the amount of PPFD per hour and the amount of solar radiation data per hour from JMA. In this way, the coefficients of PPFD transmittance into each high tunnel were determined using the total solar radiation data from JMA.

2.2. Destructive Measurements and Growth Analysis

Initial total biomass and total leaf area were measured destructively in ten plants before transplanting on 6 August 2020. Twenty-four plants (8 replicates; three plants per replication) grown under NR or PO were also destructively harvested at 42 and 92 DAT. Leaves, stems, and fruits were separated, weighed to determine the fresh weight, dried at 105 °C for at least 3 days, and reweighed to determine the dry weight of each organ. Total leaf area was determined using an area meter (AAC-400, Hayashi Denko Co. Ltd., Tokyo, Japan). Crop growth rate (CGR; DM production per unit ground area per unit time) and net assimilation rate (NAR; DM production per unit leaf area per unit time) were calculated between the time points of the destructive measurements [26,27]. The leaf area index (LAI; leaf area per cultivated area) was calculated using the total leaf area by destructive measurement.

2.3. Measurement of Temperature and Fruit Quality

The number of mature fruits, as well as fresh and dry fruit weight, were measured twice a week from 48 DAT onwards. All fruits with cracks that were detectable by eye were counted as cracked. Marketable fruit yield was defined as the total fruit yield minus disordered fruits (cracking, blossom-end rot, or catface). Fruit temperature was measured on 28 September 2020 (54 DAT) from 1.40 p.m., and the measurements were completed within 10 min. The fruit temperature was measured at the equatorial plane for all fruits in three randomly selected plants from the first to the fourth truss, using an infrared thermometer (IT-545, Horiba Ltd., Kyoto, Japan). Weather conditions did not change during the measurements.

2.4. Determination of Chl and Total N per Leaf Area

Two leaf disks (8 mm in diameter) from the leaves were punched and chl was extracted with dimethylformamide for 48 h in the dark. Chl content was determined as described by Porra et al. [28]. The remainder of the leaf was dried at 105 °C for at least 3 days and total N content was measured with a CN analyzer (JM1000N, J-Science Lab Co., Ltd., Kyoto, Japan).

2.5. Statistical analysis

Destructive measurements, growth analysis, and fruit yield and quality were based on eight replicates. The chl and total N per leaf area were assessed using four replicates (one leaf per plant, four plants). To assess the statistical significance of the differences, all the data were analyzed using an independent Student's *t*-test with a level of significance of $p < 0.05$, using Microsoft Excel for Microsoft 365 MSO.

A Pearson correlation between mean daytime (7 a.m. to 5 p.m.) air temperature and the rate of fruit cracking under each high tunnel was conducted using the R statistical software (version 4.1.1, https://www.R-project.org/, accessed on 7 October 2021). The flowering day was defined as the day when the cumulative daily mean air temperature reached approximately 1100 °C from the harvest date [29]. The mean daytime temperature from flowering to harvest was calculated and used as a variable. The rate of fruit cracking was also used as a variable. The values were calculated by the number of cracked fruits, divided by the total number of harvested fruits for each harvest date.

3. Results

3.1. The Climate in High Tunnel

The daily mean air temperature and cumulative PPFD during the experiments in each high tunnel are shown in Figure 2 (panels A and B, respectively). The daily mean air temperature throughout the growing periods was 23.1 °C under NR and 23.4 °C under PO. In particular, from August to mid-September, the daily mean temperature under NR was lower than that under PO (Figure S1). The cumulative hours of air temperature above 28.0 °C during the daytime (7 a.m. to 5 p.m.) and the ratio of each categorical temperature in cumulative hours under NR or PO are shown in Figure 3. Because the side vent was fully open, a fog cooling system was used when the inside air temperature was above 28.0 °C. Each instantaneous air temperature (recorded every minute) was categorized every 2.0 degrees. The cumulative hours of instantaneous temperatures above 28.0 °C were approximately 418 h under NR and approximately 448 h under PO. The ratio of cumulative hours of air temperature above 34.0 °C to the cumulative hours of air temperature above 28.0 °C under PO was about 1.6 times the ratio of that under NR. The maximum air temperature was 38.3 °C under NR and 39.6 °C under PO throughout the experiment. The ability of NIR reflective film to suppress the increase in air temperature has been reported [4–6,30,31] and our data is consistent with these reports. The air temperature did not increase excessively because of a decrease in solar radiation from 17 DAT (23 August) (Figure 2B). Under PO, the rate of shading screen operation during the daytime was 51.4% in August, 2.4% in September, and 1.6% in October. Thus, the total transmitted PPFD was approximately 15.6% lower under NR compared with PO (1346.5 mol m^{-2} experiment^{-1} under NR and 1556.5 mol m^{-2} experiment^{-1} under PO).

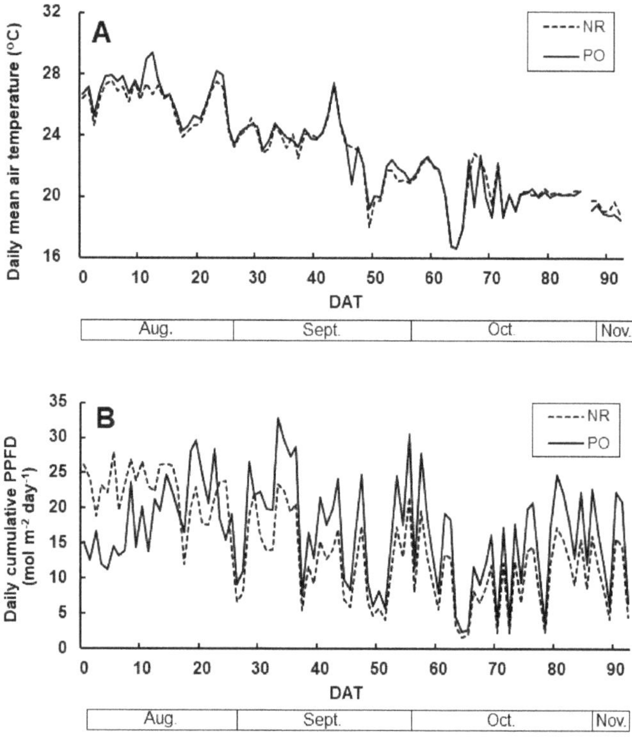

Figure 2. Daily mean air temperature (A) and daily cumulative PPFD (B) in each high tunnel. NR, NIR reflective film; PO, polyolefin film.

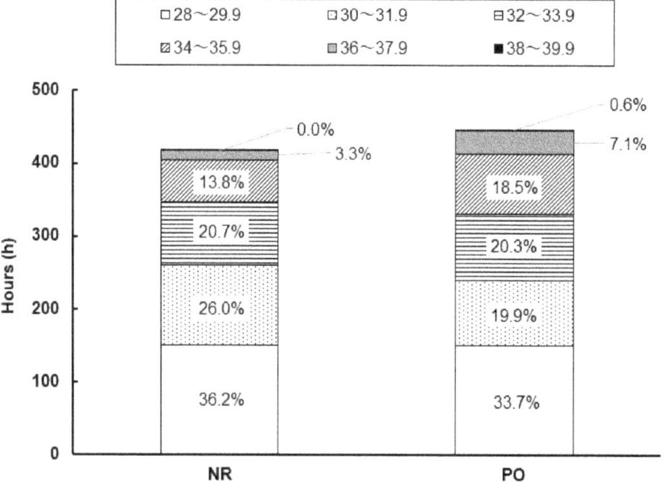

Figure 3. The cumulative hours of air temperature above 28.0 °C and the ratio of each air temperature hour to that of cumulative hours. NR, NIR reflective film; PO, polyolefin film. Each instantaneous air temperature (recorded every minute) was categorized by every 2.0 degrees from 28.0 °C. The values in the stacked bars are the percentages of each temperature category in the total cumulative hours of air temperature above 28.0 °C.

3.2. Fruit Yield and Quality

No significant differences were detected in total yield, the number of fruits, or marketable fruit yield between NR and PO (Table 1). DM content, fresh fruit weight and cracking fruit rate were significantly lower under NR compared with PO ($p < 0.01$, $p < 0.05$, $p < 0.01$, respectively; Table 1).

Table 1. Effects of high tunnel-covering film on total yield, number of fruits, fresh fruit weight, dry matter (DM) content, marketable fruit yield, and fruit cracking rate.

Film	Total Yield (kg m^{-2})	No. Fruits (Fruit m^{-2})	Fresh Fruit Weight (g Fruit^{-1})	DM Content (%)	Marketable Fruit Yield (kg m^{-2})	Fruit Cracking Rate (%)
NR	6.0 ± 0.3	74.0 ± 2.4	80.9 ± 3.0	4.76 ± 0.03	3.9 ± 0.1	13.4 ± 2.2
PO	6.6 ± 0.4	74.0 ± 3.5	89.6 ± 2.7	5.13 ± 0.02	3.8 ± 0.5	29.6 ± 3.6
p-value	0.18	0.68	<0.05 *	<0.01 **	0.87	<0.01 **

Values are the means of eight replicates $\pm$ SE. The *p*-values from an independent Student's *t*-test are shown. Asterisks (*, **) indicate a significant difference between NR and PO at $p < 0.05$ or $p < 0.01$, respectively.

As shown in Figure 4, there was a significant positive correlation between mean daytime air temperature from flowering to harvest and the rate of fruit cracking under PO ($r = 0.788$, $p < 0.001$), whereas the correlation between these factors was weak and the relationship was not significant under NR ($r = 0.200$, $p = 0.53$).

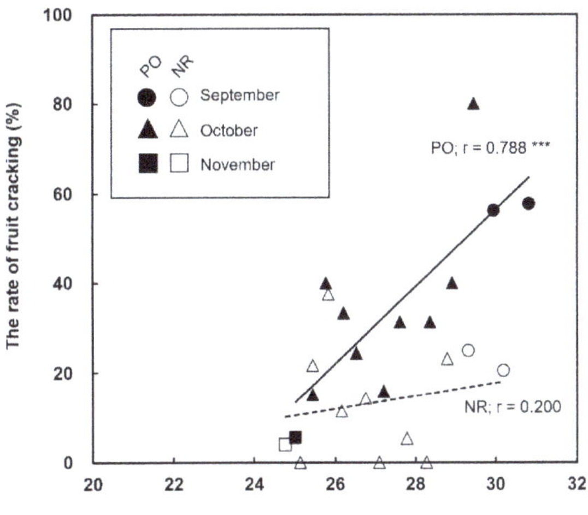

Figure 4. Pearson correlation coefficient (r) and *p*-value between the mean daytime temperature (7 a.m. to 5 p.m.) from flowering to harvest and the rate of fruit cracking. The circles (○●), triangles (△▲), and squares (□■) indicate the rate of fruit cracking when the tomato fruit was harvested in September, October, and November, respectively. The rate of fruit cracking was calculated as the number of cracked fruits divided by the total number of harvested fruits for each harvest date ($n = 12$). Asterisks (***) indicate significant correlation at $p < 0.001$.

Fruit temperatures in the first to third trusses were significantly lower under NR compared with PO ($p < 0.05$), whereas that of the fourth truss tended to be lower, although this difference was not statistically significant ($p = 0.06$; Figure 5). The difference in fruit temperature increased with truss number. It was 1.6 °C in the first truss, 2.1 °C in the second truss, 2.9 °C in the third truss and 4.8 °C in the fourth truss.

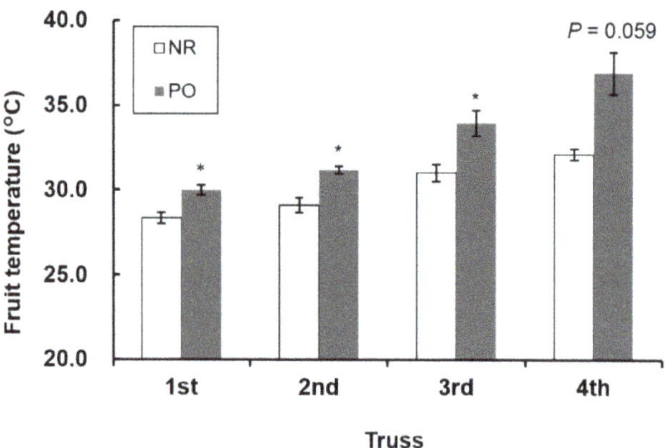

Figure 5. Fruit temperature of the first to fourth trusses of tomatoes grown in high tunnels covered with NR or PO. Mean values at each truss number from three plants ± SE are shown. * Significantly different from PO, as determined by an independent Student's *t*-test: $p < 0.05$.

On the measurement day of fruit temperature (52 DAT), the shading screen did not operate in PO. The air temperature in this high tunnel reached 30 °C at around 8 a.m. and was maintained until 3 p.m. (Figure 6). In NR, the air temperature increased gradually and reached 30 °C at approximately 2 p.m. (Figure 6).

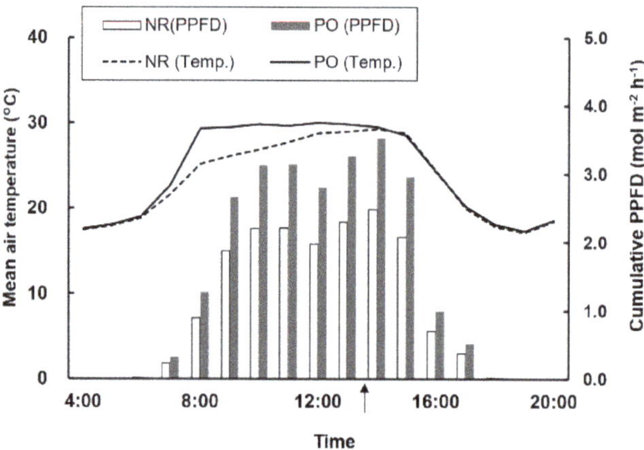

Figure 6. Daily change in microclimate in the high tunnels covered with NR or PO at 52 DAT. The arrow below the X-axis indicates the time of fruit temperature measurement.

3.3. Plant Growth and Leaf Properties

TDM was significantly lower under NR compared with PO at 42 and 92 DAT (both $p < 0.01$; Table 2). No significant difference was detected in DM allocation to the organs (Table 2) or in the number of leaves (data not shown).

For plants between 0 and 42 DAT, the CGR was slightly, but significantly, lower ($p < 0.01$) and the NAR tended to be lower ($p = 0.08$) under NR, compared with PO, whereas the LAI was not significantly different. For plants between 42 and 92 DAT, the CGR ($p < 0.01$), NAR ($p < 0.01$), and LAI ($p < 0.05$) were significantly lower under NR, compared with PO (Table 3).

Total N and chl content per leaf area were significantly lower under NR (both $p < 0.01$), but no significant difference was observed in the chl a/b ratio (Table 4).

Table 2. Effects of high tunnel-covering film on total dry matter (TDM) and dry matter (DM) allocation to each organ at 42 and 92 DAT.

Film	42DAT					92DAT				
	TDM (g m^{-2})	DM Allocation to Each Organ (%)				TDM (g m^{-2})	DM Allocation to Each Organ (%)			
		Stem	Leaves	Fruits	Others		Stem	Leaves	Fruits	Others
NR	409.2 ± 7.2	30.6 ± 0.4	42.7 ± 0.7	4.7 ± 0.2	3.4 ± 0.2	720.9 ± 26.7	20.3 ± 0.6	31.9 ± 0.6	40.6 ± 1.2	7.1 ± 0.3
PO	447.1 ± 4.8	31.7 ± 0.4	42.0 ± 1.1	4.7 ± 0.3	2.9 ± 0.2	891.2 ± 21.0	20.2 ± 0.4	32.0 ± 0.5	38.3 ± 1.0	9.5 ± 0.7
p-value	<0.01 **	0.08	0.63	1.00	0.09	<0.01 **	0.81	0.88	0.14	<0.01 **

Values are the means of eight replicates ± SE. *p*-values from an independent Student's *t*-test are shown. Asterisks (**) indicate a significant difference between NR and PO at $p < 0.01$.

Table 3. Effects of high tunnel-covering film on the CGR, NAR, and LAI for plants between 0 and 42 DAT and between 42 and 92 DAT.

Film	0-42DAT			42-92DAT		
	CGR (g m^{-2} Day^{-1})	NAR (g m^{-2} Day^{-1})	LAI (m^2 m^{-2})	CGR (g m^{-2} Day^{-1})	NAR (g m^{-2} Day^{-1})	LAI (m^2 m^{-2})
NR	9.57 ± 0.17	6.80 ± 0.17	3.17 ± 0.10	6.11 ± 0.52	1.74 ± 0.14	3.86 ± 0.18
PO	10.50 ± 0.12	7.36 ± 0.23	3.35 ± 0.16	8.71 ± 0.41	2.26 ± 0.08	4.37 ± 0.13
p-value	<0.01 **	0.08	0.38	<0.01 **	<0.01 **	<0.05 *

Values are the means of eight replicates ± SE. *p*-values from an independent Student's *t*-test are shown. Asterisks (*, **) indicate a significant difference between NR and PO at $p < 0.05$ or $p < 0.01$, respectively.

Table 4. Effects of high tunnel-covering film on total nitrogen (N) and chlorophyll (chl) content per leaf area and on the chl a/b ratio.

Film	Total N (mmol m^{-2})	chl (mmol m^{-2})	chl *a/b*
NR	87.6 ± 1.2	0.187 ± 0.006	3.19 ± 0.05
PO	107.4 ± 2.0	0.229 ± 0.006	3.11 ± 0.10
p-value	<0.01 **	<0.01 **	0.49

Values are the means of four replicates ± SE. *p*-values from an independent Student's *t*-test are shown. Asterisks (**) indicate a significant difference between NR and PO at $p < 0.01$.

4. Discussion

4.1. NIR Reflective Film Contributes to Fruit Quality

The mean air temperature in the high tunnel was lower and the cumulative hours of air temperature above 28.0 °C were shorter under NR, compared with that under PO, throughout the experiment (Figures 2A and 3). However, there was a significant co-relationship between mean daytime temperature and the rate of fruit cracking under PO, whereas there was no relationship under NR. The relative humidity was maintained above 78% under both PO and NR throughout the experiment, and the relative humidity in both high tunnels was controlled in almost the same range (data not shown). These results suggest that air temperature inside the greenhouse is not the only cause of the reduced rate of fruit cracking under NR.

The fruit temperature in different trusses during the afternoon was 1.6–4.8 °C lower under NR compared with PO and these differences increased with truss number (Figure 5). Air temperature on the measurement day increased more slowly under NR and reached 30 °C almost 6 h after it did so under PO (Figure 6). Such a moderate increase in air and fruit temperature resulted from the suppression of a rapid rise in temperature by a lower transmittance of NIR radiation under NR. Lang and Düring [13] reported that a higher temperature in grape berries dramatically increased the pressure exerted by the pulp on the skin and decreased skin stiffness and strength, thus increasing the rate of fruit cracking. Similarly, for tomatoes, Corey and Tan [32] postulated that the positive pressure inside the fruit stretches the CO_2-impermeable pericarp outward and increases fruit volume with increased fruit temperature. Higher fruit temperature increases fruit sink strength [33] and promotes fruit expansion [34,35]. Furthermore, Peet [14] suggested that higher soluble solids in fruit also cause an increase in fruit cracking because a higher photosynthesis rate relative to the translocation site is considered to promote rapid fruit expansion [36,37]. Accordingly, fruit cracking is more likely to occur at times of peak temperatures and solar radiation in the field and in high tunnels [12,14]. In the present study, the DM content was significantly decreased under NR. There was a positive correlation between fruit DM content and Brix (soluble solids content) [38,39]. Therefore, the reason for the reduction in fruit cracking was associated with a decrease in soluble solid content in the fruit, accompanied by decreasing DM content in the fruit. In addition, the decrease in fruit DM content under NR was associated with a decrease in the TDM, but not as the result of a difference in the DM allocation to the fruits (Table 2). In other words, the decrease in DM content in fruit under NR was accompanied by a reduction in whole-plant DM production. Furthermore, NR reduced the transmittance of UV radiation into high tunnels compared to PO (Figure 1). Kimura at al. [40] showed that tunnels covered with UV-cut film tended to reduce the incidence of fruit cracking. Overall, these results suggest that the reduction in fruit cracking under NR resulted from a decrease in fruit DM content caused by lower fruit temperature and decreased TDM, and the effects of the wavelength properties of NR.

4.2. Changes in Plant Growth under NR

At 42 DAT, TDM was significantly lower under NR (Table 2). The CGR is derived from NAR and LAI. The NAR is considered to represent the amount of apparent photosynthesis per plant and is known as a physiological index of whole-plant growth. For plants between 0 and 42 DAT, there were no significant differences in NAR or LAI, although NAR tended to be lower under NR (Table 3). Thus, the significant decrease in CGR under NR was caused primarily by a decrease in NAR. No differences were detected in total N content or N allocation to leaves and stems between the high tunnels (Table S1). At 92 DAT, TDM was also significantly lower under NR. For plants between 42 and 92 DAT, NAR, and LAI were significantly lower under NR; thus, a significant decrease in CGR was caused by a lower NAR and LAI. Total N content in the vegetative organs (leaves and stems) was significantly lower under NR (Table S1), whereas N allocation to the leaves or stems did not differ significantly. Makino et al. [41] reported that an increase in N allocation to leaves increased in rice grown under low light intensity. We did not observe such changes in tomatoes grown under NR; thus, these tomatoes did not show much of a compensatory response to reduced PPFD. According to Kaneko et al. [42] and Higashide [43], cumulative intercepted light positively correlates with DM production. The decrease in PAR and LAI resulted in a reduction in cumulative intercepted light, which was thought to be a factor in the decrease of TDM under NR.

To clarify the cause of decreased TDM under NR from physiological factors, we measured total N and chl content per leaf area. The results indicated that total N and chl content per leaf area were significantly decreased under NR ($p < 0.01$, respectively). Total N content per leaf area was positively correlated with photosynthesis-related factors, such as ribuloses-1,5-bisphosphate carboxylase (Rubisco), chl, cytochrome *f*, and sucrose phosphate synthase activity [41,44]. These results suggest that the photosynthetic capacity of individual tomato leaves was lower under NR compared with that under PO. These results support the presence of physiological factors for reduced NAR under NR. Therefore, the significantly lower TDM was caused not only by the lower transmitted PPFD and LAI but also by the reduced photosynthetic capacity of individual leaves under NR.

5. Conclusions

NIR reflective film suppressed the extreme increase in air and fruit temperatures in the high tunnel, and total yield did not differ between the high tunnels. Moreover, the rate of fruit cracking was reduced significantly under NR. These results suggest that the reduction in fruit cracking under NR resulted from decreased fruit DM content, caused by lower fruit temperature and decreased TDM. In contrast, TDM was decreased under NR. It was caused not only by lower transmitted PPFD and LAI but also by the reduced photosynthetic capacity of individual leaves, resulting from a decrease in both total N and chl content. We conclude that NIR reflective film can reduce the incidence of fruit cracking in exchange for a slight reduction in TDM.

Supplementary Materials: The following supporting information can be downloaded at: https://www.mdpi.com/article/10.3390/horticulturae8010051/s1, Table S1: Effects of high tunnel-covering film on total nitrogen (N) and total N allocation to each vegetative organ at 42 and 92 DAT. Figure S1: The difference in daily mean air temperature under NR compared to PO throughout the experiment.

Author Contributions: Conceptualization, M.I. and Y.I.; methodology, H.Y. and K.K.; formal analysis, H.Y.; investigation, H.Y., S.F., N.T., I.A., Y.W. and T.A.; resources, T.A., I.A. and Y.W.; writing—original draft preparation, H.Y.; writing—review and editing, K.K., Y.N., I.A., Y.W., T.A. and Y.I.; supervision, H.Y., Y.I., K.K. and M.I.; project administration, M.I.; funding acquisition, M.I. All authors have read and agreed to the published version of the manuscript.

Funding: This research was supported by grants from the Project of the Bio-oriented Technology Research Advancement Institution (R&D Matching funds on the field for Knowledge Integration and innovation).

Institutional Review Board Statement: Not applicable.

Informed Consent Statement: Not applicable.

Data Availability Statement: The data that support the findings of this study are restrictions apply to the availability of these data, which were used under license for the current study, and so are not publicly available. Data are however available from the corresponding author (M.I.) upon reasonable request and with permission.

Acknowledgments: We thank T. Kurisaki and T. Nakayama for technical assistance and Y. Murakami, J. Okubo, J. Ono, N. Nakane, A. Okano, M. Yamaguchi, and T. Kunifuda for their great help in supporting the management of this project.

Conflicts of Interest: The authors declare no conflict of interest.

References

1. Faostat. 2021. Available online: https://www.fao.org/faostat/ (accessed on 17 November 2021).
2. Kinet, J.M.; Peet, M.M. 1997 Tomato. In *The Physiology of Vegetable Crops*; Wien, H.C., Ed.; C.A.B. International: Wallingford, UK, 1997; pp. 207–258. ISBN 9780851991467.
3. Suzuki, K. Physiological disorders and their management in greenhouse tomato cultivation at high temperatures. In *Adaptation to Climate Change in Agriculture*; Izumi, T., Hirata, R., Matsuda, R., Eds.; Springer: Singapore, 2019; pp. 81–96. ISBN 978-981-13-9234-4. [CrossRef]
4. Hemming, S.; Kempkes, F.; Van der Braak, N.; Dueck, T.; Marissen, N. Greenhouse cooling by NIR-reflection. *Acta Hortic.* **2006**, *719*, 97–106. [CrossRef]
5. Mutwiwa, U.N.; Von Elsner, B.; Tantau, H.J.; Max, J.F.J. Cooling naturally ventilated greenhouses in the tropics by near-infra red reflection. *Acta Hortic.* **2008**, *801*, 259–266. [CrossRef]
6. López-Marín, J.; González, A.; García-Alonso, Y.; Espí, E.; Salmerón, A.; Fontecha, A.; Real, A.I. Use of cool plastic films for greenhouse covering in Southern Spain. *Acta Hortic.* **2008**, *801*, 181–186. [CrossRef]
7. Stanghellini, C.; Dai, J.; Kempkes, F. Effect of near-infrared-radiation reflective screen materials on ventilation requirement, crop transpiration and water use efficiency of a greenhouse rose crop. *Biosyst. Eng.* **2011**, *110*, 261–271. [CrossRef]
8. Mutwiwa, U.N.; Von Elsner, B.; Tantau, H.J.; Max, J.F.J. Effects of near infrared reflection greenhouse cooling on blossom-end rot and fruit cracking in tomato (*Solanum lycopersicum* L.). *Afr. J. Hortic. Sci.* **2008**, *1*, 33–43.
9. Ikeda, T.; Ishigami, Y.; Goto, E. The effect of CO_2 enrichment in a closed greenhouse equipped with NIR-reflecting film and EHP cooling on the yield and quality of tomato fruits during the summer season. *J. Agric. Meteorol.* **2020**, *76*, 104–110. [CrossRef]

10. Nakayama, M.; Fujita, S.; Watanabe, Y.; Ando, T.; Isozaki, M.; Iwasaki, Y. The effect of greenhouse cultivation under a heat insulation film covering on tomato growth, yield, and fruit quality in a subtropical area. *Hort. J.* **2021**, *90*, 304–313. [CrossRef]

11. Dorais, M.; Papadopoulos, A.; Gosselin, A. Greenhouse tomato fruit quality. *Hort. A Rev.* **2001**, *26*, 239–319.

12. Frazier, W.A.; Bowers, J.L. A report on studies of tomato fruit cracking in Maryland. *Proc. Soc. Hortic. Sci.* **1947**, *49*, 241–255.

13. Lang, A.; Düring, H. Grape berry splitting and some mechanical properties of the skin. *Vitis* **1990**, *29*, 61–70. [CrossRef]

14. Peet, M.M. Fruit cracking in tomato. *Horttechnology* **1992**, *2*, 216–223. [CrossRef]

15. Blanchard, M.G.; Runkle, E.S. Influence of NIR-reflecting shading paint on greenhouse environment, plant temperature, and growth and flowering of bedding plants. *Trans. ASABE* **2010**, *53*, 939–944. [CrossRef]

16. Lobos, G.A.; Acevedo-Opazo, C.; Guajardo-Moreno, A.; Valdés-Gómez, H.; Taylor, J.; Laurie, V.F. Effects of kaolin-based particle film and fruit zone netting on Cabernet Sauvignon physiology and fruit quality. *J. Int. Sci. Vigne Vin* **2015**, *49*, 137–144. [CrossRef]

17. Spicher, L.; Glauser, G.; Kessler, F. Lipid antioxidant and galactolipid remodeling under temperature stress in tomato plants. *Front. Plant Sci.* **2016**, *7*, 167. [CrossRef] [PubMed]

18. Haque, M.S.; Husna, M.T.; Uddin, M.N.; Hossain, M.S.; Ali, A.K.M.G.; Abdel, O.M.; Latef, A.A.H.A.; Hossain, A. Heat Stress at Early Reproductive Stage Differentially Alters Several Physiological and Biochemical Traits of Three Tomato Cultivars. *Horticulturae* **2021**, *7*, 330. [CrossRef]

19. Pan, T.; Wang, Y.; Wang, L.; Ding, J.; Cao, Y.; Qin, G.; Yan, L.; Xi, L.; Zhang, J.; Zou, Z. Increased CO_2 and light intensity regulate growth and leaf gas exchange in tomato. *Physiol. Plant* **2020**, *168*, 694–708. [CrossRef] [PubMed]

20. Heuvelink, E. Influence of day and night temperature on the growth of young tomato plants. *Sci. Hortic.* **1989**, *38*, 11–22. [CrossRef]

21. Fan, X.X.; Xu, Z.G.; Liu, X.Y.; Tang, C.M.; Wang, L.W.; Han, X.L. Effects of light intensity on the growth and leaf development of young tomato plants grown under a combination of red and blue light. *Sci. Hortic.* **2013**, *153*, 50–55. [CrossRef]

22. Gent, M.P.N. Carbohydrate level and growth of tomato plants: II. The effect of irradiance and temperature. *Plant Physiol.* **1986**, *81*, 1075–1079. [CrossRef]

23. Bruggink, G.T.; Heuvelink, E. Influence of light on the growth of young tomato, cucumber and sweet pepper plants in the greenhouse: Effects on relative growth rate, net assimilation rate and leaf area ratio. *Sci. Hortic.* **1987**, *31*, 161–174. [CrossRef]

24. Abdel-Ghany, A.M.; Al-Helal, I.M.; Alzahrani, S.M.; Alsadon, A.A.; Ali, I.M.; Elleithy, R.M. Covering materials incorporating radiation-preventing techniques to meet greenhouse cooling challenges in arid regions: A review. *Sci. World J.* **2012**, *2012*, 906360. [CrossRef] [PubMed]

25. Japan Meteorological Agency. 2020 Past Weather Data Search. Available online: https://www.jma.go.jp/jma/index.html (accessed on 19 February 2021).

26. Saeki, T. Growth analysis of plants. *Shokubutsugaku Zasshi* **1965**, *78*, 111–119. [CrossRef]

27. Watson, D.J. The dependence of net assimilation rate on leaf-area index. *Ann. Bot.* **1958**, *22*, 37–54. [CrossRef]

28. Porra, R.J.; Thompson, W.A.; Kriedemann, P.E. Determination of accurate extinction coefficients and Simultaneous equations for assaying chlorophylls a and b extracted with four different solvents: Verification of the concentration of chlorophyll standards by atomic absorption spectroscopy. *Biochim. Biophys. Acta* **1989**, *975*, 384–394. [CrossRef]

29. Yasuba, K.; Suzuki, K.; Sasaki, H.; Higashide, T.; Takaichi, M. Fruit yield and environmental condition under integrative environment control for high yielding production at long-time culture of tomato. *Bull. Naro Veg. Flor. Sci.* **2011**, *3*, 19–28. (In Japanese)

30. Hemming, S.; Kempkes, F.; van der Braak, N.; Dueck, T.; Marissen, N. Filtering natural light at the greenhouse covering—Better greenhouse climate and higher production by filtering out NIR? *Acta Hortic.* **2006**, *711*, 411–416. [CrossRef]

31. Kempkes, F.L.K.; Stanghellini, C.; Hemming, S. Cover materials excluding near infrared radiation: What is the best strategy in mild climates? *Acta Hortic.* **2009**, *807*, 67–72. [CrossRef]

32. Corey, K.A.; Tan, Z.Y. Induction of changes in internal gas pressure of bulky plant organs by temperature gradients. *JASHS* **1990**, *115*, 308–312. [CrossRef]

33. Walker, A.J.; Ho, L.C. Carbon translocation in the tomato: Carbon import and fruit growth. *Ann. Bot.* **1977**, *41*, 813–823. [CrossRef]

34. Pearce, B.D.; Grange, R.I.; Hardwick, K. The growth of young tomato fruit. I. Effects of temperature and irradiance on fruit grown in controlled environments. *J. Hortic. Sci.* **1993**, *68*, 1–11. [CrossRef]

35. Pearce, B.D.; Grange, R.I.; Hardwick, K. The growth of young tomato fruit. II. Environmental influences on glasshouse crops grown in rock wool or nutrient film. *J. Hortic. Sci.* **1993**, *68*, 13–23. [CrossRef]

36. Bakker, J.C.; Janse, J. Lage etmaaltemperatuur geeft meer kans op zwelscheuren bij tomaat. *Weekbl. Groenten Fruit* **1988**, *26*, 30–31.

37. Schilstra-van Veelen, I.M.; Bakker, J.C. Cracking of Tomato Fruits. In *Annual Report 1985 Glasshouse Crops Research Station Naaldwijk*; Glasshouse Crops Research Station: Naaldwijk, The Netherlands, 1985; p. 39.

38. Higashide, T.; Yasuba, K.; Suzuki, K.; Nakano, A.; Ohmori, H. Yield of Japanese tomato cultivars has been hampered by a breeding focus on flavor. *HortScience* **2012**, *47*, 1408–1411. [CrossRef]

39. Itoh, M.; Goto, C.; Iwasaki, Y.; Sugeno, W.; Ahn, D.; Higashide, T. Production of high soluble Solids fruits without reducing dry matter production in tomato plants grown in salinized nutrient solution controlled by electrical conductivity. *Hort. J.* **2020**, *89*, 403–409. [CrossRef]

40. Kimura, M.; Fujitani, S.; Itimanda, K. Mitigation techniques on fruit cracking in tomato cultivation under rain shelter in summer and autumn. *Bull. Oita Pref. Agric. Forest. Fish. Res. Cent.* **2012**, *2*, 23–42. (In Japanese)

41. Makino, A.; Sato, T.; Nakano, H.; Mae, T.T. Leaf photosynthesis, plant growth and nitrogen allocation in rice under different irradiances. *Planta* **1997**, *203*, 390–398. Available online: https://link.springer.com/content/pdf/10.1007/s004250050205.pdf (accessed on 17 January 2021). [CrossRef]

42. Kaneko, S.; Higashide, T.; Yasuba, K.; Ohmori, H.; Nakano, A. Effects of planting stage and density of tomato seedlings on growth and yield component in low-truss cultivation. *Hort. Res.* **2015**, *14*, 163–170. (In Japanese) [CrossRef]

43. Higashide, T. Review of dry matter production and light interception by plants for yield improvement of greenhouse tomatoes in Japan. *Hort. Res. (Jpn.)* **2018**, *17*, 133–146. (In Japanese) [CrossRef]

44. Nakano, H.; Makino, A.; Mae, T. The effects of elevated partial pressures of CO_2 on the relationship between photosynthetic capacity and N content in rice leaves. *Plant Physiol.* **1997**, *115*, 191–198. [CrossRef] [PubMed]

 *horticulturae*

Article

Effects of Plastic Shed Cultivation System on the Properties of Red Paddy Soil and Its Management by Reductive Soil Disinfestation

Liangliang Liu [1,2,3,4,*], Sha Long [1], Baoping Deng [1], Jiali Kuang [1], Kexin Wen [1], Tao Li [1], Zurong Bai [1] and Qin Shao [1,*]

[1] College of Life Science and Environmental Resources, Yichun University, Yichun 336000, China; wsycwhzwdnr10086@163.com (S.L.); dengbaoping_0626@163.com (B.D.); jialikuang@126.com (J.K.); wenkexin0812@163.com (K.W.); litao3311860974@163.com (T.L.); bai13872650272@163.com (Z.B.)
[2] School of Geography Science, Nanjing Normal University, Nanjing 210023, China
[3] Engineering Technology Research Center of Jiangxi Universities and Colleges for Selenium Agriculture, Yichun University, Yichun 336000, China
[4] Zhongke Clean Soil (Guangzhou) Technology Service Co., Ltd., Guangzhou 510000, China
* Correspondence: 190030@jxycu.edu.cn (L.L.); shaoqin2013@126.com (Q.S.)

Abstract: Red paddy soil is widely distributed in the south of China and has become an important production system for food and cash crops. However, the key factors limiting the quality of this soil type under the plastic shed cultivation system and the effective management strategies are still unclear. In the present study, the physicochemical and microbial properties of red paddy soil in a plastic shed (PS-Soil) and open-air (OA-Soil) cultivation systems were compared. Subsequently, reductive soil disinfestation (RSD) and organic fertilizer treatment (OF) were used to improve the soil properties in a representative PS-Soil. Results showed that the physicochemical and microbial properties in PS-Soil were significantly altered compared with those in the nearby OA-Soil, and those differences were primarily dominated by the cultivation system rather than the sampling site. Specifically, the electrical conductivity (EC) and available nutrients (NO_3^--N, NH_4^+-N, available K, and available P) contents, as well as the abundances of fungi, potential fungal soil-borne pathogens (*F. oxysporum* and *F. solani*), and fungi/bacteria were significantly increased in PS-Soil. In addition, the OF treatment could not effectively improve the above-mentioned soil properties, which was mainly reflected by that soil EC and the abundances of potential fungal soil-borne pathogens were considerably increased in the OF-treated soil. In contrast, soil EC and NO_3^--N content, the abundances of fungi, *F. oxysporum*, *F. solani*, and fungi/bacteria were remarkably decreased by 76%, 99%, 98%, 92%, 73%, and 85%, respectively. Moreover, soil pH, the abundance of bacteria, total microbial activity, metabolic activity, and carbon source utilization were significantly increased in the RSD-treated soil. Collectively, red paddy soil is significantly degraded under the plastic shed cultivation system, and RSD rather than OF can effectively improve the quality of this soil type.

Keywords: microbial activity; organic fertilizer; plastic shed cultivation system; reductive soil disinfestation; soil-borne pathogens; soil quality

Citation: Liu, L.; Long, S.; Deng, B.; Kuang, J.; Wen, K.; Li, T.; Bai, Z.; Shao, Q. Effects of Plastic Shed Cultivation System on the Properties of Red Paddy Soil and Its Management by Reductive Soil Disinfestation. *Horticulturae* **2022**, *8*, 279. https://doi.org/10.3390/horticulturae8040279

Academic Editor: Giovanni Bubici

Received: 1 March 2022
Accepted: 23 March 2022
Published: 24 March 2022

Publisher's Note: MDPI stays neutral with regard to jurisdictional claims in published maps and institutional affiliations.

1. Introduction

Paddy soil developed from the parent red soil material is one of the most important soil resources in China. It is widely distributed in the tropical and subtropical regions and has become an important production system for food and cash crops [1,2]. Therefore, the rational utilization of red paddy soil to maintain its quality is a major task in China.

Plastic shed cultivation, as a popular soil utilization strategy, can effectively avoid the loss of soil chemical elements and nutrients during rain leaching [3]. However, driven by economic benefits, plastic shed cultivation has gradually developed into an intensive cultivation system in the red paddy soil on a large scale, which is mainly characterized by continuous monoculture and over-fertilization, and thus often leads to soil degradation [4,5]. Many studies [4–7] have demonstrated that the severity of acidification, salinization, accumulation of plant soil-borne pathogens, such as *Fusarium oxysporum*, *Fusarium solani*, and *Ralstonia solanacearum*, and imbalance of microbial community are the common characteristics of the degraded soil under the plastic shed cultivation system, and ultimately lead to plant disease outbreaks and economic losses. For example, our previous study [7] has demonstrated that the abundance of soil-borne pathogens (*Fusarium oxysporum*) was significantly increased in the plastic shed cultivation system and caused severe *Fusarium* wilt disease in lisianthus.

Notably, the contributions of these degraded properties on plant health status may be different across diverse soil conditions. A previous study showed that the degradation of soil microbial properties, such as the accumulation of pathogens and imbalance of microbial community, rather than physicochemical properties, determined the plant health status [8]. However, other studies revealed that the degradation of soil physicochemical properties, such as acidification and salinization, can significantly affect the plant's healthy growth [9,10]. Reflecting on these findings, it is necessary to investigate the main properties that are degraded in red paddy soil under the plastic shed cultivation system and then adopt reasonable management strategies to improve its quality.

Organic amendments, such as organic composts, wastes, and crop residues, have been documented as effective strategies to improve soil quality by increasing soil organic matter and improving soil structure [11–14]. However, their effects on the suppression of soil-borne pathogens remain inconsistent. For example, previous studies have reported that soil with organic amendments stimulated the proliferation of soil-borne pathogens and increased the incidence of plants disease [12,14]. In contrast, soil incorporated with the above-mentioned organic materials, irrigated to saturation, and covered with a plastic film can effectively suppress many soil-borne pathogens by creating a strongly reductive and anaerobic soil environment and producing diverse organic acids [15,16]. This management strategy is called reductive soil disinfestation (RSD), anaerobic soil disinfestation, or biological soil disinfestation in different countries [8,17,18]. Many studies further revealed that RSD can also significantly improve soil physicochemical properties and microbial communities, such as eliminating soil acidification and salinization and increasing soil organic carbon content and microbial diversities [5,8,19]. These results imply that RSD is a promising strategy in the improvement of soil quality.

Here, the physicochemical (soil pH, electrical conductivity, organic carbon and available nutrients contents) and microbial (the abundances of bacteria, fungi, and common soil-borne pathogens of wilt diseases, total microbial activity, and microbial metabolic activity) properties were used in this study as the indicators of soil quality. To reveal the main properties that degraded in red paddy soil under the plastic shed cultivation system (defined as PS-Soil), we collected 24 PS-Soils located in Yichun City, Jiangxi Province, China, and the nearby open-air cultivated soils (OA-Soil) were used as a control group. Subsequently, RSD and organic amendment (application of organic fertilizer) were employed as the management strategies for a representative PS-Soil, to clarify the improvement effects of RSD and organic amendment on the above properties that degradated in the PS-Soil.

2. Materials and Methods

2.1. Sampling Site Description and Soil Collection

Three sampling sites, located at Nanmiao (114°42′ E, 28°09′ N), Xicun (114°21′ E, 28°13′ N), and Xinfang Towns (114°47′ E, 28°09′ N) in Yichun City, Jiangxi Province, China were selected to collect field soil samples (Figure S1). Each sampling site consisted of two cultivation system soils, PS-Soil and OA-Soil. The distance between each sampling site varied from 5 to 30 km. All of the sampling sites harbored the typical characteristics of red paddy soil and were classified as the Ferralic Cambisol based on the FAO classification system [20]. The soil in each cultivation system has a similar planting history. PS-Soil was primarily used for limited rotation of vegetables and melons. Chemical fertilizer combined with organic fertilizer was the most important fertilization pattern in the planting process. The soil-borne diseases in this soil type have exceeded 30% in recent years, and the yield of crops has also largely decreased. OA-Soil was used for the cultivation of grain crops, such as rice and rapeseed. Specifically, PS-Soils were converted from the open-air fields that are similar to the above-mentioned OA-Soil 4 years ago. Thus, the PS-Soils and OA-Soils used in this study were comparable.

The tillage layer soils (0–20 cm depth) were collected from eight greenhouses and three open-air cultivation fields nearby in each sampling site using sampling method according to Huang et al. [21], representing each sampling site contained 8 replicates of PS-Soil and 3 replicates of OA-Soil (regarded as the control group). These soils were then stored at 4 °C to measure physicochemical properties, total microbial activity, and microbial metabolic activity as well as −20 °C for DNA extraction and microbial quantification.

2.2. Field Experiment Design

After obtaining the main properties that degraded in PS-Soils, we selected a representative PS-Soil located in Nanmiao to conduct a management experiment. The initial properties of this soil were as follows: pH 4.11, electrical conductivity (EC) 406 μS cm^{-1}, total organic carbon (TOC) 26 g kg^{-1}, and available K and P 329 and 179 mg kg^{-1}, respectively. Three treatments were performed in this soil: CK, soil without any treatment; OF, soil amended with 7.5 t ha^{-1} organic fertilizer (TOC 380 g kg^{-1}, TN 8 g kg^{-1}, and C/N 48) that comprised chicken manure and rice husks and fully mixed by a rotary cultivator; RSD [5,8], soil firstly incorporated with 15 t ha^{-1} grain chaff (size < 1 mm, TOC 348 g kg^{-1}, TN 5 g kg^{-1}, and C/N 70) and fully mixed by a rotary cultivator, and then irrigated to saturation and covered with a plastic film (transparent, thickness 0.08 mm) (Figure S2). These treatments were incubated in place for 18 days with an average temperature ranging from 28 °C to 40 °C. There were three replicates for each treatment, and each replicate covered an area of 60 m^2 in a randomized complete block design. After treatment, the plastic film in the RSD treatment was removed, and the soil was drained. The soil samples (0–20 cm) were collected from each replicate and stored as described above. In addition to the above-mentioned soil parameters, microbial activity, and metabolic activity were further detected to clarify the effect of RSD treatment.

2.3. Physicochemical Properties Analysis

A total of 42 soil samples [(8 PS-Soils + 3 OA-Soils) × 3 sampling sites + 3 treatments × 3 replicates)] were used for physicochemical properties analysis. Soil pH and EC were detected in slurries (water: soil of 2.5:1 and 5:1) using the S220k and S230 electrode (Mettler, Greifense, Switzerland), respectively. Soil TOC was measured using wet digestion with H_2SO_4-$K_2Cr_2O_7$ [22]. Soil $NO_3{}^-$-N and $NH_4{}^+$-N were extracted using the 2 mol L^{-1} KCl solution (solution: Soil of 5:1) and then detected using a continuous flow analyzer (Sann++; Skalar, Breda, The Netherlands). Available soil phosphorus (AP) and potassium (AK) were extracted using 0.5 mol L^{-1} NaHCO$_3$ and 1 mol L^{-1} ammonium acetate, respectively [23,24]. Their concentrations were then determined by molybdenum-antimony resistance colorimetry and flame photometry, respectively.

2.4. DNA Extraction and Microbial Properties Measurement

Microbial NDA was extracted from 0.5 g soil of each replicate using the FastDNA® Spin Kit (MP Biomedicals, Santa Ana, CA, USA) according to the instruction and finally dissolved in 100 μL of elution buffer. The quality of DNA was determined using a DS-11 spectrophotometer (Denovix, Wilmington, DE, USA). The abundances of soil bacteria, fungi, and potential soil-borne pathogens (*Fusarium oxysporum*, *Fusarium solani*, and *Ralstonia solanacearum*) were quantified using the QuanStudio 3 Real-Time PCR system (Applied

Biosystems, Waltham, MA, USA). The amplification mixtures in each gene were prepared using 10 μL of SYBR Green premix Taq ($2\times$), 6 μL of sterile distilled water, 2 μL of soil DNA, and 1 μL of forward and reverse primers. The amplification primers (Eub338/Eub518 for bacterial 16S rDNA, Flic-F/Flic-R for *R. solanacearum*, ITS1f/ITS2R for fungal ITS, ITS1f/AFP346 for *F. solani*, ITS1f/AFP308 for *F. oxysporum*) and protocols are listed in Table S1. The standard curves and amplification specificity values were generated as previously described [7,8].

Total microbial activity of the soil (3 treatments $\times$ 3 replicates) was determined using the fluorescein diacetate hydrolysis method [25]. Microbial metabolic activity (CK and RSD $\times$ 3 replicates) was determined by measuring the carbon source utilization pattern using Biolog EcoPlates™ (Biolog, Inc., Hayward, CA, USA) [26]. Briefly, 10 g of fresh soil sample in each treatment was shaken in 90 mL of sterile saline for 30 min at 200 rpm and then diluted to 10^{-3}. Subsequently, 125 μL aliquots of 10^{-3} dilution were inoculated into each well of the ECO MicroPlate and incubated in the dark at 25 °C for 6 days. The optical density (OD) at 590 nm was measured at 0, 24, 48, 72, 96, 120, and 144 h by using an automated microplate reader (BioTek Instruments Inc., Winooski, VT, USA). The average well color development (AWCD) of all carbon sources was calculated as described by the previous study [27], and then clustered as carboxylic acids, carbohydrates, phenolic acids, amino acids, polymers, and amines. This carbon source utilization pattern was used to indicate soil microbial metabolic activity.

2.5. Data Analysis

To compare the differences in physciochemical and microbial properties (Tables S2 and S3) between PS-Soil and OA-Soil, nonmetric multidimensional scaling (NMDS) analysis based on Bray–Curtis distance matrices was performed using the R "phyloseq" package [28]. Relative contributions of sampling site and cultivation system on the differences in physicochemical and microbial properties were calculated using the variance partitioning analysis (VPA) and permutational multivariate analysis of variance (PERMANOVA) with R "vegan" package [29]. Relationships between physicochemical and microbial properties were analyzed using the redundancy analysis and Pearson correlation of heatmap analysis through Wekemo Bioincloud (https://www.bioincloud.tech, accessed date: 24 February 2022). The statistical differences in physicochemical properties and microbial abundance ($\log_{10}$-transformed) between PS-Soil and OA-Soil, as well as metabolic activity and microbial activity among different treatments, were determined using the two independent-samples *t*-test and Duncan's test, respectively.

3. Results

3.1. Differences in Physicochemical and Microbial Properties between PS-Soil and OA-Soil and Their Contributors

The physicochemical and microbial properties of PS-Soil and OA-Soil collected from each sampling site were listed in Tables S2 and S3, which showed that these properties of the soils between different cultivation systems were largely different. Except for pH, TOC, the abundance of *R. solanacearum*, *R. solanacearum*/bacteria, and *F. solani*/fungi, the rest of physicochemical and microbial properties exhibited significant ($p < 0.05$) differences between all of the PS-Soils and OA-Soils ($p < 0.05$) (Table 1). Compared with that in OA-Soil, the abundance of bacteria was significantly ($p < 0.05$) reduced by 53.34% in the PS-Soil (Table 1). However, soil EC and contents of NH_4^+-N, NO_3^--N, AK, AP, and the abundances of fungi, *F. oxysporum* and *F. solani*, as well as the fungi/bacteria and *F. oxysporum*/fungi were significantly ($p < 0.05$) increased in the PS-Soil (Table 1). Moreover, the potential fungal pathogens *F. oxysporum* and *F. solani* were significantly ($p < 0.05$) increased by 89.23% and 97.87% in the PS-Soil, respectively.

Table 1. The compositions of physicochemical and microbial properties between PS-Soil and OA-Soil.

Soil Properties	PS-Soil	OA-Soil
pH	5.11 ± 1.30 a	5.30 ± 0.41 a
EC (μS cm^{-1})	476.6 ± 202.3 a	51.02 ± 29.07 b
NH$_4^+$-N (mg kg^{-1})	13.00 ± 11.89 a	2.67 ± 0.47 b
NO$_3^-$-N (mg kg^{-1})	148.7 ± 85.33 a	5.61 ± 6.50 b
TOC (g kg^{-1})	34.41 ± 4.83 a	36.62 ± 6.90 a
AK (mg kg^{-1})	410.7 ± 103.10 a	132.0 ± 40.37 b
AP (mg kg^{-1})	85.14 ± 32.49 a	15.49 ± 7.42 b
Moisture (%)	19.88 ± 2.77 b	27.32 ± 2.52 a
Bacterial abundance (log$_{10}$ 16S rDNA copies g^{-1} soil)	9.51 ± 0.23 b	9.74 ± 0.09 a
Fungal abundance (log$_{10}$ ITS copies g^{-1} soil)	8.34 ± 0.21 a	7.84 ± 0.35 b
F. oxysporum abundance (log$_{10}$ ITS copies g^{-1} soil)	6.96 ± 0.28 a	5.66 ± 0.68 b
F. solani abundance (log$_{10}$ ITS copies g^{-1} soil)	7.24 ± 0.48 a	5.19 ± 0.92 b
R. solanacearum abundance (log$_{10}$ *flic* copies g^{-1} soil)	5.78 ± 0.31 a	5.97 ± 0.17 a
Fungi/bacteria (%)	8.95 ± 6.37 a	1.57 ± 1.23 b
R. solanacearum/bacteria (%)	0.02 ± 0.01 a	0.02 ± 0.01 a
F. oxysporum/fungi (%)	4.88 ± 2.66 a	1.33 ± 1.40 b
F. solani/fungi (%)	18.72 ± 30.47 a	0.77 ± 0.82 a

Values (means $\pm$ standard deviations) within the same line followed by different letters are significantly different ($p < 0.05$) according to the independent-samples t-test. EC: electrical conductivity; TOC: total organic carbon; AK: available potassium; AP: available phosphorus. PS-Soil and OA-Soil represent the plastic shed and open-air cultivation soils, respectively.

NMDS results also showed that the compositions of physicochemical and microbial properties between PS-Soil and OA-Soil were significantly ($p < 0.05$) different (Figure 1a,b). The differences in physicochemical and microbial properties between PS-Soil and OA-Soil were remarkably ($p < 0.05$) affected by sampling site and cultivation system, while the relative contributions of cultivation system on the differences in soil physicochemical and microbial properties were higher than that of the sampling site (Figure 1c,d).

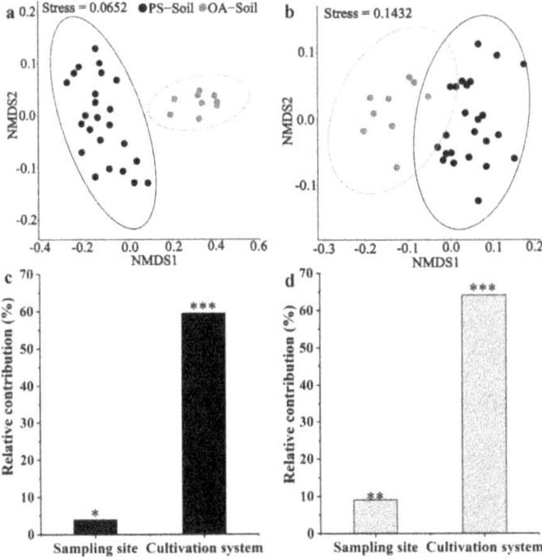

Figure 1. The dissimilarities in physicochemical (**a**) and microbial (**b**) properties between plastic shed and open-air cultivation soils and their contributors (**c,d**). Nonmetric Multidimensional Scaling (NMDS) analyses were conducted using the Bray-Curtis matrices of the physicochemical and microbial properties among different soils. The ellipses indicate confidence intervals of 0.95. The relative contributions of the sampling site and cultivation system on the dissimilarities in physicochemical (**c**) and microbial (**d**) properties were analyzed by variance partitioning analyses (VPA). * $p < 0.05$, ** $p < 0.01$, and *** $p < 0.001$, which were calculated with 999 permutations using PERMANOVA. PS-Soil and OA-Soil represent the plastic shed and open-air cultivation soils, respectively.

3.2. Relationships between Physicochemical and Microbial Properties in the PS-Soil and OA-Soil

RDA result indicated that the microbial properties in the PS-Soil were significantly and negatively ($p < 0.05$) correlated with the pH and moisture content, whereas there was a significant ($p < 0.05$) and positive correlation with the soil EC and contents of NH_4^+-N, NO_3^--N, AK, and AP (Figure 2a). Furthermore, the abundances of soil bacteria and *R. solanacearum* were significantly ($p < 0.05$) and positively correlated with the soil pH and moisture content, whereas the abundances of fungi, *F. oxysporum*, and *F. solani*, and fungi/bacteria, *F. solani*/fungi were considerably ($p < 0.05$) and positively correlated with soil EC and contents of NO_3^--N, AK, and AP (Figure 2b).

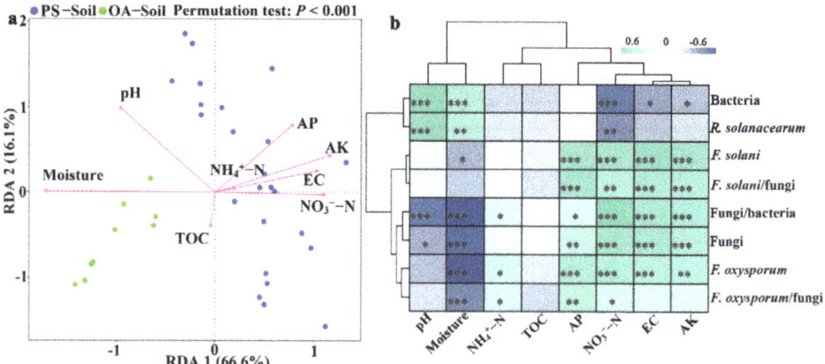

Figure 2. RDA (redundancy analysis) ordination plot (**a**) and correlation heatmap (**b**) showing the relationships between physicochemical and microbial properties in different soils. * $p < 0.05$, ** $p < 0.01$, *** $p < 0.001$, and a blank represents a nonsignificant ($p > 0.05$) correlation. The key from green (positive correlation) to blue (negative correlation) indicates the changes of Pearson correlation coefficients (r). EC: electrical conductivity; TOC: total organic carbon; AK: available potassium; AP: available phosphorus. PS-Soil and OA-Soil represent the plastic shed and open-air cultivation soils, respectively.

3.3. Effects of Different Treatments on Soil Physicochemical Properties

Compared with that in the CK soil, soil pH was significantly increased ($p < 0.05$) by 0.54 and 1.12 after OF and RSD treatment, respectively (Table 2). Soil EC and content of NO_3^--N were significantly ($p < 0.05$) decreased by 76.04% and 99.47% after RSD treatment in comparison to CK soil, respectively, while which showed an increasing trend after OF treatment (Table 2).

Table 2. Soil physicochemical properties among different treatments.

Treatments	pH	EC (μS cm^{-1})	NO_3^--N (mg kg^{-1})
CK	4.11 ± 0.11 c	406.30 ± 33.06 b	135.00 ± 2.25 a
OF	4.65 ± 0.14 b	536.13 ± 42.90 a	154.37 ± 32.48 a
RSD	5.23 ± 0.02 a	97.34 ± 6.60 c	0.71 ± 0.37 b

Values (means $\pm$ standard deviations) within the same column followed by different letters are significantly different ($p < 0.05$) according to the Duncan test. EC: electrical conductivity. CK: soil without any treatment; OF: soil incorporated with organic fertilizer; RSD: soil incorporated with grain chaff, irrigated to saturation, and covered with plastic film.

3.4. Effects of Different Treatments on Soil Microbial Abundances

After treatment, the abundances of bacteria in OF-treated (3.23×10^{10} copies g^{-1} soil) and RSD-treated (4.42×10^{10} copies g^{-1} soil) soils were significantly ($p < 0.05$) increased by 0.29 and 0.77 times compared with that in the CK (2.49×10^{10} copies g^{-1} soil) soil, respectively (Table 3). Compared with CK, the abundance of fungi (1.17×10^9 copies g^{-1} soil) and the ratio of fungi/bacteria in OF-treated soil were increased ($p < 0.05$) by 6.79

and 4.97 times, respectively, whereas the abundance of fungi (4.01×10^7 copies g^{-1} soil) and the ratio of fungi/bacteria in RSD-treated soil were decreased by 73.47% and 84.80%, respectively ($p < 0.05$) (Table 3).

Table 3. Soil microbial abundances among different treatments.

Microbial Abundance	CK	OF	RSD
Bacteria (log$_{10}$ 16S rDNA copies g^{-1} soil)	10.39 ± 0.00 c	10.51 ± 0.03 b	10.64 ± 0.04 a
Fungi (log$_{10}$ ITS copies g^{-1} soil)	8.16 ± 0.14 b	9.06 ± 0.09 a	7.57 ± 0.19 c
Fungi/bacteria (%)	0.61 ± 0.20 b	3.63 ± 0.67 a	0.09 ± 0.05 c
F. oxysporum (log$_{10}$ ITS copies g^{-1} soil)	6.67 ± 0.26 b	7.24 ± 0.11 a	4.33 ± 0.11 c
F. solani (log$_{10}$ ITS copies g^{-1} soil)	6.51 ± 0.07 b	7.11 ± 0.12 a	4.82 ± 0.03 c
F. oxysporum/fungi	4.02 ± 2.88 a	1.54 ± 0.23 ab	0.06 ± 0.03 b
F. solani/fungi (%)	2.38 ± 1.15 a	1.13 ± 0.35 ab	0.19 ± 0.08 b

Values (means $\pm$ standard deviations) within the same column followed by different letters are significantly different ($p < 0.05$) according to the Duncan test. CK: soil without any treatment; OF: soil incorporated with organic fertilizer; RSD: soil incorporated with grain chaff, irrigated to saturation, and covered with plastic film.

The abundances of *F. oxysporum* (1.81×10^7 copies g^{-1} soil) and *F. solani* (1.31×10^7 copies g^{-1} soil) in OF-treated soil were significantly ($p < 0.05$) increased by 2.53 and 3.03 times, respectively, compared with those in the CK (5.13×10^6 and 3.25×10^6 copies g^{-1} soil) soil, whereas they considerably ($p < 0.05$) decreased by 98.35% and 91.95%, respectively, in the RSD-treated soil (2.19×10^4 and 6.71×10^4 copies g^{-1} soil) (Table 3). Furthermore, the ratios of *F. oxysporum*/fungi and *F. solani*/fungi were remarkably ($p < 0.05$) decreased in the RSD-treated soil in comparison to CK, while the non-significant differences of these parameters were observed between CK and OF-treated soil (Table 3).

3.5. Effects of Different Treatments on Soil Microbial Activity and Metabolic Activity

Compared with that in CK soil, the total microbial activity was significantly ($p < 0.05$) increased in RSD-treated soil, whereas the OF treatment did not affect microbial activity (Figure 3a). Moreover, RSD also significantly ($p < 0.05$) increased the microbial metabolic activity and changed the carbon source utilization pattern compared to that of CK soil (Figure 3b,c). The AWCD values of the carbon sources, such as carbohydrates, carboxylic acids, amino acids, phenolic acids, amines, and polymers, were considerably ($p < 0.05$) increased in the RSD-treated soil (Figure 3c).

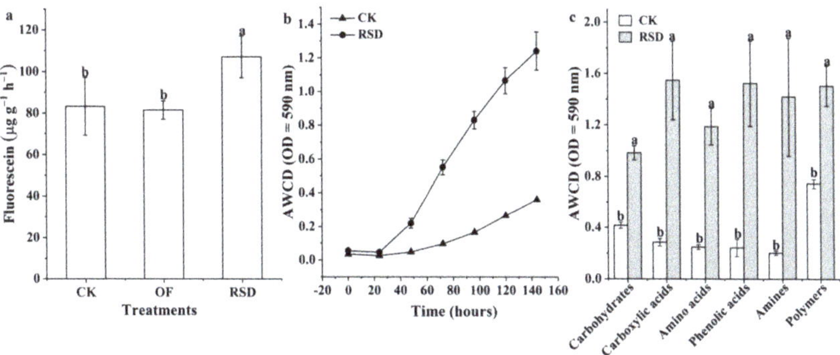

Figure 3. Changes in the microbial activity (**a**), average well color development (AWCD) values (**b**), and categorized carbon utilization pattern at the end of incubation time (**c**). Error bars indicate the standard deviations. Different letters represent significant differences at $p < 0.05$ according to the Duncan test and independent-samples *t*-test. OD: optical density. CK: soil without any treatment; OF: soil incorporated with organic fertilizer; RSD: soil incorporated with grain chaff, irrigated to saturation, and covered with plastic film.

4. Discussion

The primary soil properties restricting its quality are often inconsistent under different conditions [30,31]. In the present study, we found that the physicochemical and microbial properties between PS-Soil and OA-Soil were significantly different, while the differences were mainly dominated by the cultivation system rather than the sampling site. This result was primarily due to the dissimilarities in micro-environments, such as temperature, crop type, and cultivation intensity, as well as water and fertilizer regime, between these two cultivation systems [32–34]. Specifically, many studies have revealed that soil pH and EC are the crucial physicochemical properties affecting soil quality and plant health status [8,9,30]. For example, the soil disease resistance could be significantly inhibited at pH below 5.0, and meanwhile, the crops were susceptible to the toxic effects of aluminum [35]. Furthermore, soil with high EC would increase its osmotic potential and thereby hinder crop uptake of water needed for healthy growth [36]. In this study, we found that the plastic shed cultivation system significantly increased soil EC but had an insignificant effect on soil pH. The increase in soil EC might be primarily caused by overfertilization and lack of rain leaching, which lead to the accumulation of NO_3^- and SO_4^{2-} [37]. In addition, previous studies have reported that the red paddy soils distributed in Jian City, Jiangxi Province, and Hangzhou City, Zhejiang Province, China also had a low pH value (5.03 and 5.97, respectively) [38,39], which is similar to our study. However, the soil organic carbon contents (5.50 and 2.64 g kg^{-1}, respectively) in these studies were obviously lower than our study, which might be related to the differences in artificial management between these areas.

In addition, microbial properties, such as microbial abundance, diversity, and community composition, are the key indicators for evaluating soil quality and are often used to predict plant health status [7,21]. Previous studies have indicated that healthy soils were often dominated by a beneficial bacterial microbiome, such as plant growth-promoting rhizobacteria [40,41], whereas the opposite result was found in the long-term continuous cropping soils that mainly controlled by a harmful fungal microbiome, such as the potential fungal soil-borne pathogens [21,42]. Although the microbial communities were not investigated in this study, we found that the ratio of fungi to bacteria in PS-Soil was significantly increased compared to that in OA-Soil. This result is consistent with the above-mentioned studies and indirectly indicated that the microbial communities could be considerably changed in the PS-Soil. In particular, the potential plant soil-borne pathogens in PS-Soil were dominated by fungal pathogens, such as *F. oxysporum* and *F. solani*, whereas the abundance of the bacterial pathogen (*R. solanacearum*) between PS-Soil and OA-Soil had no significant differences. This result further confirmed our speculation and this difference may be related to the root exudates that are released by melons or some vegetables rather than *solanaceae* crops. As we know, the melons, such as watermelon and cucumber, are the main host crops of *F. oxysporum* and *F. solani*, while *solanaceae* crops, such as tomato and eggplant, are the host crops of *R. solanacearum*. Consistently, the PS-Soils used in this study were primarily characterized by limited rotations of melons and vegetables. Collectively, these results indicated that the quality of red paddy soil has been significantly degraded under the plastic shed cultivation system, especially soil salinization (EC and NO_3^--N), dominant microbiome alteration, and fungal soil-borne pathogens accumulation were the key factors that degraded in the PS-Soil.

Improving soil quality and maintaining soil health by reliable management strategies is essential for the sustainable development of agriculture. In this study, we found that soil amended with organic fertilizer significantly increased the soil EC, abundances of fungi, *F. oxysporum* and *F. solani*, and fungi/bacteria, indicating that the soil quality could be reduced by organic fertilizer treatment. This result was unexpected because the application of organic fertilizer is a common soil management strategy by local farmers. Moreover, although a few similar results with us have been reported in recent years [12,14], the overall evaluation of organic fertilizers are positive [11,13,43]. The results of this study may be related to the following: (1) this organic fertilizer may carry more salt and a larger number of potential soil-borne pathogens or (2) the nutrients, such as carbon sources, of this organic fertilizer were easily utilized by these indigenous soil-borne pathogens. As a result, the salinization and proliferation of soil-borne pathogens could be accelerated in the soil after applying such organic fertilizers. These results further indicate that the application of suitable organic amendments, such as bio-organic fertilizers, is important for improving soil quality.

In contrast, soil salinization, acidification, and the accumulation of soil-borne pathogens, were significantly improved in the RSD-treated soil. This result indicated that RSD is an effective strategy for the improvement of soil quality and is also consistent with many previous studies [8,16–19]. Specifically, the elimination of soil salinization and acidification by RSD was primarily due to the formation of a reductive and anaerobic environment, which can effectively convert the soil nitrate and sulfate into gases of N_2O, N_2, and H_2S and consume a large amount of H^+ [37,44]. The anaerobic environment and organic acids (acetic, propionic, butyric, and isovaleric acids) produced during the RSD process can significantly kill the soil-borne pathogens [18,19]. Meanwhile, we found that the total microbial activity was significantly increased after RSD treatment, while the ratio of fungi to bacteria showed an opposite trend. This result showed that the soil microbial community was changed significantly and may be dominated by a bacterial microbiome. Moreover, previous studies have demonstrated that the dominant microbes during the RSD treatments included many carbon sources decomposers, such as *Clostridium*, *Cytophaga*, *Ruminococcus*, *Chaetomium*, and *Caulobacteracere* [45–47]. It is consistent with our study that the utilization abilities of carbohydrates, carboxylic acids, amino acids, phenolic acids, amines, and polymers by soil microbes were significantly increased in the RSD treatment. Therefore, RSD rather than organic fertilizer can significantly and effectively increase the quality of red paddy soil.

5. Conclusions

In conclusion, our study revealed that the quality of red paddy soil decreased significantly in the plastic shed cultivation system, reflected by severe soil salinization and fungal pathogens accumulation. Organic fertilizer could not effectively remediate soil degradation. In contrast, RSD is an effective tool to eliminate soil salinization and fungal pathogens, increase soil microbial activity and metabolic activity, and improve the quality of red paddy soil under the plastic shed cultivation system.

Supplementary Materials: The following supporting information can be downloaded at: https://www.mdpi.com/article/10.3390/horticulturae8040279/s1, Table S1: Primers list, Table S2: Physicochemical properties of each PS-Soil and OA-Soil, Table S3: Microbial properties of each PS-Soil and OA-Soil. References [48–54] are cited in the supplementary materials. Figure S1: Map of soil sampling sites. PS-Soil and OA-Soil represent the plastic shed and open air cultivation soils, respectively. Figure S2: The operation processes of RSD treatment. (a), soil incorporated with organic material; (b), the fully mixed of soil and organic material using a rotary cultivator; (c), soil irrigated to saturation; and (d), soil covered with a plastic film. This illustration was taken from our previous RSD experiment rather than from this study, but it was not published anywhere.

Author Contributions: Conceptualization, L.L. and S.L.; methodology, B.D.; software, B.D. and J.K.; validation, K.W., T.L. and Z.B.; formal analysis, J.K. and Z.B.; investigation, S.L.; resources, L.L.; data curation, S.L.; writing—original draft preparation, L.L.; writing—review and editing, Q.S.; visualization, L.L.; supervision, Q.S.; project administration, L.L. and Q.S.; funding acquisition, L.L. and Q.S. All authors have read and agreed to the published version of the manuscript.

Funding: This research was funded by the Key-Area Research and Development Program of Guangdong Province (2020B0202010006), the National Natural Science Foundation of China (32160748, 32160716), the China Postdoctoral Science Foundation (2021M691625), and the Key Research and Development Project (Agriculture) of Yichun City, Jiangxi Province (20211YFN4240).

Institutional Review Board Statement: Not applicable.

Informed Consent Statement: Not applicable.

Data Availability Statement: Not applicable.

Conflicts of Interest: The authors declare no conflict of interest.

References

1.   Gao, S.J.; Zhang, R.G.; Cao, W.D.; Fan, Y.Y.; Gao, J.S.; Huang, J.; Bai, J.S.; Zeng, N.H.; Chang, D.N.; Shimizu, K.Y.; et al. Long-term rice-rice-green manure rotation changing the microbial communities in typical red paddy soil in south china. *J. Integr. Agric.* **2015**, *14*, 2512–2520.
2.   Huang, B.; Li, Z.W.; Li, D.Q.; Yuan, Z.J.; Nie, X.D.; Huang, J.Q.; Zhou, Y.Y. Effect of moisture condition on the immobilization of cd in red paddy soil using passivators. *Environ. Technol.* **2018**, *40*, 2705–2714. [PubMed]
3.   Chen, X.M.; Zhang, Q.; Zeng, S.M.; Chen, Y.; Guo, Y.Y.; Huang, X.Z. Rain-shelter cultivation affects fruit quality of pear, and the chemical properties and microbial diversity of rhizosphere soil. *Can. J. Plant Sci.* **2020**, *100*, 683–691.
4.   Runia, W.T.; Molendijk, L.P.G. Physical methods for soil disinfestation in intensive agriculture: Old methods and new approaches. *Acta Hortic.* **2010**, *883*, 249–258.
5.   Meng, T.Z.; Ren, G.D.; Wang, G.F.; Ma, Y. Impacts on soil microbial characteristics and their restorability with different soil disinfestation approaches in intensively cropped greenhouse soils. *Appl. Microbiol. Biotechnol.* **2019**, *103*, 6369–6383.
6.   Zhao, J.; Li, Y.L.; Wang, B.Y.; Huang, X.Q.; Yang, L.; Lan, T.; Zhang, J.B.; Cai, Z.C. Comparative soil microbial communities and activities in adjacent sanqi ginseng monoculture and maize-sanqi ginseng systems. *Appl. Soil Ecol.* **2017**, *120*, 89–96.
7.   Liu, L.L.; Yan, Y.Y.; Ding, H.X.; Zhao, J.; Cai, Z.C.; Dai, C.C.; Huang, X.Q. The fungal community outperforms the bacterial community in predicting plant health status. *Appl. Microbiol. Biotechnol.* **2021**, *105*, 6499–6513.
8.   Liu, L.L.; Huang, X.Q.; Zhao, J.; Zhang, J.B.; Cai, Z.C. Characterizing the key agents in a disease-suppressed soil managed by reductive soil disinfestation. *Appl. Environ. Microb.* **2019**, *85*, e02992-18.
9.   Li, S.L.; Liu, Y.Q.; Wang, J.; Yang, L.; Zhang, S.T.; Xu, C.; Ding, W. Soil Acidification Aggravates the Occurrence of Bacterial Wilt in South China. *Front. Microbiol.* **2017**, *8*, 703.
10.   Xun, W.B.; Huang, T.; Zhao, J.; Ran, W.; Wang, B.R.; Shen, Q.R.; Zhang, R.F. Environmental conditions rather than microbial inoculum composition determine the bacterial composition, microbial biomass and enzymatic activity of reconstructed soil microbial communities. *Soil Biol. Biochem.* **2015**, *90*, 10–18.
11.   Coventry, E.; Noble, R.; Mead, A.; Whipps, J.M. Suppression of allium white rot (*Sclerotium cepivorum*) in different soils using vegetable wastes. *Eur. J. Plant Pathol.* **2005**, *111*, 101–112.
12.   Tilston, E.L.; Pitt, D.; Groenhof, A.C. Composted recycled organic matter suppresses soil-borne diseases of field crops. *New Phytol.* **2002**, *154*, 731–740. [PubMed]
13.   Bonanomi, G.; Antignani, V.; Capodilupo, M.; Scala, F. Identifying the characteristics of organic soil amendments that suppress soilborne plant diseases. *Soil Biol. Biochem.* **2010**, *42*, 136–144.
14.   Mazzola, M.; Granatstein, D.M.; Elfving, D.C.; Mullinix, K. Suppression of specific apple root pathogens by brassica napus seed meal amendment regardless of glucosinolate content. *Phytopathology* **2007**, *91*, 673–679.
15.   Blok, W.J.; Lamers, J.G.; Termorshuizen, A.J.; Bollen, G.J. Control of soilborne plant pathogens by incorporating fresh organic amendments followed by tarping. *Phytopathology* **2000**, *90*, 253–259.
16.   Momma, N.; Momma, M.; Kobara, Y. Biological soil disinfestation using ethanol: Effect on *Fusarium oxysporum* f. sp. *lycopersici* and soil microorganisms. *J. Gen. Plant Pathol.* **2010**, *76*, 336–344.
17.   Butler, D.M.; Kokalis-Burelle, N.; Muramoto, J.; Shennan, C.; McCollum, T.G.; Rosskopf, E.N. Impact of anaerobic soil disinfestation combined with soil solarization on plant–parasitic nematodes and introduced inoculum of soilborne plant pathogens in raised-bed vegetable production. *Crop Prot.* **2012**, *39*, 33–40.
18.   Momma, N. Biological soil disinfestation (BSD) of soilborne pathogens and its possible mechanisms. *Jpn. Agr. Res. Q.* **2008**, *42*, 7–12.
19.   Zhu, R.; Huang, X.Q.; Zhang, J.B.; Cai, Z.C.; Li, X.; Wen, T. Efficiency of Reductive Soil Disinfestation Affected by Soil Water Content and Organic Amendment Rate. *Horticulturae* **2021**, *7*, 559.
20.   IUSS Working Group. *World Reference Base for Soil Resources 2006*; World Soil Resources Reports; FAO: Rome, Italy, 2007; First Update 2007; NO. 103.
21.   Huang, X.Q.; Zhou, X.; Zhang, J.B.; Cai, Z.C. Highly connected taxa located in the microbial network are prevalent in the rhizosphere soil of healthy plant. *Biol. Fert. Soils* **2019**, *55*, 299–312.
22.   Bremner, J.M.; Jenkinson, D.S. Determination of organic carbon in soil. I. Oxidation by dichromate of organic matter in soil and plant materials. *Eur. J. Soil Sci.* **1960**, *11*, 394–402.
23.   Robertson, J.A. Comparison of an acid and an alkaline extracting solution for measuring available phosphorus in alberta soils. *Can. J. Soil Sci.* **1962**, *42*, 115–121.
24.   Peck, N.H.; Macdonald, G.E. Table beet responses to residual and band-applied phosphorus and potassium. *Agron. J.* **1981**, *73*, 1037–1041.

25. Adam, G.; Duncan, H. Development of a sensitive and rapid method for the measurement of total microbial activity using fluorescein diacetate (FDA) in a range of soils. *Soil Biol. Biochem.* **2001**, *33*, 943–951.
26. Garland, J.L.; Mills, A.L. Classification and characterization of heterotrophic microbial communities on the basis of patterns of community-level sole-carbon-source utilization. *Appl. Environ. Microbiol.* **1991**, *57*, 2351–2359.
27. Garland, J.L. Analytical approaches to the characterization of samples of microbial communities using patterns of potential C source utilization. *Soil Biol. Biochem.* **1996**, *28*, 213–221.
28. McMurdie, P.J.; Holmes, S.; Michael, W. Phyloseq: An R package for reproducible interactive analysis and graphics of microbiome census data. *PLoS ONE* **2013**, *8*, e61217.
29. Oksanen, J.; Blanchet, F.G.; Friednly, M.; Kindt, R.; Legendre, P.; McGlinn, D.; Minchin, P.R.; O'Hara, R.B.; Simpson, G.L.; Solymos, P.; et al. Vegan: Community Ecology Package. R Package Version 2.5–7. The R Journal. Published 28 November 2020. Available online: https://cran.r-project.org/web/packages/vegan/index.html (accessed on 28 February 2022).
30. Janvier, C.; Villeneuve, F.; Alabouvette, C.; Edel-Hermann, V.; Mateille, T.; Steinberg, C. Soil health through soil disease suppression: Which strategy from descriptors to indicators? *Soil Biol. Biochem.* **2007**, *39*, 1–23.
31. Raaijmakers, J.M.; Paulitz, T.C.; Steinberg, C.; Alabouvette, C.; Moenneloccoz, Y. The rhizosphere: A playground and battlefield for soilborne pathogens and beneficial microorganisms. *Plant Soil* **2009**, *321*, 341–361.
32. Liu, L.L.; Huang, X.Q.; Zhang, J.B.; Cai, Z.C.; Jiang, K.; Chang, Y.Y. Deciphering the relative importance of soil and plant traits on the development of rhizosphere microbial communities. *Soil Biol. Biochem.* **2020**, *148*, 107909.
33. Preem, J.K.; Truu, J.; Truu, M.; Mander, Ü.; Oopkaup, K.; Lõhmus, K.; Helmisaari, H.S.; Uri, V.; Zobel, M. Bacterial community structure and its relationship to soil physico-chemical characteristics in alder stands with different management histories. *Ecol. Eng.* **2012**, *49*, 10–17.
34. Stres, B.; Danevcic, T.; Pal, L.; Fuka, M.M.; Resman, L.; Leskovec, S.; Hacin, J.; Stopar, D.; Mahne, I.; Mandic-Mulec, I. Influence of temperature and soil water content on bacterial, archaeal and denitrifying microbial communities in drained fen grassland soil microcosms. *FEMS Microbiol. Ecol.* **2010**, *66*, 110–122.
35. Watanabe, K.; Matsui, M.; Honjo, H.; Becker, J.O.; Fukui, R. Effects of soil pH on *rhizoctonia* damping-off of sugar beet and disease suppression induced by soil amendment with crop residues. *Plant Soil* **2011**, *347*, 255–268.
36. Nachmias, A.; Kaufman, Z.; Livescu, L.; Tsror, L.; Meiri, A.; Caligari, P.D.S. Effects of salinity and its interactions with disease incidence on potatoes grown in hot climates. *Phytoparasitica* **1993**, *21*, 245–255.
37. Meng, T.Z.; Zhu, T.B.; Zhang, J.B.; Cai, Z.C. Effect of liming on sulfate transformation and sulfur gas emissions in degraded vegetable soil treated by reductive soil disinfestation. *J. Environ. Sci.* **2015**, *36*, 112–120.
38. Zhong, W.H.; Cai, Z.C.; Zhang, H. Effects of Long-Term Application of Inorganic Fertilizers on Biochemical Properties of a Rice-Planting Red Soil. *Pedosphere* **2007**, *17*, 419–428.
39. Dong, W.; Zhang, X.; Wang, H.; Dai, X.; Sun, X.; Qiu, W.; Yang, F. Effect of Different Fertilizer Application on the Soil Fertility of Paddy Soils in Red Soil Region of Suothern China. *PLoS ONE* **2012**, *7*, e44504.
40. Wang, T.T.; Hao, Y.W.; Zhu, M.Z.; Yu, S.T.; Ran, W.; Xue, C.; Ling, N.; Shen, Q.R. Characterizing differences in microbial community composition and function between *fusarium* wilt diseased and healthy soils under watermelon cultivation. *Plant Soil* **2019**, *438*, 421–433.
41. Wei, Z.; Gu, Y.A.; Friman, V.; Kowalchuk, G.; Xu, Y.C.; Shen, Q.R.; Jousset, A. Initial soil microbiome composition and functioning predetermine future plant health. *Sci. Adv.* **2019**, *5*, eaaw0759.
42. Huang, X.Q.; Liu, S.Z.; Liu, X.; Zhang, S.R.; Li, L.; Zhao, H.T.; Zhao, J.; Zhang, J.B.; Cai, Z.C. Plant pathological condition is associated with fungal community succession triggered by root exudates in the plant-soil system. *Soil Biol. Biochem.* **2020**, *151*, 108046.
43. Li, B.Y.; Zhou, D.M.; Cang, L.; Zhang, H.L.; Fan, X.H.; Qin, S.W. Soil micronutrient availability to crops as affected by long-term inorganic and organic fertilizer applications. *Soil Tillage Res.* **2007**, *96*, 166–173.
44. Zhu, T.B.; Dang, Q.; Zhang, J.B.; Müller, C.; Cai, Z.C. Reductive soil disinfestation (RSD) alters gross N transformation rates and reduces NO and N2O emissions in degraded vegetable soils. *Plant Soil* **2014**, *382*, 269–280.
45. Huang, X.Q.; Liu, L.L.; Zhao, J.; Zhang, J.B.; Cai, Z.C. The families *Ruminococcaceae*, *Lachnospiraceae*, and *Clostridiaceae* are the dominant bacterial groups during reductive soil disinfestation with incorporated plant residues. *Appl. Soil Ecol.* **2019**, *135*, 65–72.
46. Huang, X.Q.; Zhao, J.; Zhou, X.; Zhang, J.B.; Cai, Z.C. Differential responses of soil bacterial community and functional diversity to reductive soil disinfestation and chemical soil disinfestation. *Geoderma* **2019**, *348*, 124–134.
47. Liu, L.L.; Kong, J.J.; Cui, H.L.; Zhang, J.B.; Wang, F.H.; Cai, Z.C.; Huang, X.Q. Relationships of decomposability and C/N ratio in different types of organic matter with suppression of *Fusarium oxysporum* and microbial communities during reductive soil disinfestation. *Biol. Control* **2016**, *101*, 103–113.
48. Lane, D.J. 16S/23S rRNA sequencing. In *Nucleic Acid Techniques in Bacterial Systematics*, 2nd ed.; Stackenbrandt, E., Goodfellow, M., Eds.; John Wiley and Sons, Inc.: Chichester, UK, 1991; pp. 115–175.
49. Muyzer, G.; de Waal, E.C.; Uitterlinden, A.G. Profiling of complex microbial populations by denaturing gradient gel electrophoresis analysis of polymerase chain reaction-amplified genes coding for 16S rRNA. *Appl. Environ. Microb.* **1993**, *59*, 695–700.

50. Schonfeld, J.; Heuer, H.; van Elsas, J.D.; Smalla, K. Specific and Sensitive Detection of *Ralstonia solanacearum* in Soil on the Basis of PCR Amplification of fliC Fragments. *Appl. Environ. Microb.* **2003**, *69*, 7248–7256.
51. Gardes, M.; Bruns, T.D. TS primers with enhanced specificity for basidiomycetes–application to the identification of mycorrhizae and rusts. *Mol. Ecol.* **1993**, *2*, 113–118.
52. Vilgalys, R.; Hester, M. Rapid genetic identification and mapping of enzymatically amplified ribosomal DNA from several *Cryptococcus* species. *J. Bacteriol.* **1990**, *172*, 4238–4246.
53. Lievens, B.; Brouwer, M.; Vanachter, A.C.R.C.; Cammue, B.P.A.; Thomma, B.P.H.J. Real-time PCR for detection and quantification of fungal and oomycete tomato pathogens in plant and soil samples. *Plant Sci.* **2006**, *171*, 155–165.
54. Lievens, B.; Brouwer, M.; Vanachter, A.C.R.C.; Lévesque, C.A.; Cammue, B.P.A.; Thomma, B.P.H.J. Quantitative assessment of phytopathogenic fungi in various substrates using a DNA macroarray. *Environ. Microbiol.* **2005**, *7*, 1698–1710. [PubMed]

horticulturae

Article

Fruit Quality Response to Different Abaxial Leafy Supplemental Lighting of Greenhouse-Produced Cherry Tomato (*Solanum lycopersicum* var. *Cerasiforme*)

Chengyao Jiang [1,*,†], Jiahui Rao [1,†], Sen Rong [1], Guotian Ding [1], Jiaming Liu [2], Yushan Li [3] and Yu Song [3,*]

[1] College of Horticulture, Sichuan Agricultural University, Chengdu 611130, China; raojiahui1205@163.com (J.R.); rongsen1232022@163.com (S.R.); dingguotian2022@163.com (G.D.)
[2] Laboratory of Crop Immune Gene Editing Technology, Chengdu NewSun Crop Science Co., Ltd., Chengdu 611600, China; liujiaming@nwafu.edu.cn
[3] Research Institute of Crop Germplasm Resources, Xinjiang Academy of Agricultural Sciences, Urumqi 830091, China; lys8302@163.com
* Correspondence: catherinejiang@126.com (C.J.); songyu150@163.com (Y.S.)
† These authors contributed equally to this work.

Abstract: Insufficient light supply for canopies is a constant issue during greenhouse production in most areas of Northern China. Applying supplemental lighting to plant canopies is an efficient method of solving this problem. Several studies were conducted to identify the optimal, economically efficient abaxial leafy supplemental lighting mode to produce high-quality greenhouse tomatoes. In this experiment, no supplemental treatment was used as a blank control (CK), while three supplemental lighting modes were used as treatments: T1, continuous supplemental lighting from 8:00–9:00 (at GMT+8, which is 6:00–7:00 local time, before the thermal insulation covers, abbreviated as TIC below, opening), and 20:00–22:00 (after TIC closing) with photosynthetic photon fluxion density (PPFD) of 200 $\mu mol \cdot m^{-2} \cdot s^{-1}$; T2, dynamic altered supplemental lighting with PPFD rising from 100 $\mu mol \cdot m^{-2} \cdot s^{-1}$ to 200 $\mu mol \cdot m^{-2} \cdot s^{-1}$ before TIC opening and falling from 200 $\mu mol \cdot m^{-2} \cdot s^{-1}$ to 100 $\mu mol \cdot m^{-2} \cdot s^{-1}$ after TIC closing; and T3, intermittent supplemental lighting which was automatically conducted with PPFD of 100 $\mu mol \cdot m^{-2} \cdot s^{-1}$ when indoor PPFD below 150 $\mu mol \cdot m^{-2} \cdot s^{-1}$ from 8:00–22:00. The results demonstrated that abaxial leafy supplemental lighting treatment could improve both fruit yield and quality. The total yield in the T1 and T2 treatments was higher than in other treatments, though there was no significant difference. Differences in leaf carbon exportation showed the possibility of determining fruit yield from the 3rd leaf under the fruit. The overall appearance, flavor quality, nutrient indicators, and aroma of cherry tomato fruits under T1 and T2 treatments were generally higher than in other treatments. Correlation analysis of fruit yield and quality parameters suggested that they produce relatively high yield and fruit quality. Combined with a cost-performance analysis, dynamic altered supplemental lighting (T2) is more suitable for high-valued greenhouse cherry tomato production.

Keywords: aromatic substances; carbon exportation; flavor quality; nutrient and function indicator; overall appearance

Citation: Jiang, C.; Rao, J.; Rong, S.; Ding, G.; Liu, J.; Li, Y.; Song, Y. Fruit Quality Response to Different Abaxial Leafy Supplemental Lighting of Greenhouse-Produced Cherry Tomato (*Solanum lycopersicum* var. *Cerasiforme*). *Horticulturae* **2022**, *8*, 423. https://doi.org/10.3390/horticulturae8050423

Academic Editors: Xiaohui Hu, Shiwei Song and Xun Li

Received: 18 April 2022
Accepted: 6 May 2022
Published: 10 May 2022

Publisher's Note: MDPI stays neutral with regard to jurisdictional claims in published maps and institutional affiliations.

1. Introduction

Light makes it possible for plants to conduct photosynthesis and generate energy and also triggers numerous physiological processes, such as seed germination [1,2], leaf development [3,4], flowering [5,6], stomatal regulation [7,8], and membrane transport of cells [9]. All of this extensively regulates the growth and development of plants and determines greenhouse crop productivity. However, insufficient canopy lighting during greenhouse production often occurs to recurring snowy or rainy weather during winter and early spring in most Northern China areas. In Southern Xinjiang, this problem is even more serious in the early spring due to the prevalence of sand and dust, which leads to low-light stress. Low-light stress usually causes low temperatures and high humidity indoors, which decreases tomato yield and quality and frequently causes diseases.

Supplemental lighting is an artificial light resource that is applied to plant canopies and is considered an efficient method of solving the problem of indoor light insufficiency. Numerous studies have been conducted on the canopy layer [10,11], light source [12], light intensity [13,14], and light period [15] to assess the effects of supplemental lighting. At the same time, studies on light and other environmental factors [16–19] have recently become a popular research topic. Research on effective positioning of supplemental lighting is increasing, and selecting an efficient irradiated surface of the functional blade is a good starting point [14,20–24]. These results demonstrate that abaxial leafy supplemental lighting can significantly improve the photosynthetic efficiency of leaves, improve the yield and quality of tomato fruit, and produce higher economic returns.

However, previous research on abaxial leafy supplemental lighting has primarily focused on tomato yield, while the fruit quality only involves limited indicators, such as soluble solids and ascorbic acid. Only a few experimental results have assessed fruit volatile matter contents. In [24], the researcher found that a relatively high PPFD abaxial leafy supplemental lighting mode (before TIC opening and after TIC closing) could improve the total content of aromatic components in common tomatoes [24]. As the high-end consumer market has rapidly developed, cherry tomatoes, with their attractive appearance, unique flavor, and tempting aroma, have become popular and produced higher profits for greenhouse growers [25]. Therefore, it is important to explore methods of increasing the quality of cherry tomatoes and implementing economically efficient abaxial leafy supplemental lighting, which produces high-quality fruit. Therefore, this study systematically investigates and outlines how different intensities and times of abaxial leafy supplemental lighting affect tomato quality, including appearance indicators, nutritional parameters, flavor indexes, and aromatic contents.

2. Materials and Methods

2.1. Plant Material and Growth Conditions

Tomato plants ('Meiying', Institute of Vegetables and Flowers, China Academy of Agricultural Science, Beijing, China) were substrate cultivated (Peilei No. 2, Peile Organic Fertilizer Co., Zhenjiang, China) in a Venlo-type intelligently controlled greenhouse in Atushi City, Xinjiang, China ($39°43'48''$ N, $76°7'12''$ E) from October 2020 to May 2021. Seeds were sown in plug trays filled with the same substrate and germinated in a germination-nursery greenhouse for 24 days (3 days in warm darkness and 21 days in a nursery chamber with environment parameters described in Jiang et al., 2017) [21]. Plants were transplanted at a density of 12 plants $\cdot m^{-2}$, and the variation of the thermal environment index is shown in Figure 1. During the experiment, the daily indoor amount of solar radiation was typically below 10 MJ$\cdot m^{-2}$, except for October. Air conditioners were used to maintain the thermal environment, making the average daily air temperature around 18–25 °C, while the daily mean relative humidity was kept above 65%, and the CO_2 concentration was 400–600 μmol$\cdot$mol^{-1}.

Each treatment included 2 rows and 40 plants, while 1 row and 20 plants mutually settled to avoid inter-group interference in each experimental area. In contrast, 66 plants were set at the boundary to prevent boundary effects. A total of 3 replicate test areas were used.

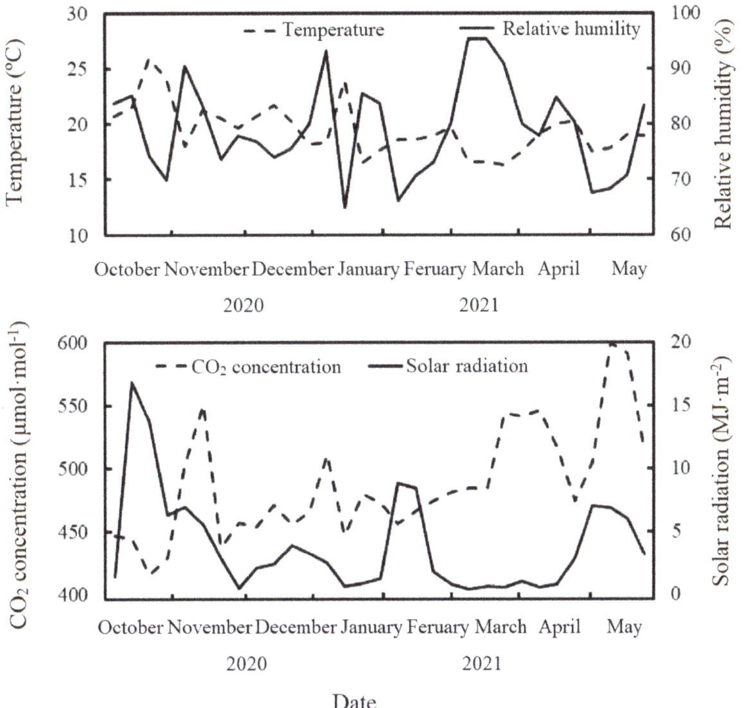

Figure 1. Variation in daily mean air temperature, relative humidity, CO_2 concentration, and integral amount of solar radiation inside the greenhouse from October 2020 to May 2021.

2.2. Supplemental Lighting Treatment

Light-emitting diodes (LEDs; Chenhua lighting Co. Ltd., Guangzhou, China), with a light wavelength of white (W, all wavelength), red (R, peak 635 nm), blue (B, peak 480 nm) were applied at a ratio of 3:2:1 (Figure 2). Lighting began from the fruit setting to the mature green stage [24] and was orientated from underneath the canopy and maintained 10 cm away from the abaxial epidermis of the lowest functional leaf [21]. A no-supplement treatment was used as a blank control (CK), while three supplemental lighting mode treatments were used (Table 1). The lighting system, linked with a time and a TIC time module, was operated automatically from late November 2020 to early April 2021.

Table 1. The arrangement of different abaxial leafy supplemental lighting treatments.

Treatment	Supplemental Lighting Arrangement
CK	No supplement treatment, a blank control
T1	Continuous supplemental lighting during 8:00–9:00 (before TIC opening), and 20:00–22:00 (after TIC closing) with PPFD of 200 $\mu mol \cdot m^{-2} \cdot s^{-1}$
T2	Dynamic altered supplemental lighting: PPFD rising from 100 $\mu mol \cdot m^{-2} \cdot s^{-1}$ to 200 $\mu mol \cdot m^{-2} \cdot s^{-1}$ (8:00–9:00) and falling from 200 $\mu mol \cdot m^{-2} \cdot s^{-1}$ to 100 $\mu mol \cdot m^{-2} \cdot s^{-1}$ (20:00–22:00).
T3	Intermittent supplemental lighting, which automatically used PPFD of 100 $\mu mol \cdot m^{-2} \cdot s^{-1}$ when indoor PPFD is below 150 $\mu mol \cdot m^{-2} \cdot s^{-1}$ from 8:00–22:00.

Damping coil with a computer used software that simulated sunrise and sunset and was connected in T2 to simulate a natural solar period.

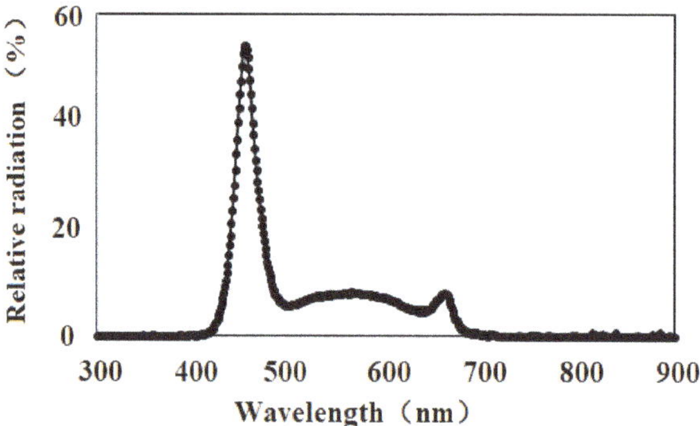

Figure 2. Relative spectral photon flux of polychromatic LEDs was used for the supplemental lighting treatment in this experiment. The wavelengths of the light sources were recorded at 240–800 nm with a spectrometer (SR9910-v7, Irradiant Ltd., Tranent, UK). A digital timer, dimmer, and transformer were used to maintain the light period and light intensity.

2.3. Fruit Yield and Carbon Sequestration Analyses

The fresh weight of each fruit truss was measured to calculate the total yield, while 15 fruit trusses treatment-1 were used for dry weight determination. Plant ^{13}C feeding treatment was conducted on the 15th day when the third fruit truss began lighting treatments for 3 h [14]. Three plants were randomly selected from each treatment for sampling, which was repeated 3 times. The ^{13}C content in the fruit and the 1st to 3rd leaves under the fruit [14] were analyzed by a stable isotope ratio mass spectrometer (Integra CN, SerCon Ltd., Cheshire, UK).

2.4. Fruit Appearance and Flavor Quality Analyses

Fruit quality analyses were conducted on the third and fourth trusses [26], and 15 fruits with the same maturity were randomly selected from each treatment for overall appearance index analyses. The transverse and longitudinal diameters were measured with vernier calipers (605A-05, Harbin Measuring & Cutting Tools Group Co., Ltd., Harbin, China), and the fruit shape index was calculated by Formula (1). Tomato peel color, including peel lightness L^*, peel red saturation a^*, and peel yellow saturation b^*, were determined with a color measuring instrument (YQ-Z-48A, Hangzhou Qingtong Boke Co., Hangzhou, China). The color saturation C and the chromaticity angle H were calculated using Formulas (2) and (3), respectively.

$$\text{Fruit shape index} = \frac{\text{Longitudinal Diameter}}{\text{Transverse Diameter}} \tag{1}$$

$$\text{Color saturation } C = a^{*2} + b^{*2} \tag{2}$$

$$\text{Chromaticity angle } H = tanh^{-1}\frac{b^*}{a^*} \tag{3}$$

Fifteen fruits with the same maturity were randomly selected from each treatment, and the soluble solids content, acidity, and sugar–acid ratio of the tomato fruits were determined with a sugar–acid integrated machine (PAL-BXIACID3, ATAGO Co. Ltd., Tokyo, Japan); another 15 fruits with the same maturity were randomly selected for soluble sugar content determination using the anthrone colorimetric method [27].

2.5. Fruit Nutrient and Function Indicator Analyses

Fifteen fruits with the same maturity were randomly selected from each treatment, and the content of soluble protein in the tomato fruit was determined using the Coomassie brilliant blue method. In contrast, the ascorbic acid content was determined by the molybdenum blue colorimetric method [27]. The lycopene content was determined by the spectrophotometric method [28], and the nitrate–nitrogen content was determined by the salicylic acid colorimetric method [27].

2.6. Fruit Aromatic Substances Analyses

Fifteen fruits with the same maturity were randomly selected from each treatment, and a gas chromatography-mass spectrometry (ISQ 7000 GC-MS, Thermo Fisher Scientific. In. Co. Ltd., Waltham, MA, USA) was used to determine the volatile substances. The volatile substances in each fruit were calculated using a standard internal method with Formula (4).

$$\text{Volatile substances} = \frac{\text{Single peak area of the substance}}{\text{Peak area of internal standard}} \times \frac{\text{Mass of internal standard}}{\text{Mass of total sample}} \quad (4)$$

2.7. Economic Performance Analyses

Each lighting treatment was applied with an ammeter to determine the electricity consumption. Electricity efficiency ($kg \cdot kWh^{-1}$) and cost performance (return/cost) were calculated with Formulas (5) and (6). The average wholesale price of common cherry tomatoes was 10 $CNY \cdot kg^{-1}$, (January to May 2021, Urumqi Wholesale Market), and the electricity unit price was 0.186 $CNY \cdot kWh^{-1}$ (electricity for agricultural usage, State Grid Xinjiang Electric Power Co. Ltd., Xinjiang, China). Each LED unit covered 1 m^2, and the purchase prices were 75 $CNY \cdot unit^{-1}$ (100 $\mu mol \cdot m^{-2} \cdot s^{-1}$) and 95 $CNY \cdot unit^{-1}$ (200 $\mu mol \cdot m^{-2} \cdot s^{-1}$), with a theoretical service life of 50,000 h.

$$\text{Electricity efficiency} = \frac{\text{Yield enhancement with supplemental lighting treatment } \left(kg \cdot m^{-2}\right)}{\text{Electricity consumption } \left(kWh \cdot m^{-2}\right)} \quad (5)$$

$$\text{Cost performance} = \frac{\text{Price of cherry tomato } \left(CNY \cdot kg^{-1}\right) \times \text{Yield enhancement } \left(kg \cdot m^{-2}\right)]}{\text{Price of electricity } \left(CNY \cdot kg^{-1}\right) \times \text{Electricity consumption } \left(kWh \cdot m^{-2}\right) + \text{LED input cost } \left(CNY \cdot m^{-2}\right)} \quad (6)$$

The simplified calculation of LED input cost = (LED purchase price × usage time) / (lighting range × product life)

2.8. Statistical Analyses

Statistical software (SPSS 11.0, SPSS Inc., Chicago, IL, USA) was used for all analyses. Mean separations were conducted using a Tukey's HSD test protected by ANOVA (analysis of variance) at $p < 0.05$.

3. Results

3.1. Fruit Yield and Carbon Sequestration of Function Leaves

Table 2 shows that the abaxial leafy supplemental lighting treatments could significantly enhance the fruit yield. There was a similar trend in the fresh weight of single fruit and the fresh weight of single fruit trusses; the highest appeared in T2. However, the total yield in the T1 and T2 treatments was higher than in other treatments, although there was no significant difference between these two.

Table 2. Effects of different abaxial leafy supplemental lighting treatments on fruit yield of cherry tomato.

Treatment	Fresh Weight of Single Fruit (g)	Fresh Weight of Single Fruit Truss (g·truss^{-2})	Dry Weight of Single Fruit Truss (g·truss^{-2})	Total Yield (kg·m^{-2})
CK	18.03 ± 0.11 c	208.03 ± 5.11 c	2.21 ± 0.03 c	15.41 ± 4.61 c
T1	25.88 ± 2.14 ab	265.18 ± 2.14 ab	3.29 ± 0.03 a	22.12 ± 4.14 a
T2	26.81 ± 0.87 a	281.83 ± 4.87 a	3.32 ± 0.02 a	23.35 ± 5.36 a
T3	20.79 ± 1.02 b	254.79 ± 1.02 b	3.28 ± 0.02 a	18.01 ± 4.61 b

Means ± standard deviation (n = 15) with different letters within each row are significantly different using Tukey's HSD test at $p < 0.05$.

The 3 leaves under the fruit were used as the ^{13}C carbon source to evaluate the contribution to fruit material accumulation of the leaves. The distribution patterns of exported ^{13}C in fruit fixed by different source leaves showed that, except for the 3rd leaf under the fruit, the fixed ^{13}C exported 15.3% more to the fruit than remained in the leaf (Figure 3). When the 3rd leaf under the fruit was treated as a ^{13}C source leaf, the values of the ^{13}C ratio in leaf or fruit were lower than the 1st and 2nd source leaves. Data on fixed ^{13}C remaining in plant leaves under CK treatment were significantly higher than the others (Figure 3C).

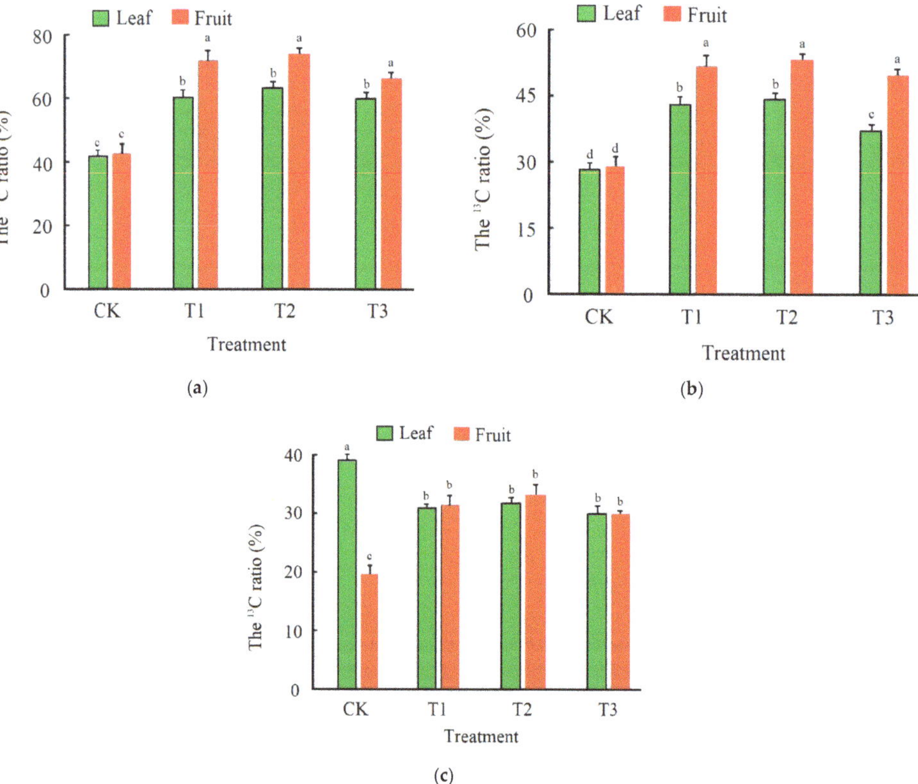

Figure 3. Distribution patterns of exported ^{13}C in fruit fixed by different source leaves. (**a**) The 1st leaf as ^{13}C source leaf, (**b**) the 2nd leaf as ^{13}C source leaf, and (**c**) the 3rd leaf as ^{13}C source leaf. ^{13}CO$_2$ was applied to leaves for 3 h and left to sit for 48 h; the effect of photosynthetic products on fruit development was determined by a stable isotope ratio mass spectrometer. Tukey's HSD test was used, and vertical bars represent standard errors of the means (n = 15). Different letters indicate significant differences at $p < 0.05$, according to Tukey's HSD test.

3.2. Fruit Appearance Characters

Table 3 shows the effects of different abaxial leafy supplemental light treatments on the shape of the tomato fruit. Compared with CK, the transverse diameter of the fruit significantly extended as the supplemental light treatment was applied. However, there was no significant difference in the longitudinal diameter of the fruit among the treatments. The results of the fruit shape index calculation demonstrated that the value under the T1 treatment was the smallest, indicating that the tomato fruit under this treatment was oval-shaped, compared with other treatments.

Table 3. Effect of different abaxial leafy supplemental lighting treatments on fruit shape of cherry tomato.

Treatment	Transverse Diameter (mm)	Longitudinal Diameter (mm)	Fruit Shape Index
CK	27.58 ± 0.17 c	26.36 ± 1.21 a	0.96 ± 0.01 a
T1	32.89 ± 1.11 a	30.92 ± 1.25 a	0.94 ± 0.01 b
T2	35.16 ± 3.48 a	33.47 ± 2.98 a	0.95 ± 0.00 a
T3	30.23 ± 2.36 a	30.09 ± 0.56 a	0.99 ± 0.00 a

Means $\pm$ standard deviation (n = 15) with different letters within each row are significantly different by Tukey's HSD test at $p < 0.05$.

Abaxial leafy supplement lighting treatment could significantly improve the brightness of tomato fruit, the saturation of the red and yellow peel, and the C values of the peel, indicating that deep-colored fruits were produced under these treatments (Table 4). Though the H values were significantly higher in the T1 and T3 treatments, there was no significant difference between the T2 and the CK group.

Table 4. Effect of different abaxial leafy supplemental lighting treatments on fruit color of cherry tomato.

Treatment	Luminosity (L^*)	Pericarp Chroma		Color Saturation (C)	Hue (H)
		Red Saturation (a^*)	Yellow Saturation (b^*)		
CK	45.23 ± 0.02 b	20.13 ± 0.06 b	12.24 ± 0.16 b	23.56 ± 0.61 b	31.30 ± 0.45 b
T1	52.15 ± 1.01 a	26.18 ± 0.13 a	19.04 ± 0.13 a	32.38 ± 0.03 a	36.02 ± 0.12 a
T2	51.40 ± 0.77 a	26.65 ± 0.16 a	18.50 ± 0.22 a	32.44 ± 0.15 a	34.76 ± 0.97 ab
T3	51.67 ± 0.98 a	26.05 ± 0.68 a	18.41 ± 0.59 a	31.90 ± 0.94 a	35.25 ± 1.09 a

Means $\pm$ standard deviation (n = 15) with different letters within each row are significantly different by Tukey's HSD test at $p < 0.05$.

3.3. Fruit Flavor Quality Analyses

Figure 4 shows that the soluble sugar content in tomato plants under CK was significantly lower than under supplemental lighting treatments. The solid soluble content of tomato plants was highest under the T1 treatment, followed by the T2 and T3 treatments. Compared to CK treatment, the values of the T1, T2, and T3 treatments increased by 22.0%, 21.2% and 15.4%, respectively. A similar trend was observed in the contents of total acids and the ratio of sugar to acid in the fruit: the highest values were observed under the T1 and T2 treatments, while the CK treatment had the lowest values.

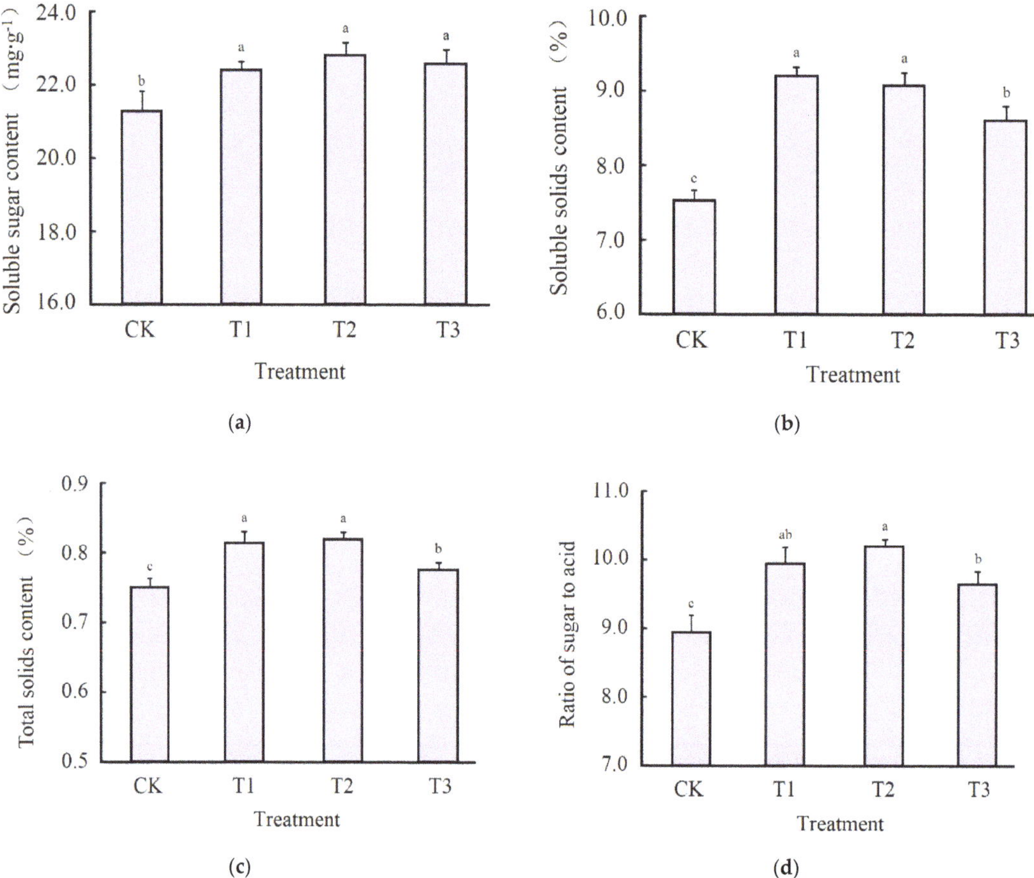

Figure 4. Effect of different abaxial leafy supplemental lighting treatments on flavor quality of cherry tomato. (**a**) Soluble sugar; (**b**) soluble solids; (**c**) total acid; (**d**) ratio of sugar to acid. Tukey's HSD test was used, and vertical bars represent standard errors of the means (*n* = 15). Different letters indicate significant differences at $p < 0.05$ according to Tukey's HSD test.

3.4. Fruit Nutrient and Function Indicator Analyses

Abaxial leafy supplemental lighting treatments positively affected the fruit nutrient and function indicators. The highest value of soluble protein, ascorbic acid, and lycopene content was found in the T1 and T2 treatments, followed by the T3 treatment. Compared to the CK group, the nitrate–nitrogen content in fruits under T1, T2, and T3 treatments decreased by 42.8%, 44.1.3%, and 45.4%, respectively (Figure 5).

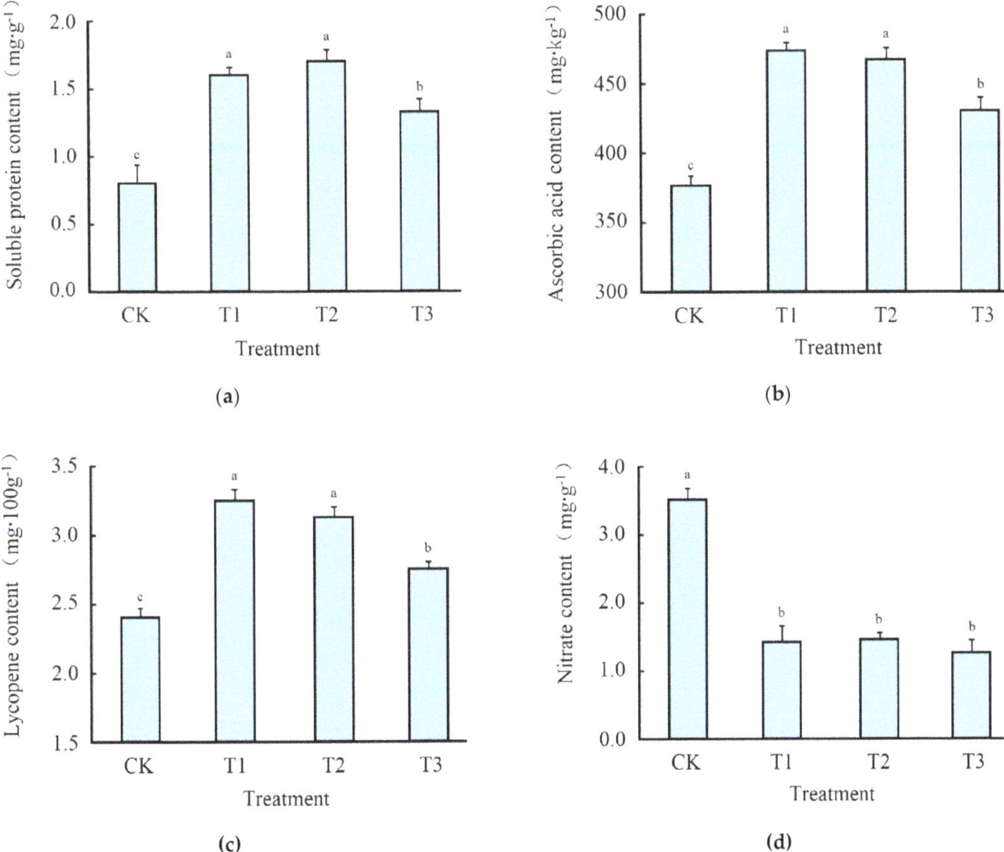

Figure 5. Effect of different abaxial leafy supplemental lighting treatments on nutrient and function quality of cherry tomato. (**a**) Soluble protein; (**b**) ascorbic acid; (**c**) lycopene; (**d**) nitrate. Tukey's HSD test was used, and vertical bars represent the standard error of the means (n = 15). Different letters indicate significant differences at $p < 0.05$ according to Tukey's HSD test.

3.5. Fruit Aromatic Substances Analyses

Forty-five kinds of volatile substances were detected, and the types and contents of volatile substances in each treatment were different (Figures 6 and 7). Of them, the types of aromatic substances in the tomato fruit under the T3 treatment were the richest, with 35 species. The total content of the volatile substances in cherry tomato fruit under the T2 treatment was the largest, 10,266.26 μg·kg^{-1}, which was 52.6% more than in the CK treatment. Nine main aromatic components were detected, including the hexanal, trans-2-heptenal, trans-2-hexenal, trans-2-heptenal, 1-pentene-3-ketone, 6-methyl Base-5-heptene-2-one, β-ionone, 2-phenylethanol, methylsalicylic acid. Of them, eigt were detected in the T2 treatment, and only seven were detected in the CK, T1, and T3 treatments. Hexenal is related to cherry tomato sweetness, and it had the highest value under the T2 treatment. It increased by 215.5%, 8.7%, and 32.9% compared with the CK, T1, and T3 treatments, respectively. The content of 6-methyl-5-heptene-2-ketone is related to the fruit flavor and was highest under the CK treatment. It decreased by 36.1%, 63.5%, and 64.8% compared to the T1, T2, and T3 treatments, respectively. The content of β-ionone, which is related to the sour taste of fruit, was highest in the T3 treatment. Aldehydes and ketones accounted for more than 80% of the total volatile substances, and the relative percentage of alcohols, esters, and hydrocarbons was less than 15%.

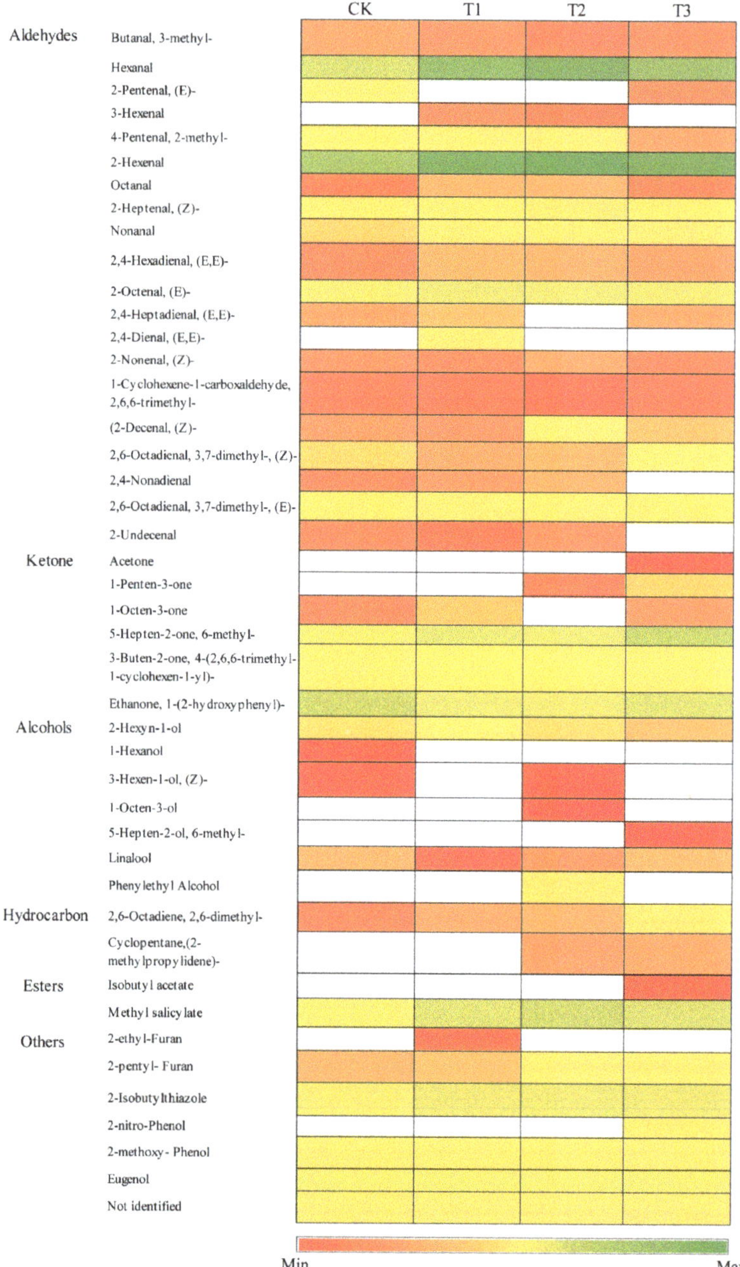

Figure 6. Heat map of fruit aroma components of cherry tomato under different abaxial leafy supplemental lighting treatments. White indicates no data, and the red minimum and green maximum values for the heat map are shown in the color scale bar.

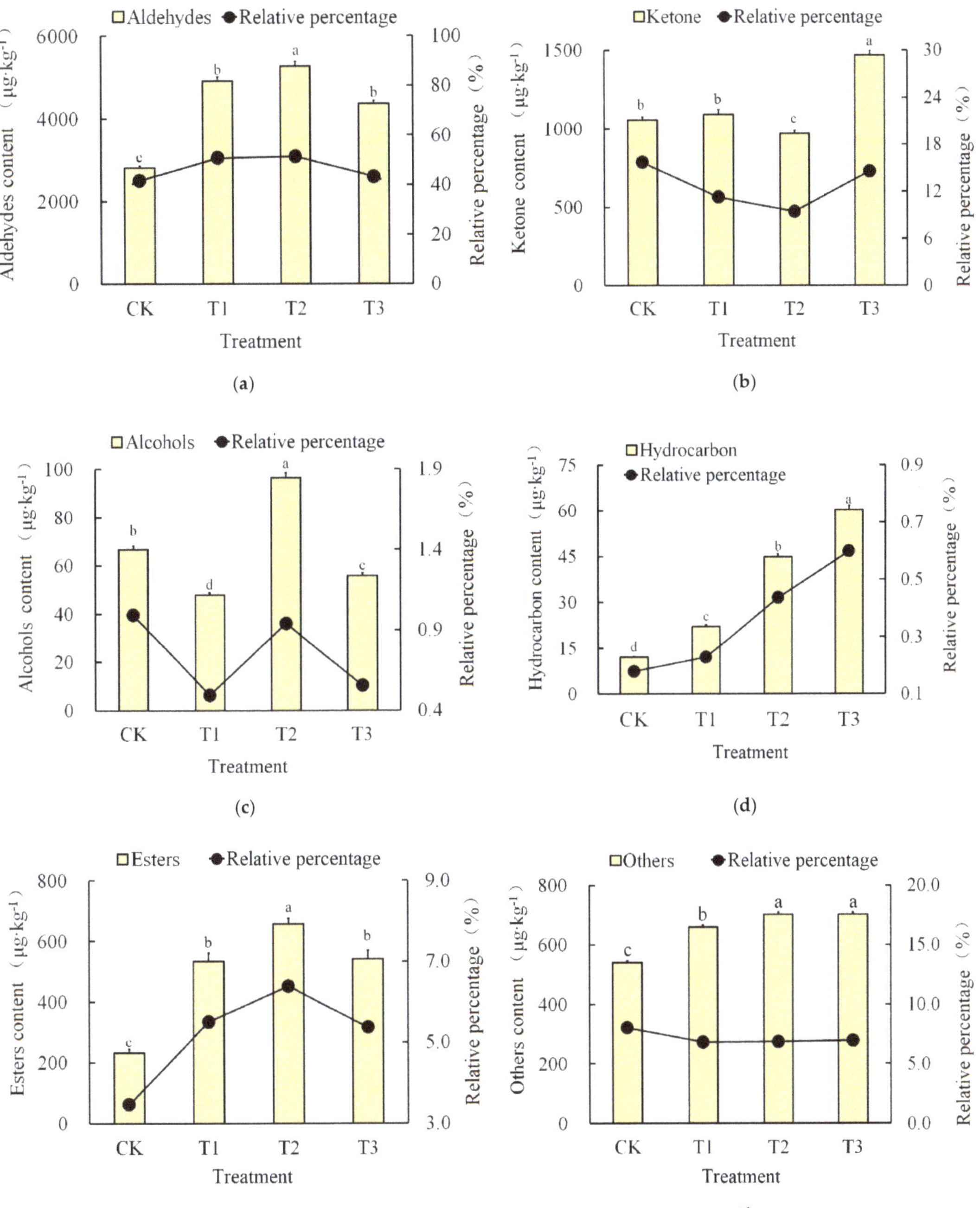

Figure 7. Effect of different abaxial leafy supplemental lighting treatments on the content and relative percentage of various volatile substances of cherry tomato fruit. (**a**) Aldehydes; (**b**) ketone; (**c**) alcohols; (**d**) hydrocarbon; (**e**) esters; (**f**) others. Tukey's HSD test was used, and vertical bars represent standard errors of the means ($n = 15$). Different letters indicate significant differences at $p < 0.05$ according to Tukey's HSD test.

3.6. Correlation Analysis of Fruit Yield and Quality Parameters

The correlation analysis of the selected fruit yield and quality parameters demonstrated that the fresh weight of a single fruit truss and total yield were highly significantly correlated to the relative percentage of aldehydes and the relative ketone percentage of tomato plants ($p < 0.05$; Table 5). Fruit shape index and relative ketone percentage were negatively correlated with other parameters but not with each other. Lycopene content was significantly positively correlated with fruit appearance characteristics, while the ascorbic acid content showed a reverse correlation between fruit shape index and fruit nutrient indicators.

Table 5. Correlation analysis of selected fruit yield, appearance characteristics, flavor quality, nutrients, and the aromatic substance of cherry tomatoes under different abaxial leafy supplemental lighting treatments.

Parameter [y]	FWSFT	TY	FSI	L^*	C	SSC	LC	AAC	ARP	KRP
FWSFT	1									
TY	0.97 *[z]	1								
FSI	0.64	0.83	1							
L^*	0.88	0.95 *	−0.94 *	1						
C	0.66	−0.68	−0.94 *	0.90	1					
SSC	0.83	0.92 *	−0.85	0.95 *	0.95 *	1				
LC	0.87	0.86	−0.87	0.93 *	0.94 *	0.92 *	1			
AAC	0.86	0.78	−0.89 *	0.58	0.62	0.95 *	0.90 *	1		
ARP	0.91 *	0.93 *	−0.40	0.84	0.86	0.91 *	0.81	0.94 *	1	
KRP	−0.96 *	−0.91 *	0.82	−0.95 *	−0.90	−0.88	−0.71	−0.68	−0.94 *	1

[y] Fresh weight of single fruit truss (FWSFT), total yield (TY), fruit shape index (FSI), luminosity (L^*), color saturation (C), soluble solids content (SSC), lycopene content (LC), ascorbic acid content (AAC), aldehydes relative percentage (ARP), ketone relative percentage (KRP). [z] *, significant by t-test at $p < 0.05$.

3.7. Economic Performance Analyses

Statistics on the electricity consumption of each supplementary light treatment (Table 6) demonstrate that the power consumption of the T1 treatment significantly exceeds that of the T2 and T3 treatments, while the yield enhancement was highest in T2, followed by the T1 and T3 treatments. The electricity efficiency results showed a trend similar to yield enhancement. However, after substituting the wholesale price of tomato and the input cost of LED units, the T2 to T3 treatments significantly decreased the value of the final cost performance. In contrast, the T1 treatments were in the middle and did not significantly differ from either. This demonstrates that T2 could produce higher economic benefits.

Table 6. Economic analysis of different abaxial leafy supplemental lighting treatments.

Treatment	Electricity Consumption (kWh·m^{-2})	Yield Enhancement (kg·m^{-2})	Electricity Efficiency (g·kWh^{-1})	Cost Performance
T1	52.09 ± 2.11 a	6.71 ± 1.12 b	128.81 ± 3.65 b	2.03 ± 0.62 ab
T2	45.16 ± 1.88 b	7.94 ± 0.48 a	175.82 ± 2.88 a	2.64 ± 0.33 a
T3	40.99 ± 1.16 c	2.60 ± 0.56 c	63.43 ± 2.12 c	1.57 ± 0.41 b

The average wholesale price of common cherry tomatoes was 10 CNY·kg^{-1} (January to May 2021, Urumqi Wholesale Market), and the electricity unit price was 0.186 CNY·kWh^{-1} (electricity for agricultural usage, State Grid Xinjiang Electric Power Co. Ltd., Xinjiang, China). Each LED unit covered 1 m^2 and the purchase price was 75 CNY·unit^{-1} (100 μmol·m^{-2}·s^{-1}) and 95 CNY·unit^{-1} (200 μmol·m^{-2}·s^{-1}), with a theoretical service life of 50,000 h. Means ($n = 15$) with different letters within each row are significantly different by Tukey's HSD test at $p < 0.05$.

4. Discussion

The overall appearance of cherry tomato fruit, including its shape, size, and color, directly affects whether it will be purchased by consumers and therefore determines its economic potential. Our results demonstrate that abaxial leafy supplemental lighting can significantly increase the single fruit weight and horizontal and vertical diameters of tomato fruits, which is consistent with the results of previous research [14,23,24]. Jiang et al. [23] showed that under abaxial leafy supplemental lighting treatment, the time it takes tomato fruit to change color shortened, the brightness and color of the fruit improved, and the single fruit weight significantly increased, which was consistent with the results of our research. However, variation in fruit yield indicates a relationship between fruit development, with fixed carbon exported in fruit and left in source leaves (Table 2; Figure 3). Previous research results demonstrated that when plants are in the reproductive stage, their carbon exportation pattern changes, and more photosynthetic products are transported to the fruit, while concentrations in different leaves vary [14,23,29]. Since carbon assimilates in functional leaves, three leaves below the fruit largely determine fruit production [12,23], and different photosynthetic products from the 3rd leaf below the fruit under different supplemental lighting treatments (Figure 3) could determine fruit yield. This demonstrates that more research is needed to determine how to develop the potential of this leaf.

Cherry tomato fruit mark indicators typically refer to appearance and peel color. Fruit diameters of this variety usually range from 25 mm to 30 mm and have an oval shape, while environmental factors such as irrigation method [30], cultivation schedule [31], and thermal conditions [32] can largely determine fruit size, which affects fruit quality and shelf life [33]. In this study, fruit under supplemental lighting treatments was larger and had brighter peels than fruit under CK treatment (Tables 3 and 4). This was consistent with previous research [23]. Flavor quality indicators typically include soluble sugar, soluble solids, total acids, and the ratio of sugar to acid. Although the changes in various indicators under supplemental lighting treatments were inconsistent, the contents of soluble solids, total solids, and the ratio of sugar to acid were largest under the T2 treatment (Figure 4). This indicates that this supplemental lighting mode contributed to the accumulation of sweet substances in fruit, which typically sell well. The nutritional quality of tomato fruit is primarily measured by the content of soluble protein, ascorbic acid, and lycopene. Studies have shown that the content of soluble protein and ascorbic acid in tomatoes can be significantly increased by increasing the lighting intensity [12] and the number of fruit truss [14]. In this experiment, the content of soluble protein and ascorbic acid in the T1 and T2 treatments were significantly higher than that for other treatments. This indicates that the total light intensity that the plant received could be higher in these two treatments (Figure 5a,b). Lycopene is a strong antioxidant, which affects scavenging inner free radicals, delaying aging, and preventing cancer and cardiovascular diseases [33]. It is typically used as an important indicator to measure whether the tomato is a functional food. Previous results [24] demonstrated that supplemental lighting could significantly increase the lycopene concentration in tomato fruit. In this experiment, the lycopene content in the T2 and T1 treatments was also higher (Figure 5c), suggesting that these two supplemental lighting modes could promote the healthy properties of tomatoes. However, compared with other parameters, there is a relative lack of research on lycopene response under abaxial supplemental lighting methods, and further study is needed. The tomato fruit aroma is composed of various volatile substances, which can directly affect its flavor. There are currently more than 400 kinds of aromatic substances in tomatoes, including aldehydes, ketones, esters, alcohols, and hydrocarbons [34]. Only nine main aromatic components were detected in this experiment, which could be due to different cultivation conditions and tomato varieties. The T2 process detected the most species (eight) (Figures 6 and 7). Some volatile aromatic substances have a strong odor, and some have a mild odor, though all substances must be assessed to reflect the aromatic characteristics of the fruit [34]. Our results showed that abaxial leafy supplemental lighting could increase the types and total amounts of volatile substances in tomato fruit and that T2 had the highest increase in the total amount of aromatic substances.

The results of our correlation analysis assessing fruit yield and quality parameters in cherry tomatoes demonstrated that fruit yield parameters (FWSFT and TY) were highly significantly correlated to aromatic fruit indicators (ARP and KRP) and were not significantly correlated to the appearance, flavor, or nutrient characteristics (FSI, L^*, C, SSC, LC, and AAC) (Table 6). In general, better yields are typically correlated with poorer fruit quality [30–33]. Our results did not reflect this trend (i.e., FWSFT and TY are positively correlative to ARP), indicating that plants treated with abaxial supplemental lighting could have relatively high yield and high quality. We also previously conducted an economic analysis on abaxial leafy supplemental lighting and classic adaxial leafy supplemental lighting. Our results showed that the electricity efficiency of abaxial leafy supplemental lighting was 46.6% more energy efficient than plants exposed to classic adaxial leafy supplemental lighting. This resulted in a 45.0% higher cost performance for abaxial leafy supplemental lighting, compared to classic adaxial leafy supplemental lighting [14,21]. In this study, the economic analysis still used a uniform unit price rather than step prices in our calculations, which was determined according to different fruit quality. This method should be used in future studies. While it was not a rigorous analysis, the economic performance analysis suggested that T2 had a higher cost return.

5. Conclusions

Abaxial leafy supplemental lighting treatment can optimize both fruit yield and quality, and the leaf carbon exportation showed the possibility of fruit yield determination. The overall appearance, flavor quality, nutrient indicators, and aroma of cherry tomato fruits under the T1 and T2 treatments were higher than in the other treatments. The correlation analysis of fruit yield and quality parameters suggested that this lighting could produce high-yield and high-quality tomato fruit. Combined with the cost performance analysis, the dynamic altered supplemental lighting (T2) is more suitable for high-valued greenhouse cherry tomato production.

Author Contributions: Investigation, analysis, and validation, J.R., S.R., G.D. and Y.L.; formal analysis and writing—original draft preparation; C.J. and J.L.; conceptualization, resources, writing—review and editing, and funding acquisition, C.J. and Y.S. All authors have read and agreed to the published version of the manuscript.

Funding: This research was funded by the National Natural Science Foundation of China, grant number 31960622; the China Postdoctoral Science Foundation, grant number 2021MD703889; the Natural Science Basic Research Program of Shaanxi, grant number 2020JQ-249; the Shaanxi Postdoctoral Research Project Support, grant number 2018BSHYDZZ69.

Institutional Review Board Statement: Not applicable.

Informed Consent Statement: Not applicable.

Data Availability Statement: The datasets generated for this study are available on request to the corresponding author.

Acknowledgments: We would like to thank Xiaoting Zhou for her information support and Jue Wang for her support in laboratory work. We also wish to thank the anonymous reviewers for their valuable comments and suggestions that helped improve the manuscript.

Conflicts of Interest: The authors declare no conflict of interest.

References

1. De Villers, A.J.; van Rooyen, M.W.; Theron, G.K.; van de Venter, H.A. Germination of three Namaqualand pioneer species, as influenced by salinity, temperature and light. *Seed Sci. Technol.* **1994**, *22*, 427–433.
2. Zohar, Y.; Waisel, Y.; Karschon, R. Effects of light, temperature and osmotic stress on seed germination of Eucalyptus occidentalis Endl. *Aust. J. Bot.* **1975**, *23*, 391–397. [CrossRef]

3. Evans, J.; Poorter, H.R. Photosynthetic acclimation of plants to growth irradiance: The relative importance of specific leaf area and nitrogen in maximizing carbon gain. *Plant Cell Environ.* **2001**, *24*, 755–767. [CrossRef]

4. Fu, Q.; Zhao, B.; Wang, X.; Wang, Y.; Ren, S.; Guo, Y. The responses of morphological trait, leaf ultrastructure, photosynthetic and biochemical performance of tomato to differential light availabilities. *Agric. Sci. China* **2011**, *10*, 1887–1897. [CrossRef]

5. Cerdán, P.D.; Chory, J. Regulation of flowering time by light quality. *Nature* **2003**, *423*, 881–885. [CrossRef] [PubMed]

6. Goto, N.; Kumagai, T.; Koornneef, M. Flowering responses to light-breaks in photomorphogenic mutants of Arabidopsis thaliana, a long-day plant. *Physiol. Plantarum.* **1991**, *83*, 209–215. [CrossRef]

7. O'Carrigan, A.; Hinde, E.; Lu, N.; Xu, X.; Duanc, H.; Huang, G.; Mak, M.; Bellotti, B.; Chen, Z. Effects of light irradiance on stomatal regulation and growth of tomato. *Environ. Exp. Bot.* **2014**, *98*, 65–73. [CrossRef]

8. Shimazaki, K.; Doi, M.; Assmann, S.M.; Kinoshita, T. Light regulation of stomatal movement. *Annu. Rev. Plant Biol.* **2007**, *58*, 219–247. [CrossRef]

9. Mullineaux, P.; Karpinski, S. Signal transduction in response to excess light getting out of the chloroplast. *Curr. Opin. Plant Biol.* **2002**, *5*, 43–48. [CrossRef]

10. Hovi, T.; Nakkila, J.; Tahvonen, R. Intra-canopy lighting improves production of year-round cucumber. *Sci. Hortic.* **2004**, *102*, 283–294. [CrossRef]

11. Pettersen, R.I.; Torre, S.; Gislerod, H.R. Effects of intera-canopy lighting on photosynthesis characteristics in cucumber. *Sci. Hortic.* **2010**, *125*, 77–81. [CrossRef]

12. Lu, N.; Maruo, T.; Johkan, M.; Hohjo, M.; Tsukagoshi, S.; Ito, Y.; Ichimura, T.; Shinohara, Y. Effect of supplemental lighting within the canopy at different developing stages on tomato yield and quality of single-truss tomato plant grown at high density. *Environ. Control Biol.* **2012**, *50*, 1–11. [CrossRef]

13. Dorais, M. The use of supplemental lighting for vegetable crop production: Light intensity, crop response, nutrition, crop management, cultural practices. In Proceedings of the Canadian Greenhouse Conference, Toronto, QN, Canada, 1 August 2003.

14. Song, Y. Study on the Relationship between LED Supplementary Light and Greenhouse Tomato Light Utilization Characteristics and Growth and Development. Ph.D. Thesis, China Agricultural University, Beijing, China, 2017. (In Chinese with English Abstract).

15. Tewolde, F.T.; Lu, N.; Shiina, K.; Maruo, T.; Takagaki, M.; Kozai, T.; Wataru, Y. Nighttime supplemental LED inter-lighting improves growth and yield of single-truss tomatoes by enhancing photosynthesis in both winter and summer. *Front. Plant Sci.* **2016**, *7*, 448. [CrossRef] [PubMed]

16. Dou, H.; Niu, G.; Gu, M.; Masabni, J.G. Effects of Light Quality on Growth and Phytonutrient Accumulation of Herbs under Controlled Environments. *Horticulturae* **2017**, *3*, 36. [CrossRef]

17. Rezazadeh, A.; Harkess, R.L.; Telmadarrehei, T. The Effect of Light Intensity and Temperature on Flowering and Morphology of Potted Red Firespike. *Horticulturae* **2018**, *4*, 36. [CrossRef]

18. Zha, L.; Liu, W. Effects of light quality, light intensity, and photoperiod on growth and yield of cherry radish grown under red plus blue LEDs. *Hortic. Environ. Biotechnol.* **2018**, *59*, 511–518. [CrossRef]

19. Ahmed, H.A.; Tong, Y.; Li, L.; Sahari, S.Q.; Almogahed, A.M.; Cheng, R. Integrative Effects of CO_2 Concentration, Illumination Intensity and Air Speed on the Growth, Gas Exchange and Light Use Efficiency of Lettuce Plants Grown under Artificial Lighting. *Horticulturae* **2022**, *8*, 270. [CrossRef]

20. Song, Y.; Jiang, C.; Gao, L. Polychromatic supplemental lighting from underneath canopy is more effective to enhance tomato plant development by improving leaf photosynthesis and stomatal regulation. *Front. Plant Sci.* **2016**, *7*, 1832. [CrossRef]

21. Jiang, C.; Johkan, M.; Hohjo, M.; Tsukagoshi, S.; Ebihara, M.; Nakaminami, A.; Maruo, T. Photosynthesis, plant growth, and fruit production of single-truss tomato improves with supplemental lighting provided from underneath or within the inner canopy. *Sci. Hortic.* **2017**, *222*, 221–229. [CrossRef]

22. Jiang, C.; Johkan, M.; Hohjo, T.; Tsukagoshi, S.; Ebihara, M.; Nakaminami, A.; Maruo, M. Supplemental lighting applied within or underneath the canopy enhances leaf photosynthesis, stomatal regulation and plant development of tomato under limiting light conditions. *Acta Hortic.* **2018**, *1227*, 645–652. [CrossRef]

23. Jiang, C.; Song, Y.; Li, Y. Effects of different supplemental lightning modes on photosynthetic performance and carbon sequestration of tomato leaves in Gobi greenhouse. *China Veg.* **2019**, *10*, 32–38, (In Chinese with English Abstract).

24. Song, Y.; Jiang, C.; Li, Y. Effects of different abaxial leaf supplemental lightning modes on fruit quality of tomato produced in Gobi greenhouses. *Xinjiang Agric. Sci.* **2021**, *58*, 294–303, (In Chinese with English Abstract).

25. National Bureau of Statistics. *China Statistical Yearbook-2021*; China Statistics Press: Beijing, China, 2021.

26. Aguirre, N.C.; Cabrera, F.A.V. Evaluating the fruit production and quality of cherry tomato (Solanum lycopersicum var. cerasiforme). *Rev. Fac. Natl. Agric. Medellín* **2012**, *65*, 6593–6604.

27. Sun, Q.; Zhang, N.; Wang, J.; Zhang, H.; Li, D.; Shi, J.; Li, R.; Weeda, S.; Zhao, B.; Ren, S.; et al. Melatonin promotes ripening and improves quality of tomato fruit during postharvest life. *J. Exp. Bot.* **2015**, *66*, 657–668. [CrossRef] [PubMed]

28. Davis, A.R.; Fish, W.W.; Perkins, P. A rapid spectrophotometric method for analyzing lycopene content in tomato products. *Postharvest Biol. Tec.* **2003**, *28*, 425–430. [CrossRef]

29. Reich, P.B.; Walters, M.B.; Tjoelker, M.G.; Vanderklein, D.; Buschena, C. Photosynthesis and respiration rates depend on leaf and root morphology and nitrogen concentration in nine boreal tree species differing in relative growth rate. *Func. Ecol.* **2010**, *12*, 395–405. [CrossRef]
30. Petrovic, I.; Savic, S.; Jovanovic, Z.; Stikić, R.; Brunel, B.; Sérino, S.; Bertin, N. Fruit quality of cherry and large fruited tomato genotypes as influenced by water deficit. *Zemdirbyste* **2019**, *106*, 123–128. [CrossRef]
31. Dias, N.D.S.; Diniz, A.A.; de Morais, P.L.D.; Pereira, G.D.S.; Sá, F.V.D.S.; Souza, B.G.D.A.; Cavalcante, L.F.; Neto, M.F. Yield and quality of cherry tomato fruits in hydroponic cultivation. *Biosci. J.* **2019**, *35*, 1470–1477. [CrossRef]
32. Gautier, H.; Rocci, A.; Buret, M.; Grasselly, D.; Causse, M. Fruit load or fruit position alters response to temperature and subsequently cherry tomato quality. *J. Sci. Food Agric.* **2005**, *85*, 1009–1016. [CrossRef]
33. Islam, M.Z.; Lee, Y.T.; Mele, M.A.; Choi, I.L.; Kang, H.M. Effect of fruit size on fruit quality, shelf life and microbial activity in cherry tomatoes. *AIMS Agric. Food* **2019**, *4*, 340–348. [CrossRef]
34. Selli, S.; Kelebek, H.; Ayseli, M.T.; Tokbas, H. Characterization of the most aroma-active compounds in cherry tomato by application of the aroma extract dilution analysis. *Food Chem.* **2014**, *165*, 540–546. [CrossRef] [PubMed]

 *horticulturae*

Article

Reductive Soil Disinfestation Enhances Microbial Network Complexity and Function in Intensively Cropped Greenhouse Soil

Yuanyuan Yan [1,†], Ruini Wu [1,†], Shu Li [1], Zhe Su [1], Qin Shao [2], Zucong Cai [1,3,4], Xinqi Huang [1,3,4,5] and Liangliang Liu [1,2,*]

[1] School of Geography, Nanjing Normal University, Nanjing 210023, China; yyy18336222721@163.com (Y.Y.); rainywu0000@163.com (R.W.); lishu0003@163.com (S.L.); lwhbjslh@163.com (Z.S.); zccai@njnu.edu.cn (Z.C.); xqhuang@njnu.edu.cn (X.H.)

[2] Engineering Technology Research Center of Jiangxi Universities and Colleges for Selenium Agriculture, College of Life Science and Environmental Resources, Yichun University, Yichun 336000, China; shaoqin2013@126.com

[3] Jiangsu Center for Collaborative Innovation in Geographical Information Resource Development and Application, Nanjing 210023, China

[4] Jiangsu Engineering Research Center for Soil Utilization & Sustainable Agriculture, Nanjing Normal University, Nanjing 210023, China

[5] Zhongke Clean Soil (Guangzhou) Technology Service Co., Ltd., Guangzhou 510000, China

* Correspondence: 190030@jxycu.edu.cn

† These authors contributed equally to this work.

Citation: Yan, Y.; Wu, R.; Li, S.; Su, Z.; Shao, Q.; Cai, Z.; Huang, X.; Liu, L. Reductive Soil Disinfestation Enhances Microbial Network Complexity and Function in Intensively Cropped Greenhouse Soil. *Horticulturae* **2022**, *8*, 476. https://doi.org/10.3390/horticulturae 8060476

Academic Editors: Xiaohui Hu, Shiwei Song and Xun Li

Received: 2 May 2022
Accepted: 25 May 2022
Published: 27 May 2022

Publisher's Note: MDPI stays neutral with regard to jurisdictional claims in published maps and institutional affiliations.

Abstract: Reductive soil disinfestation (RSD) is an effective practice to eliminate plant pathogens and improve the soil microbial community. However, little is known about how RSD treatment affects microbial interactions and functions. Previous study has shown that RSD-regulated microbiomes may degenerate after re-planting with former crops, while the effect of planting with different crops is still unclear. Here, the effects of both RSD treatment and succession planting with different crops on microbial community composition, interactions, and functions were investigated. Results showed that RSD treatment improves the soil microbial community, decreases the relative abundance of plant pathogens, and effectively enhances microbial interactions and functions. The microbial network associated with RSD treatment was more complex and connected. The functions of hydrocarbon (C, H), nitrogen (N), and sulfur (S) cycling were significantly increased in RSD-treated soil, while the functions of bacterial and fungal plant pathogens were decreased. Furthermore, the bacterial and fungal communities present in the RSD-treated soil, and soil succession planted with different crops, were found to be significantly different compared to untreated soil. In summary, we report that RSD treatment can improve soil quality by regulating the interactions of microbial communities and multifunctionality.

Keywords: reductive soil disinfestation; microbial community; microbial function; co-occurrence network; soil ecosystem

1. Introduction

Plastic shed production systems (PSPSs) play an important role in meeting the increasing food demands of a growing population [1]. However, primarily due to economic factors, PSPSs tend to be dominated by both continuous mono-cropping and over-fertilization [2]. As a result of these poor management practices, soil quality in PSPSs can become severely degraded, and PSPS soils often suffer from imbalanced soil microbiota and elevated plant pathogen populations, resulting in increased plant disease pressure and economic losses [3,4].

Microbial communities and their functional properties are critical for maintaining soil health [5], as microbes are the dominant contributors to both soil ecosystem succession and plant disease suppression [6,7]. For example, soil microbes are heavily involved in carbon, nitrogen, and sulfur cycling, helping to maintain soil nutrient balance and promote healthy plant growth [8]. Furthermore, a highly-connected microbial community network is a key indicator of soil ecosystem stability and functions to guard against invasion by soil-borne pathogens [9]. Microbial communities in healthy soils are often dominated by beneficial microbes which can release a variety of antimicrobial compounds to suppress plant pathogens [10–12]. Additionally, microbial network interactions in healthy soils are more complex than in diseased soils, suggesting that microbes in healthy soils can limit pathogen invasion through niche competition [13,14]. Therefore, how to improve microbial network interactions and functions is a major question in agronomic soil ecology.

Reductive soil disinfestation (RSD), also known as anaerobic soil disinfestation (ASD) or biological soil disinfestation (BSD), is a soil management practice consisting of (1) incorporating organic materials (e.g., crop residues, fresh cover crops, manure, molasses, etc.), (2) irrigating to maximum field capacity, and (3) covering the soil surface with plastic film to produce a reductive and anaerobic soil environment [14,15]. Previous studies have demonstrated that the combination of organic acids, ammonia, hydrogen sulfide, and metal ions produced during RSD treatment can successfully eliminate a broad spectrum of soil-borne pathogens [16–18]. Moreover, in practice, RSD has been shown to produce positive impacts on both soil physicochemical properties and microbial communities, including alleviating soil acidification and salinization, and optimizing the soil microbiome [14,18].

Despite such promising results, we still lack a holistic understanding of how and why RSD results in soil improvement. Most studies on RSD have focused on the effect of this management practice on the above-mentioned soil characteristics, leaving many unanswered questions regarding its effect on soil microbial network interactions and functions. Alarmingly, the results of several studies indicated that the microbiomes of soils treated by RSD can degenerate to their previously diseases state when subjected to prior management practices, such as replanting with the previous crops [19–21]. Liu et al. [20,22] have suggested that this phenomenon can be explained by further degradation of the soil abiotic environment and the reintroduction of root exudates produced by the former crops. However, whether and how disparate crops differentially drive soil microbial community succession remains unclear.

Here, we used RSD to treat diseased, *Fusarium*-infected strawberry cultivation soil. Additionally, we succession-planted this soil with both cabbage and tomato. We sought to answer the following questions: (1) What is the influence of RSD treatment on microbial community interactions? (2) How the microbial community composition and function change when RSD-treated soil is succession planted with different crops?

2. Materials and Methods

2.1. Field Site Description

The experimental field is located at a plastic shed greenhouse in Suzhou City, Jiangsu Province, China (31°17′ N, 120°49′ E). This region has a marine subtropical monsoon climate, with an average annual temperature of 15.7 °C and precipitation of 1100 mm. Strawberry has been planted continuously in this greenhouse for nearly 5 years, and the plant disease incidence has exceeded 30% in recent seasons. The initial soil properties were as follows: pH, 5.26; electrical conductivity (EC), 0.16 mS cm^{-1}; NH$_4^+$-N, 28.68 mg kg^{-1}, and available potassium (AK), 256.0 mg kg^{-1}.

2.2. Field Experiment Design

Two treatments, control ("CTL") and RSD, were performed in the field. CTL was the soil without any treatment, except for its moisture content maintained at 15–20% during the incubation. RSD was the soil amended with 10 t ha^{-1} molasses (TOC, 347.8 g kg^{-1}; TN,

16.8 g kg^{-1}; C/N, 20.7), irrigated to saturation, and covered with transparent plastic film (0.08 mm). The molasses was diluted 20 times before adding to the soil. These treatments were performed for three weeks in September 2019, with the temperature ranging from 25 to 40 °C. Each treatment had three replicates and each replicate covered an area of 80 m^2. After treatment ("AT"), the plastic film was removed and the soil was drained. Subsequently, the soil was planted with cabbage during the first season ("FS", from October to November 2019) and tomato during the second season ("SS", from December 2019 to May 2020). We collected soil samples from each replicate at all three time points (AT, FS, and SS) using the 5–point sampling method [23]. Soil samples were stored at 4 °C for physicochemical analyses and −20 °C for DNA extraction.

2.3. Analysis of Soil Physicochemical Properties

Soil pH and electrical conductivity (EC) were determined at a water/soil (*v:w*) of 2.5:1 and 5:1 using the S220 and S230 metre (Mettler, Greifense, Switzerland), respectively. Soil NH$_4^+$-N was extracted using 2 mol L^{-1} KCl solution and determined by a continuous flow analyser (San + +; Skalar, Breda, The Netherlands). Available potassium (AK) was extracted with 1 mol L^{-1} ammonium acetate and determined by flame photometry.

2.4. Microbial DNA Extraction and Quantification

Microbial DNA was extracted from each 0.50 g soil replicate sample using the FastDNA SPIN Kit (MP Biomedicals, Santa Ana, CA, USA). Both the concentration and purity of DNA were assessed using a DS-11 spectrophotometer (Denovix, Wilmington, DE, USA). Bacterial and fungal abundances were detected using the CFX96^{TM} Real-Time System (Bio-Rad Laboratories Inc., Hercules, CA, USA). The PCR amplification mixtures were prepared using 10 μL of SYBR Green Premix Ex Taq^{TM} (2×, TaKaRa, Kyoto, Japan), 1 μL of each primer (Eub338 ACTCCTACGGGAGGCAGCAG and Eub518 ATTACCGCGGCTGCTGG for bacteria, ITS1f CTTGGTCATTTAGAGGAAGTAA and ITS2R GCTGCGTTCTTCATC-GATGC for fungi), 2 μL of the DNA template, and 6 μL of ddH$_2$O. The amplification conditions and standard curves were established as previously described [24].

2.5. Illumina MiSeq Sequencing and Data Processing

Illumina MiSeq sequencing was used to investigate the compositions, networks, and functions of soil microbial communities. The primers 515F (5′-adapter-MID-GTGCCAGCM GCCGCGG-3′) and 907R (5′-adapter-MID-CCGTCAATTCMTTTRAGTTT-3′) were used to amplify the V4-V5 region of the bacterial 16s rRNA gene [25]. The nuclear ribosomal internal transcribed spacer (ITS) region of fungi was amplified using the primers ITS1F (5′-adapter-MID-CTTGGTCATTTAGAGGAAGTAA-3′) and ITS2 (5′-adapter-MID-GCTGCGTTCTTCATCGATGC-3′) [14]. The reaction mixtures, amplification conditions, and PCR product purification methods were performed as previously described [13]. After quantification, the equimolar-concentration PCR products were sequenced on an Illumina MiSeq Benchtop Sequencer (Illumina Inc., San Diego, CA, USA) at Tianhao Biotechnologies, Inc. (Shanghai, China).

FASTQ sequence data were processed using the QIIME software(version 1.9.1, Colorado, USA) that developed by Caporaso et al [26]. Briefly, the paired-end sequences were merged and quality-controlled using the default arguments implemented in multiple_joined_paired_ends.py and multiple_split_libraries_fastq.py, respectively. The quality-controlled bacterial and fungal sequences were clustered to operational taxonomic units (OTUs) at a similarity of 97%, and then classified and annotated according to the Green-genes 13_8 database (for bacteria) and UNITE database (for fungi), respectively [27,28]. Finally, the bacterial and fungal sequences were rarified to 137,666 and 134,430 in each soil, respectively. Microbial alpha-diversity was analyzed using the default arguments of alpha_diversity.py.

2.6. Microbial Functional Prediction and Data Analysis

The functional compositions of bacterial and fungal communities were predicted using the FAPROTAX (script version 1.2.1, developed by Louca et al., Vancouver, Canada) and FUNGuild (v1.0, developed by Nguyen et al., Minnesota, USA) databases, respectively [29,30]. In order to avoid over-interpreting the fungal functional groups, we retained only those functional groups marked with the "probable" and "highly probable" confidence levels, and deleted those marked only as "possible". In order to compare microbial community and functional dissimilarities, principal coordinates analyses (PCoAs) of the distribution of OTUs, based on the Bray–Curtis distance matrices, were performed using the R 'phyloseq' package [31]. The relative contributions of treatment type and time point on microbial community dissimilarities were analyzed by variance partitioning analysis (VPA) and permutational multivariate analysis of variance (PERMANOVA) using the R 'vegan' package [32]. Co-occurrence network analysis was conducted to investigate the network interactions of microbial communities and functions in different treatments using R 'psych' package, and then visualized with Gephi (v0.9.2, Paris, France) that developed by Bastian et al. [33]. The relative abundances of the OTUs present at greater than 0.15% for bacteria and 0.05% for fungi were retained. Correlation coefficients $|r| < 0.6$ and p-values > 0.05 of the correlation R matrix were removed.

Statistically significant differences in microbial abundances (transformed by $\log_{10}$) between CTL and RSD soils at the same time point were determined using independent-samples t-test, and between different time points using LSD's test. The relationships between dominant genera and microbial functions were visualized by heatmap correlation analysis.

3. Results

3.1. Soil Physicochemical Properties

After treatment, soil NH_4^+-N content was significantly ($p < 0.05$) increased in RSD-treated soil, whereas soil pH, EC, and AK were not significantly different ($p > 0.05$) between RSD and CTL soils (Table S1). During the crops cultivation, except that soil NH_4^+-N and AK contents in RSD-treated soil were significantly higher ($p < 0.05$) than those in CTL soil after FS, other physicochemical properties after FS and all physicochemical properties after SS were not significant different ($p > 0.05$) between RSD and CTL soils (Table S1).

3.2. Quantification of Soil Microbes

After treatment, the abundance of bacteria in RSD-treated soil (1.07×10^{10} 16S rDNA copies g^{-1} dry soil) was significantly ($p < 0.05$) increased by 23.3% compared to CTL soil (8.31×10^9 16S rDNA copies g^{-1} dry soil), while the abundance of fungi and the ratio of fungi to bacteria in RSD-treated soil were considerably ($p < 0.05$) decreased by 75.3% and 80.6%, respectively (Figure 1a–c). When soil was succession-planted, the abundance of bacteria in CTL soil and the abundance of fungi and the ratio of fungi to bacteria in RSD and CTL soils increased during the FS of succession planting, but decreased during the SS, with no significant ($p > 0.05$) differences between CTL and RSD-treated soils (Figure 1a–c).

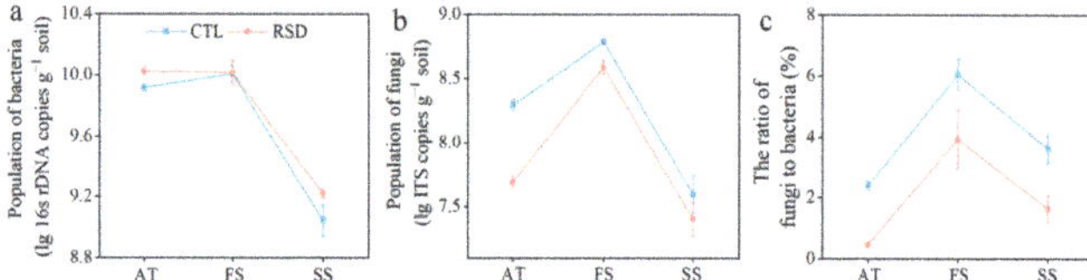

Figure 1. Populations of bacteria (**a**) and fungi (**b**) and the ratio of fungi to bacteria (**c**). Error bars represent the standard errors (SE$_S$). CTL, untreated soil; RSD, soil incorporated with molasses, irrigated to saturation, and covered with plastic film. AT, FS, and SS indicate the soils collected from after treatment, the first season, and the second season, respectively.

3.3. Soil Microbial Community and Functional Diversities

Overall, a total of 3,926,128 16S and 3,732,518 ITS sequences passed quality control. After rarefaction, 17,166 bacterial OTUs and 5447 fungal OTUs clustered at 97% similarity.

In RSD-treated soil, the fungal Shannon diversity during the AT and the species richness during the SS were significantly ($p < 0.05$) decreased in comparison to CTL soil. All other measures of bacterial and fungal diversity showed no significant ($p > 0.05$) differences between CTL and RSD-treated soils (Table S2).

PCoA plots illustrated that the microbial communities and functions were considerably distinct between CTL and RSD-treated soils across the three time points (Figure 2a). VPA and PERMANNOVA indicated that microbial communities and functions were significantly influenced ($p < 0.001$) by both treatment and time point, and the relative contribution of time point (i.e., succession planting with different crops) (24.1~47.5%) was greater than that of treatment (7.7~23.1%) (Figure 2b).

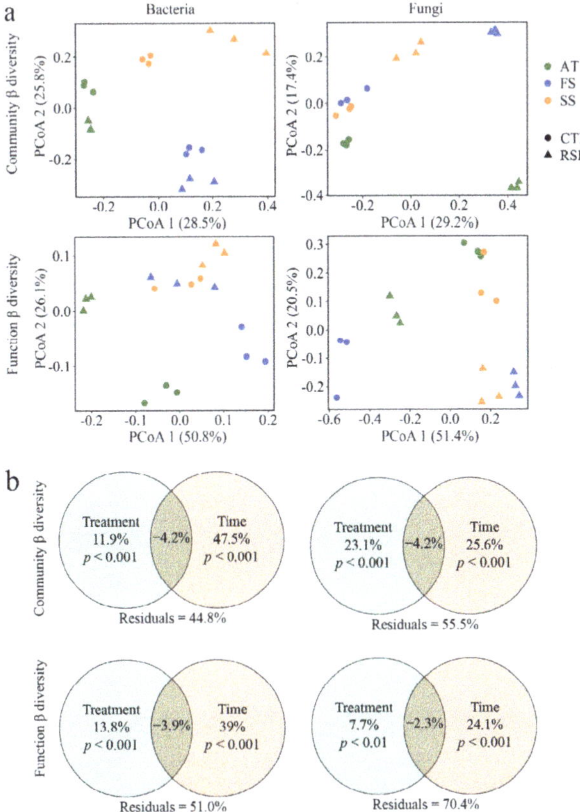

Figure 2. Dissimilarities in soil microbial communities and functions and their contributors. Principal coordinates analyses (PCoAs) (**a**) of soil microbial communities and functions were determined based on Bray−Curtis distances in different treatments and time points. The relative contributions and significant effects of treatments and time points to the dissimilarities in microbial communities and functions were calculated using VPA and PERMANOVA analyses (**b**). CTL, untreated soil; RSD, soil incorporated with molasses, irrigated to saturation, and covered with plastic film. AT, FS, and SS indicate the soils collected from after treatment, the first season, and the second season, respectively.

3.4. Co-Occurrence Networks of Microbial Communities

We found that clear differences in the bacterial and fungal community networks, and their topological characteristics, between CTL and RSD-treated soil (Figure 3, Table 1). For example, the number of nodes and edges, average connectivity, and clustering coefficient of both bacterial and fungal networks were greater in the RSD-treated soil, while modularity and average path length were lessened (Table 1). Besides, the different modules contained distinct microbial compositions, and the phylum compositions of each module were also different between CTL and RSD-treated soils (Figure S1).

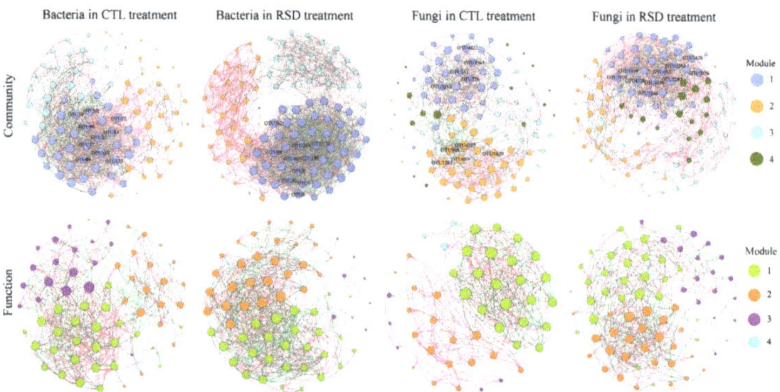

Figure 3. Co-occurrence networks of microbial communities and functions in CTL and RSD soils. The keystone taxa were marked in each network. CTL, untreated soil; RSD, soil incorporated with molasses, irrigated to saturation, and covered with plastic film.

Table 1. Topological characteristics of microbial community and function networks in the CTL and RSD soils.

Topological Characteristics	Bacteria				Fungi			
	Community		Function		Community		Function	
	CTL	RSD	CTL	RSD	CTL	RSD	CTL	RSD
Number of nodes	93	107	64	64	110	111	47	73
Number of edges	1217	1902	514	635	1060	1352	350	533
Average connectivity	26.17	35.55	16.06	20.06	19.27	24.36	14.33	14.60
Modularity	0.28	0.39	0.37	0.23	0.39	0.34	0.35	0.37
Average path length	1.99	1.87	2.07	2.01	2.31	2.10	2.06	2.29
Average clustering coefficient	0.65	0.76	0.64	0.67	0.56	0.61	0.76	0.61

In addition, the identities of the top ten bacterial and fungal keystone taxa between CTL and RSD-treated soils were strikingly different. For example, the bacterial keystone taxa *Comamonadaceae*, *Flavobacteriaceae*, *Rhizobiales*, *Bradyrhizobiaceae*, and *Xanthomonadaceae* were found in CTL soil, whereas *Rhizomicrobium*, *Amycolatopsis*, *Rhodococcus*, *Candidatus_Koribacter*, *Achromobacter*, and *Arachidicoccus* were found in RSD-treated soil (Table S3). The fungal keystone taxa *Aspergillus*, *Rhodotorula*, *Talaromyces*, *Pseudozyma*, and *Simplicillium* were found in CTL soil, while *Penicillium*, *Chaetomium*, *Purpureocillium*, and *Acremonium* were found in RSD-treated soil (Table S3).

3.5. Microbial Communities and Their Functional Compositions

The relative abundances of most microbial genera were altered by both RSD treatment and succession planting (Figure S2). After RSD treatment, the relative abundances of dominant bacterial genera *Rhizomicrobium*, *Paenibacillus*, *Ramlibacter*, *Flavisolibacter*, *Hy-*

drogenispora, *Geobacter*, and *Ruminiclostridium*, and dominant fungal genus *Zopfiella* were significantly increased ($p < 0.05$), whereas the relative abundances of dominant bacterial genera *Bacillus*, *Mizugakiibacter*, *Cladosporium*, *Aspergillus*, *Alternaria*, *Fusarium*, and *Acremonium*, and dominant fungal genus *Gibellulopsis* were significantly decreased ($p < 0.05$) compared to CTL (Figure 4a,b). After the FS of succession planting, the relative abundances of dominant bacterial genera *Rhcrobium*, *Ramlibacter*, *Flavisolibacter*, *Geobacter*, *Ruminiclostridium*, and *Zopfiella*, and dominant fungal genus *Conlarium* were significantly increased ($p < 0.05$), whereas after the SS, the relative abundances of dominant bacterial genera *Streptomyces*, *Paenibacillus*, *Ramlibacter*, and *Flavisolibacter*, and dominant fungal genus *Simplicillium* were significantly increased ($p < 0.05$), compared to CTL (Figure 4a,b).

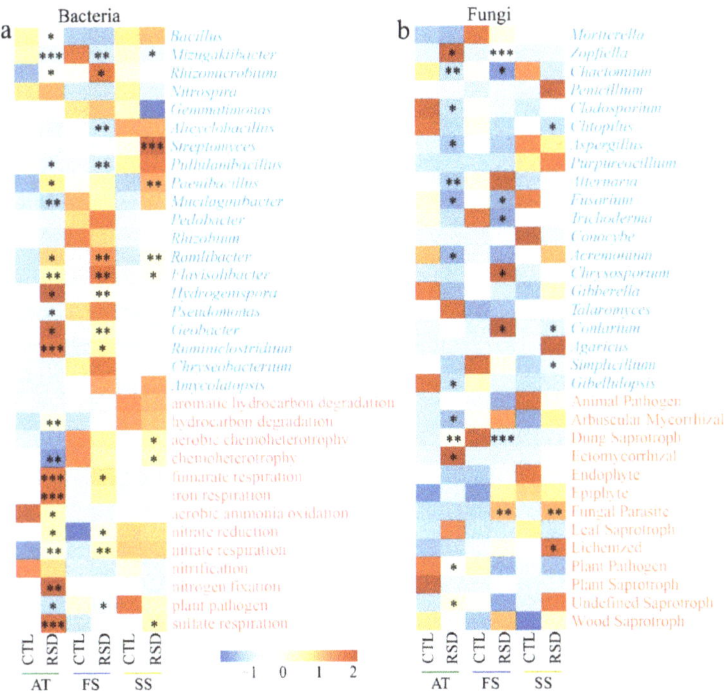

Figure 4. Heatmap displaying the relative abundances of top 20 dominant genera and functional groups in bacteria (**a**) and fungi (**b**). * ($p < 0.05$), ** ($p < 0.01$), and *** ($p < 0.001$) represent significant differences between CTL and RSD in different time points according to independent-samples *t*-test. CTL, untreated soil; RSD, soil incorporated with molasses, irrigated to saturation, and covered with plastic film. AT, FS, and SS indicate the soils collected from after treatment, the first season, and the second season, respectively.

Overall, RSD treatment had a stronger effect on bacterial functional groups than on fungal functional groups. A total of 69 bacterial function groups and 3 fungal trophic modes (pathotroph, symbiotroph, and saprotroph) were found. In RSD-treated soil, the relative abundances of bacterial functional groups related to hydrocarbon (C, H), nitrogen (N), and sulfur (S) cycling, such as hydrocarbon degradation, nitrate reduction, nitrate respiration, nitrogen fixation, sulfate respiration, and fungal functional groups associated with dung saprotrophy, ectomycorrhizal, and undefined saprotrophy were significantly ($p < 0.05$) enriched, compared to CTL. Notably, the relative abundances of bacterial aerobic ammonia oxidation, and both bacterial and fungal plant pathogens were decreased in RSD-treated soils (Figure 4a,b). After the FS of succession planting, the relative abundances of bacterial fumarate respiration, nitrate reduction, nitrate respiration, and fungal parasite, and after the SS, bacterial aerobic chemoheterotrophy, chemoheterotrophy, sulfate respiration, and fungal parasite and lichenized were significantly ($p < 0.05$) increased compared to CTL (Figure 4a,b).

3.6. Correlations between Dominant Genera and Functional Groups

Most of the dominant genera which were increased in RSD-treated soil were significantly ($p < 0.05$) correlated with functional groups and microbial β-diversity indices. For bacteria (Figure 5a), the relative abundance of *Streptomyces* was significantly ($p < 0.05$) and positively correlated with carbon and nitrogen cycling, including hydrocarbon degradation, aromatic hydrocarbon degradation, and nitrate reduction. The relative abundances of *Ramlibacter*, *Flavisolibacter*, *Geobacter*, *Ruminiclostridium*, and *Hydrogenispora* were significantly ($p < 0.05$) and positively correlated with both iron and fumarate respiration. The relative abundances of *Geobacter*, *Ruminiclostridium*, and *Hydrogenispora* were significantly ($p < 0.05$) and positively correlated with both nitrogen fixation and sulfate respiration. For fungi (Figure 5b), the relative abundance of *Zopfiella* was significantly ($p < 0.05$) and positively correlated with ectomycorrhizals, and those of *Penicillium*, *Conlarium*, *Chrysosporium*, *Alternaria*, and *Agaricus* were significantly ($p < 0.05$) and positively correlated with fungal parasites. Additionally, most of the dominant microbial genera were significantly ($p < 0.05$) and positively correlated with microbial β-diversity indices, such as *Rhizomicrobium*, *Streptomyces*, *Paenibacillus*, *Amycolatopsis*, *Zopfiella*, and *Conlarium*.

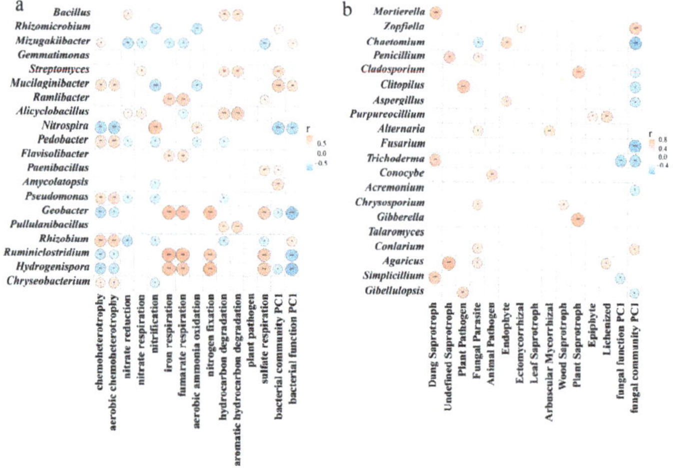

Figure 5. Pairwise spearman's correlations between the relative abundances of dominant genera and microbial functions and β diversity. (**a**) Relationships between the relative abundances of bacterial dominant genera and functional compositions. (**b**) Relationships between the relative abundances of fungal dominant genera and functional compositions. * $p < 0.05$, ** $p < 0.01$, *** $p < 0.001$, and the blank indicates $p > 0.05$. Blue and red colors indicate the negative and positive correlations, respectively.

4. Discussion

Many studies have found that RSD treatment can lead to significant changes in soil microbial communities and help resist invasion by soil-borne pathogens [14,15,34]. In agreement with these studies, we found that the relative abundances of known soil-borne pathogens, such as *Fusarium*, *Aspergillus*, *Alternaria*, and *Cladosporium*, were significantly reduced by RSD treatment. Both the bacterial and fungal communities were considerably improved by RSD treatment, showing remarkable increases in the relative abundances of known disease-suppressive agents, such as *Rhizomicrobium*, *Paenibacillus*, *Flavisolibacter*,

Geobacter, *Hydrogenispora*, and *Zopfiella*. For example, members of the genera *Rhizomicrobium* and *Paenibacillus* can inhibit a broad spectrum of plant pathogens by producing lipopeptides, polypeptides, and bacteriocins [35,36]. Additionally, *Zopfiella* and *Flavisolibacter* have been found to successfully control soil-borne diseases caused by *Fusarium oxysporum*, *Rhizoctonia solani*, and *Cladosporium* in cucumber, Sanqi ginseng, and watermelon [14,20,37].

We also observed that succession planting with different crops (cabbage and tomato) produced similarly striking changes in microbial communities. Specifically, the relative abundances of *Rhizomicrobium*, *Ramlibacter*, *Flavisolibacter*, *Hydrogenispora*, *Geobacter*, *Ruminiclostridium*, *Zopfiella*, *Chrysosporium*, and *Conlarium* after cabbage cultivation, and the relative abundances of *Paenibacillus*, *Streptomyces*, *Ramlibacter*, *Flavisolibacter*, and *Simplicillium* after tomato cultivation, were still significantly enriched in RSD-treated soil. However, the relative abundances of most dominant genera in RSD-treated soil showed a decreased trend during crops cultivation. The trends we observed in the relative abundances were found to be similar to previous studies [19,20,38,39]. That is, these previous studies found that the relative abundances of dominant microbes in RSD-treated soil could return to the same levels of untreated soil after re-planting the former season crops.

Previous studies have suggested that the microbial networks of healthy soils are more complex than those of diseased soils, indicating that the characteristics of the microbial network play an important role in predicting plant health status [9,10,12]. In the present study, we found that both the complexity and connectivity of microbial networks were greater in RSD-treated soils, indicating that RSD treatment can effectively improve the stability of the soil microbial ecosystem. Keystone nodes are generally considered network initiating components, and keystone taxa may play a critical role in maintaining biotic connectivity [40]. We observed that the keystone taxa present in RSD-treated soils included *Amycolatopsis*, *Kribbella*, *Achromobacter*, *Chaetomium*, *Purpureocillium*, *Acremonium*, *Penicillium*, *Rhizomicrobium*, *Candidatus_Koribacter*, *Achromobacter*, *Rhodococcus*, and *Arachidicoccus*. Many of these taxa have the capacity to control soil-borne pathogens and promote plant health, primarily through the production of a variety of antimicrobials. For example, members of *Amycolatopsis* and *Kribbella* produce antimicrobial compounds such as decaplanin and kribellosides to suppress a variety of pathogens [41,42]. Furthermore, *Chaetomium*, *Acremonium*, *Achromobacter*, and *Penicillium* have been widely reported to act as biological control agents effectively against plant pathogens [4,13,43,44]. Collectively, our results illustrate that RSD treatment can promote a highly connected and more complex microbial network.

Although it is clear that RSD can regulate the soil microbial community and suppress soil-borne pathogens, its effects on other soil functions are still not well-characterized. We found that RSD significantly increased microbial functions associated with nutrient cycling, ectomycorrhizal, and dung saprotrophy, and decreased both bacterial and fungal plant pathogens. These increased nutrient cycling functions, including fumarate, nitrate, sulfate, and iron respiration, nitrogen fixation, nitrate reduction, and hydrocarbon degradation, may be responsible for the decrease of NO_3^- and SO_4^{2-} and the increase of organic carbon, NH_4^+-N, and Fe^{2+}, previously reported as occurring during RSD treatment [45,46]. The application of molasses is likely responsible for the enrichment in hydrocarbon degradation and fumarate respiration noted after RSD treatment, as molasses may promote the proliferation of carbon source decomposers [4,34]. It has been reported that the removal of NO_3^- and SO_4^{2-} can decrease aluminum and iron toxicity, as well as suppress pathogenic growth, promoting the accumulation of plant biomass and increasing crop yield [38]. We found that RSD treatment enriched iron respiration functions, likely due to the increased growth of iron-reducing microbes under anaerobic conditions [47]. Additionally, ectomycorrhizal fungi, found to be enriched after RSD treatment, are able to form a large mycelial network which enhances plants tolerance to biotic and abiotic stresses [48,49]. Most of these functions were changed significantly during succession planting. For instance, the functions of hydrocarbon degradation, iron respiration, and nitrogen fixation were significantly decreased after succession planting, which may be related to the change from reducing to oxidizing soil conditions. After cabbage cultivation in particular, the abundances of fumarate, nitrate, and sulfate respiration, and nitrate reduction were still significantly higher in RSD-treated soil. This might be due to the root exudates released by and the fertilization patterns applied to distinct crops, resulting in the selective enrichment of microbiomes [50].

Most of the dominant genera which were significantly increased in RSD-treated soils were closely correlated with the above-mentioned nutrient cycling functions. For example, *Hydrogenispora*, *Ruminiclostridium*, *Geobacter*, *Flavisolibacter*, and *Ramlibacter* were associated with both iron and fumarate respiration, and *Streptomyces* was associated with both nitrate respiration and hydrocarbon degradation. Previous studies have found that *Flavisolibacter* and *Geobacter* are important iron-reducing agents, both directly and indirectly converting Fe^{3+} to Fe^{2+} under anaerobic conditions [51,52], and that the accumulation of Fe^{2+} during RSD treatment is significantly inhibitive of the growth of plant pathogens [45]. In addition, *Hydrogenispora*, *Ruminiclostridium*, and *Ramlibacter* are known to be important decomposers of various carbohydrates, and thus stimulate the soil carbon cycle [53–55]. Members of *Streptomyces* are generally considered important nitrate reducers, effectively removing soil nitrate by releasing reductases [56]. Moreover, members of *Streptomyces* have been shown to promote crop growth by decomposing organic phosphorus and producing indole acetic acid [57,58]. These results suggest that the dominant microbial taxa regulated by RSD can effectively improve soil functions, such as organic carbon degradation, nitrate and sulfate removal, and iron reduction.

5. Conclusions

Our results show that RSD treatment significantly alters the soil microbial community and improves microbial function. The ammonium and available potassium contents in RSD-treated soil were maintained at a high level, even after the planting of two crops. The microbial network associated with RSD treatment was more complex and connected. RSD treatment was able to increase microbial functions associated with hydrocarbon (C, H), nitrogen (N), sulfur (S), and iron (Fe) cycling, and reduce the functions of plant pathogens. Moreover, the improved community functions were closely associated with most of the enriched genera, including *Hydrogenispora*, *Ruminiclostridium*, *Geobacter*, *Flavisolibacter*, *Ramlibacter*, and *Streptomyces*, many of which are also known to be disease-suppressive agents. Combined with succession planting with different crops, RSD can delay the degradation of microbial communities through the recruitment of beneficial microbes to some extent. Here, we suggest that after two consecutive crops are planted in RSD-regulated soil, a new RSD treatment should be re-applied to maintain soil health.

Supplementary Materials: The following supporting information can be downloaded at: https://www.mdpi.com/article/10.3390/horticulturae8060476/s1, Table S1: Changes in soil physicochemical properties after treatment and plants cultivation; Table S2: Changes in the bacterial and fungal α diversities after treatment and plants cultivation; Table S3: Top ten keystone taxa of microbial community networks in CTL and RSD soils; Figure S1: The bacterial (a,b) and fungal (c,d) phylum composition for each module of microbial community networks in CTL and RSD soils. The relative abundance of each phylum was calculated by the average relative abundance of this phylum across all soils divided by the average total relative abundance of all phyla in each module; Figure S2: Relative abundances of bacterial (a) and fungal (b) genera. CTL, untreated soil; RSD, soil incorporated with molasses, irrigated to saturation, and covered with plastic film. AT, FS, and SS indicate the soils collected from after treatment, the first season, and the second season, respectively. "UC-" indicates that the given taxon could not be classified to genus level.

Author Contributions: Conceptualization, Y.Y.; Data curation, Y.Y. and R.W.; Funding acquisition, L.L.; Investigation, Z.S.; Methodology, S.L.; Project administration, Z.C. and L.L.; Software, S.L. and Z.S.; Supervision, Z.C.; Validation, Z.C.; Visualization, Q.S.; Writing—original draft, Y.Y. and R.W.; Writing—review and editing, X.H. and L.L. All authors have read and agreed to the published version of the manuscript.

Funding: This research was funded by the Key-Area Research and Development Program of Guangdong Province (2020B0202010006), the National Natural Science Foundation of China (Grant No. 32160748, U21A20226, 42090065), the China Postdoctoral Science Foundation (2021M691625), and the Key Research and Development Project (Agriculture) of Yichun City, Jiangxi Province (20211YFN4240).

Institutional Review Board Statement: Not applicable.

Informed Consent Statement: Not applicable.

Data Availability Statement: The raw sequencing data were deposited at the NCBI Sequence Read Archive database with the accession number of PRJNA738280.

Conflicts of Interest: The authors declare no conflict of interest.

References

1. Zhang, H.P.; Zhang, Q.K.; Song, J.J.; Zhang, Z.H.; Chen, S.Y.; Long, Z.N.; Wang, M.C.; Yu, Y.L.; Fang, H. Tracking resistomes, virulence genes, and bacterial pathogens in long-term manure-amended greenhouse soils. *J. Hazard. Mater.* **2020**, *396*, 122618. [CrossRef] [PubMed]
2. Hu, W.Y.; Zhang, Y.X.; Huang, B.; Teng, Y. Soil environmental quality in greenhouse vegetable production systems in eastern China: Current status and management strategies. *Chemosphere* **2017**, *170*, 183–195. [CrossRef] [PubMed]
3. Huang, X.Q.; Chen, L.H.; Ran, W.; Shen, Q.R.; Yang, X.M. *Trichoderma harzianum* strain SQR-T37 and its bio-organic fertilizer could control *Rhizoctonia solani* damping-off disease in cucumber seedlings mainly by the mycoparasitism. *Appl. Microbiol. Biotechnol.* **2011**, *91*, 741–755. [CrossRef] [PubMed]
4. Zhao, J.; Liu, S.Z.; Zhou, X.; Xia, Q.; Liu, X.; Zhang, S.R.; Zhang, J.B.; Cai, Z.C.; Huang, X.Q. Reductive soil disinfestation incorporated with organic residue combination significantly improves soil microbial activity and functional diversity than sole residue incorporation. *Appl. Microbiol. Biotechnol.* **2020**, *104*, 7573–7588. [CrossRef]
5. Bahram, M.; Hildebrand, F.; Forslund, S.K.; Anderson, J.L.; Soudzilovskaia, N.A.; Bodegom, P.M.; Bengtsson-Palme, J.; Anslan, S.; Coelho, L.P.; Harend, H.; et al. Structure and function of the global topsoil microbiome. *Nature* **2018**, *560*, 233–237. [CrossRef]
6. Van Agtmaal, M.; Straathof, A.; Termorshuizen, A.; Teurlincx, S.; Hundscheid, M.; Ruyters, S.; Busschaert, P.; Lievens, B.; De Boer, W. Exploring the reservoir of potential fungal plant pathogens in agricultural soil. *Appl. Soil Ecol.* **2017**, *121*, 152–160. [CrossRef]
7. Janvier, C.; Villeneuve, F.; Alabouvette, C.; Edel-Hermann, V.; Mateille, T.; Steinberg, C. Soil health through soil disease suppression: Which strategy from descriptors to indicators? *Soil Biol. Biochem.* **2007**, *39*, 1–23. [CrossRef]
8. Wagg, C.; Bender, S.F.; Widmer, F.; Van Der Heijden, M.G. Soil biodiversity and soil community composition determine ecosystem multifunctionality. *Proc. Natl. Acad. Sci. USA* **2014**, *111*, 5266–5270. [CrossRef]
9. Liu, L.L.; Yan, Y.Y.; Ding, H.X.; Zhao, J.; Cai, Z.C.; Dai, C.A.C.; Huang, X.Q. The fungal community outperforms the bacterial community in predicting plant health status. *Appl. Microbiol. Biotechnol.* **2021**, *105*, 6499–6513. [CrossRef]
10. Wang, T.T.; Hao, Y.W.; Zhu, M.Z.; Yu, S.T.; Ran, W.; Xue, C.; Ling, N.; Shen, Q.R. Characterizing differences in microbial community composition and function between *Fusarium* wilt diseased and healthy soils under watermelon cultivation. *Plant Soil* **2019**, *438*, 421–433. [CrossRef]
11. Wei, Z.; Gu, Y.; Friman, V.P.; Kowalchuk, G.A.; Xu, Y.C.; Shen, Q.R.; Jousset, A. Initial soil microbiome composition and functioning predetermine future plant health. *Sci. Adv.* **2019**, *5*, eaaw0759. [CrossRef] [PubMed]
12. Huang, X.Q.; Zhou, X.; Zhang, J.B.; Cai, Z.C. Highly connected taxa located in the microbial network are prevalent in the rhizosphere soil of healthy plant. *Biol. Fertil. Soils* **2019**, *55*, 299–312. [CrossRef]
13. Liu, L.L.; Huang, X.Q.; Zhang, J.B.; Cai, Z.C.; Jiang, K.; Chang, Y.Y. Deciphering the relative importance of soil and plant traits on the development of rhizosphere microbial communities. *Soil Biol. Biochem.* **2020**, *148*, 107909. [CrossRef]
14. Huang, X.Q.; Liu, L.L.; Wen, T.; Zhang, J.B.; Wang, F.H.; Cai, Z.C. Changes in the soil microbial community after reductive soil disinfestation and cucumber seedling cultivation. *Appl. Microbiol. Biotechnol.* **2016**, *100*, 5581–5593. [CrossRef]
15. Momma, N.; Kobara, Y.; Uematsu, S.; Kita, N.; Shinmura, A. Development of biological soil disinfestations in Japan. *Appl. Microbiol. Biotechnol.* **2013**, *97*, 3801–3809. [CrossRef]
16. Momma, N.; Momma, M.; Kobara, Y. Biological soil disinfestation using ethanol: Effect on *Fusarium oxysporum* f. sp. *lycopersici* and soil microorganisms. *J. Gen. Plant Pathol.* **2010**, *76*, 336–344. [CrossRef]
17. Huang, X.Q.; Liu, L.L.; Wen, T.; Zhu, R.; Zhang, J.B.; Cai, Z.C. Illumina MiSeq investigations on the changes of microbial community in the *Fusarium oxysporum* f.sp. cubense infected soil during and after reductive soil disinfestation. *Microbiol. Res.* **2015**, *181*, 33–42. [CrossRef]
18. Zhu, R.; Huang, X.Q.; Zhang, J.B.; Cai, Z.C.; Li, X.; Wen, T. Efficiency of Reductive Soil Disinfestation Affected by Soil Water Content and Organic Amendment Rate. *Horticulturae* **2021**, *7*, 559. [CrossRef]
19. Huang, X.Q.; Zhao, J.; Zhou, X.; Han, Y.S.; Zhang, J.B.; Cai, Z.C. How green alternatives to chemical pesticides are environmentally friendly and more efficient. *Eur. J. Soil Sci.* **2019**, *70*, 518–529. [CrossRef]

20. Liu, L.L.; Chen, S.H.; Zhao, J.; Zhou, X.; Wang, B.Y.; Li, Y.L.; Zheng, G.Q.; Zhang, J.B.; Cai, Z.C.; Huang, X.Q. Watermelon planting is capable to restructure the soil microbiome that regulated by reductive soil disinfestation. *Appl. Soil Ecol.* **2018**, *129*, 52–60. [CrossRef]
21. Mowlick, S.; Yasukawa, H.; Inoue, T.; Takehara, T.; Kaku, N.; Ueki, K.; Ueki, A. Suppression of spinach wilt disease by biological soil disinfestation incorporated with *Brassica juncea* plants in association with changes in soil bacterial communities. *Crop Prot.* **2013**, *54*, 185–193. [CrossRef]
22. Liu, L.L.; Yan, Y.Y.; Ali, A.; Zhao, J.; Cai, Z.C.; Dai, C.C.; Huang, X.Q.; Zhou, K.S. Deciphering the *Fusarium*-wilt control effect and succession driver of microbial communities managed under low-temperature conditions. *Appl. Soil Ecol.* **2022**, *171*, 104334. [CrossRef]
23. Liu, J.J.; Sui, Y.Y.; Yu, Z.H.; Yao, Q.; Shi, Y.; Chu, H.Y.; Jin, J.; Liu, X.B.; Wang, G.H. Diversity and distribution patterns of acidobacterial communities in the black soil zone of northeast China. *Soil Biol. Biochem.* **2016**, *95*, 212–222. [CrossRef]
24. Huang, X.Q.; Zhao, J.; Zhou, X.; Zhang, J.B.; Cai, Z.C. Differential responses of soil bacterial community and functional diversity to reductive soil disinfestation and chemical soil disinfestation. *Geoderma* **2019**, *348*, 124–134. [CrossRef]
25. Biddle, J.F.; Fitz-Gibbon, S.; Schuster, S.C.; Brenchley, J.E.; House, C.H. Metagenomic signatures of the Peru Margin subseafloor biosphere show a genetically distinct environment. *Proc. Natl. Acad. Sci. USA* **2008**, *105*, 10583–10588. [CrossRef]
26. Caporaso, J.G.; Kuczynski, J.; Stombaugh, J.; Bittinger, K.; Bushman, F.D.; Costello, E.K.; Fierer, N.; Peña, A.G.; Goodrich, J.K.; Gordon, J.I.; et al. QIIME allows analysis of high-throughput community sequencing data. *Nat. Methods* **2010**, *7*, 335–336. [CrossRef]
27. Mcdonald, D.; Price, M.N.; Goodrich, J.; Nawrocki, E.P.; Desantis, T.Z.; Probst, A.; Andersen, G.L.; Knight, R.; Hugenholtz, P. An improved Greengenes taxonomy with explicit ranks for ecological and evolutionary analyses of bacteria and archaea. *ISME J.* **2012**, *6*, 610–618. [CrossRef]
28. Koljalg, U.; Nilsson, R.H.; Abarenkov, K.; Tedersoo, L.; Taylor, A.F.; Bahram, M.; Bates, S.T.; Bruns, T.D.; Bengtsson-Palme, J.; Callaghan, T.M.; et al. Towards a unified paradigm for sequence-based identification of fungi. *Mol. Ecol.* **2013**, *22*, 5271–5277. [CrossRef]
29. Louca, S.; Parfrey, L.W.; Doebeli, M. Decoupling function and taxonomy in the global ocean microbiome. *Science* **2016**, *353*, 1272–1277. [CrossRef]
30. Nguyen, N.H.; Song, Z.W.; Bates, S.T.; Branco, S.; Tedersoo, L.; Menke, J.; Schilling, J.S.; Kennedy, P.G. FUNGuild: An open annotation tool for parsing fungal community datasets by ecological guild. *Fungal. Ecol.* **2016**, *20*, 241–248. [CrossRef]
31. Mcmurdie, P.J.; Holmes, S. phyloseq: An R package for reproducible interactive analysis and graphics of microbiome census data. *PLoS ONE* **2013**, *8*, e61217. [CrossRef] [PubMed]
32. Oksanen, J.; Blanchet, F.G.; Kindt, R.; Legendre, P.; Minchin, P.R.; O'hara, R.B.; Simpson, G.L.; Solymos, P.; Stevens, M.H.H.; Wagner, H. Vegan: Community Ecology Package. *R Package* **2016**, *7*, 2.
33. Bastian, M.; Heymann, S.; Jacomy, M. Gephi: An Open Source Software for Exploring and Manipulating Networks. *ICWSM* **2009**, *8*, 361–362.
34. Zhao, J.; Zhou, X.; Jiang, A.Q.; Fan, J.Z.; Lan, T.; Zhang, J.B.; Cai, Z.C. Distinct impacts of reductive soil disinfestation and chemical soil disinfestation on soil fungal communities and memberships. *Appl. Microbiol. Biotechnol.* **2018**, *102*, 7623–7634. [CrossRef]
35. Wu, S.M.; Jia, S.F.; Sun, D.D.; Chen, M.L.; Chen, X.Z.; Jin, Z.; Huan, L.D. Purification and Characterization of Two Novel Antimicrobial Peptides Subpeptin JM_4-A and Subpeptin JM_4-B Produced by *Bacillus subtilis* JM_4. *Curr. Microbiol.* **2005**, *51*, 292–296. [CrossRef]
36. Li, B.; Wang, Y.Q.; Tu, W.Q.; Wang, Z.S.; Xu, M.Q.; Xing, L.D.; Li, W.S. Improving cyclic stability of lithium nickel manganese oxide cathode for high voltage lithium ion battery by modifying electrode/electrolyte interface with electrolyte additive. *Electrochim. Acta* **2014**, *147*, 636–642. [CrossRef]
37. Li, B.; Li, Q.; Xu, Z.H.; Zhang, N.; Shen, Q.R.; Zhang, R.F. Responses of beneficial *Bacillus amyloliquefaciens* SQR9 to different soilborne fungal pathogens through the alteration of antifungal compounds production. *Front. Microbiol.* **2014**, *5*, 636. [CrossRef]
38. Meng, T.Z.; Ren, G.D.; Wang, G.F.; Ma, Y. Impacts on soil microbial characteristics and their restorability with different soil disinfestation approaches in intensively cropped greenhouse soils. *Appl. Microbiol. Biotechnol.* **2019**, *103*, 6369–6383. [CrossRef]
39. Jaiswal, A.K.; Elad, Y.; Cytryn, E.; Graber, E.R.; Frenkel, O. Activating biochar by manipulating the bacterial and fungal microbiome through pre-conditioning. *New Phytol.* **2018**, *219*, 363–377. [CrossRef]
40. Barabasi, A.L. Scale-Free Networks: A Decade and Beyond. *Science* **2009**, *325*, 412–413. [CrossRef]
41. Wink, J.; Gandhi, J.; Kroppenstedt, R.M.; Seibert, G.; Straubler, B.; Schumann, P.; Stackebrandt, E. Amycolatopsis decaplanina sp nov., a novel member of the genus with unusual morphology. *Int. J. Syst. Evol. Microbiol.* **2004**, *54*, 235–239. [CrossRef] [PubMed]
42. Igarashi, M.; Sawa, R.; Yamasaki, M.; Hayashi, C.; Umekita, M.; Hatano, M.; Fujiwara, T.; Mizumoto, K.; Nomoto, A. Kribellosides, novel RNA 5′-triphosphatase inhibitors from the rare actinomycete *Kribbella* sp. MI481-42F6. *J. Antibiot.* **2017**, *70*, 582–589. [CrossRef] [PubMed]
43. Herrero, N. A novel monopartite dsRNA virus isolated from the entomopathogenic and nematophagous fungus *Purpureocillium lilacinum*. *Arch. Virol.* **2016**, *161*, 3375–3384. [CrossRef] [PubMed]
44. Mohamadpoor, M.; Amini, J.; Ashengroph, M.; Azizi, A. Evaluation of biocontrol potential of *Achromobacter xylosoxidans* strain CTA8689 against common bean root rot. *Physiol. Mol. Plant Pathol.* **2022**, *117*, 101769. [CrossRef]

45. Momma, N.; Kobara, Y.; Momma, M. Fe^{2+} and Mn^{2+}, potential agents to induce suppression of *Fusarium oxysporum* for biological soil disinfestation. *J. Gen. Plant Pathol.* **2011**, *77*, 331–335. [CrossRef]

46. Meng, T.Z.; Zhu, T.B.; Zhang, J.B.; Xie, Y.; Sun, W.J.; Yuan, L.; Cai, Z.C. Liming accelerates the NO_3–removal and reduces N_2O emission in degraded vegetable soil treated by reductive soil disinfestation (RSD). *J. Soils Sediments* **2015**, *15*, 1968–1976. [CrossRef]

47. Ohmura, N.; Sasaki, K.; Matsumoto, N.; Saiki, H. Anaerobic respiration using Fe^{3+}, S^0, and H_2 in the chemolithoautotrophic bacterium *Acidithiobacillus ferrooxidans*. *J. Bacteriol.* **2002**, *184*, 2081–2087. [CrossRef]

48. Tedersoo, L.; Bahram, M.; Zobel, M. How mycorrhizal associations drive plant population and community biology. *Science* **2020**, *367*, a1223. [CrossRef]

49. Sa, G.; Yao, J.; Deng, C.; Liu, J.; Zhang, Y.N.; Zhu, Z.M.; Zhang, Y.H.; Ma, X.J.; Zhao, R.; Lin, S.Z.; et al. Amelioration of nitrate uptake under salt stress by ectomycorrhiza with and without a Hartig net. *New Phytol.* **2019**, *222*, 1951–1964. [CrossRef]

50. Chaparro, J.M.; Badri, D.V.; Bakker, M.G.; Sugiyama, A.; Manter, D.K.; Vivanco, J.M. Root exudation of phytochemicals in Arabidopsis follows specific patterns that are developmentally programmed and correlate with soil microbial functions. *PLoS ONE* **2013**, *8*, e55731. [CrossRef]

51. Holmes, D.E.; Nicoll, J.S.; Bond, D.R.; Lovley, D.R. Potential role of a novel psychrotolerant member of the family *Geobacteraceae*, *Geopsychrobacter electrodiphilus* gen. nov., sp nov., in electricity production by a marine sediment fuel cell. *Appl. Environ. Microbiol.* **2004**, *70*, 6023–6030. [CrossRef] [PubMed]

52. Das, S.; Liu, C.C.; Jean, J.S.; Lee, C.C.; Yang, H.J. Effects of microbially induced transformations and shift in bacterial community on arsenic mobility in arsenic-rich deep aquifer sediments. *J. Hazard. Mater.* **2016**, *310*, 11–19. [CrossRef] [PubMed]

53. Liu, Y.; Qiao, J.T.; Yuan, X.Z.; Guo, R.B.; Qiu, Y.L. *Hydrogenispora ethanolica* gen. nov., sp nov., an anaerobic carbohydrate-fermenting bacterium from anaerobic sludge. *Int. J. Syst. Evol. Microbiol.* **2014**, *64*, 1756–1762. [CrossRef] [PubMed]

54. Ravachol, J.; Borne, R.; Meynial-Salles, I.; Soucaille, P.; Pages, S.; Tardif, C.; Fierobe, H.P. Combining free and aggregated cellulolytic systems in the cellulosome-producing bacterium *Ruminiclostridium cellulolyticum*. *Biotechnol. Biofuels* **2015**, *8*, 114–128. [CrossRef]

55. Liu, W.Y.; Zhao, Q.Q.; Zhang, Z.Y.; Li, Y.; Xu, N.H.; Qu, Q.; Lu, T.; Pan, X.L.; Qian, H.F. Enantioselective effects of imazethapyr on *Arabidopsis thaliana* root exudates and rhizosphere microbes. *Sci. Total Environ.* **2020**, *716*, 137121. [CrossRef]

56. Meng, S.T.; Wu, H.; Wang, L.; Zhang, B.C.; Bai, L.Q. Enhancement of antibiotic productions by engineered nitrate utilization in actinomycetes. *Appl. Microbiol. Biotechnol.* **2017**, *101*, 5341–5352. [CrossRef]

57. Khamna, S.; Yokota, A.; Lumyong, S. Actinomycetes isolated from medicinal plant rhizosphere soils: Diversity and screening of antifungal compounds, indole-3-acetic acid and siderophore production. *World J. Microbiol. Biotechnol.* **2009**, *25*, 649–655. [CrossRef]

58. Hamdali, H.; Hafidi, M.; Virolle, M.J.; Ouhdouch, Y. Growth promotion and protection against damping-off of wheat by two rock phosphate solubilizing actinomycetes in a P-deficient soil under greenhouse conditions. *Appl. Soil Ecol.* **2008**, *40*, 510–517. [CrossRef]

 *horticulturae*

Article

Evaluation of Suitable Water–Zeolite Coupling Regulation Strategy of Tomatoes with Alternate Drip Irrigation under Mulch

Xiaolan Ju, Tao Lei *, Xianghong Guo, Xihuan Sun, Juanjuan Ma, Ronghao Liu and Ming Zhang

College of Water Resource Science and Engineering, Taiyuan University of Technology, Taiyuan 030024, China; juxiaolan0757@link.tyut.edu.cn (X.J.); guoxianghong@tyut.edu.cn (X.G.); sunxihuan@tyut.edu.cn (X.S.); mjjsxty@163.com (J.M.); liuronghao@tyut.edu.cn (R.L.); zhangming0789@link.tyut.edu.cn (M.Z.)
* Correspondence: lcsyt@126.com

Abstract: The water (W; W_{50}, W_{75}, and W_{100})–zeolite (Z; Z_0, Z_3, Z_6 and Z_9) coupling (W-Z) regulation strategy of high-quality and high-yield tomato was explored with alternate drip irrigation under mulch. Greenhouse planting experiments were used in monitoring and analyzing tomato growth, physiology, yield, quality, and water use efficiency (WUE). Suitable amounts of W and Z for tomato growth were determined through the principal component analysis (PCA) method. Results showed that tomato plant height (Ph), stem thickness (St), root indexes, leaf area index (LAI), photosynthetic rate (Pn), transpiration rate (Tr), stomatal conductance (Gs), organic acid (OA), and yield showed a positive response to W, whereas nitrate (NC), vitamin C (VC), soluble solid (SS), intercellular CO_2 concentration (Ci), fruit firmness (Ff), and WUE showed the opposite trend. The responses of Ci and Ff to Z were first negative and then positive, whereas the responses of other indexes to Z showed an opposite trend (except yield under W_{50}). The effects of W, Z, and W-Z on tomato growth, physiological, and quality indexes and yield were as follows: $W > Z > W$-Z; the effects on WUE were as follows: $Z > W > W$-Z. The two principal components of growth factor and water usage factor were extracted, and the cumulative variance contribution rate reached 93.831%. Under different treatments for tomato growth, the comprehensive evaluation score F was between -1.529 and 1.295, the highest treated with Z_6W_{100}, the lowest treated with Z_0W_{50}. The PCA method showed that under the condition of alternate drip irrigation under mulch, the most suitable W for tomato planting was 100% E (E is the water surface evaporation), and the amount of Z was 6 t·ha^{-1}.

Keywords: alternate drip irrigation under mulch; water; zeolite amount; tomato growth; water use efficiency; principal component

Citation: Ju, X.; Lei, T.; Guo, X.; Sun, X.; Ma, J.; Liu, R.; Zhang, M. Evaluation of Suitable Water–Zeolite Coupling Regulation Strategy of Tomatoes with Alternate Drip Irrigation under Mulch. *Horticulturae* **2022**, *8*, 536. https://doi.org/10.3390/horticulturae8060536

Academic Editors: Xun Li, Xiaohui Hu and Shiwei Song

Received: 1 May 2022
Accepted: 14 June 2022
Published: 16 June 2022

Publisher's Note: MDPI stays neutral with regard to jurisdictional claims in published maps and institutional affiliations.

1. Introduction

Tomatoes are popular worldwide and a high water-dependent horticultural crop, cultivated in open fields and greenhouses [1]. Water stress has negative effects on biochemical and physiological systems and affects the healthy development of crops [2]. Adding zeolite (Z) can improve soil water retention [3], improve the ability of crop roots to absorb and utilize soil water [4], and reduce the impact of water stress on crop growth and physiological development [5]. Alternate drip irrigation is an effective irrigation method that saves water and regulates soil quality. It has the advantages of low degree of soil evaporation, water conservation, high yield, and high irrigation efficiency [6]. On the basis of the coupling strategy of Z modifier and alternate irrigation, this paper is expected to further enrich the alternative irrigation theory, explore the potential of agricultural water conservation, and promote the efficient utilization of water resources and tomato yield and quality.

Previous studies mainly revealed the effects of Z on tomato growth [7], yield [8,9], and dry matter [10]. However, studies on the effects of Z on photosynthetic characteristics,

WUE, and root growth of tomato during growth period are few. The effects of *W* on the yield of tomato under drip irrigation and furrow irrigation [11–13], root growth under alternate partial root-zone irrigation [14,15], nutritional quality under drip irrigation [11], *WUE* under furrow irrigation [12], and other indicators [16,17] have been studied more. However, the effect of coupled water (*W*) regulation strategy combined with alternate drip irrigation under mulch and *Z* improvement on the growth and development of tomato throughout the growth period is still unclear. Owing to significant spatial variability in soil water between the dry and wet areas of alternate drip irrigation under mulch, the resulting physiological stimulation effect of the rhizosphere, and excellent water absorption and water retention performance of *Z*, the response of soil wetting body rhizosphere and crop growth and physiology under this coupled *W* regulation strategy must be different from that under conventional drip irrigation and furrow irrigation. Although the effects of a single factor of *W* or *Z* on crop growth have been clarified [8,12,15], the primary and secondary relationships between the effects of *W* and *Z* on tomato are unknown. The *W-Z* strategy has a synergistic effect on potato yield [18], soybean growth, and seed quality [19], and bean yield and *WUE* [20]. However, effects on the root growth, photosynthetic characteristics, and nutrient quality of tomato remain unclear.

PCA uses the idea of dimensionality reduction to convert multiple indicators with many correlations into few comprehensive indicators, so that the comprehensive performance of each indicator can be accurately determined. *PCA* has been widely applied to agriculture. For instance, it was used in establishing a comprehensive evaluation model of plant height (*Ph*), stem thickness (*St*), yield, and *WUE* for green pepper growth [21] and a comprehensive evaluation model of yield, *WUE*, and nutritional quality for potato growth [22]. However, these previous models did not involve crop root growth and physiological indicators. The root systems of plants are the main organs that absorb water and mineral nutrients, and their form and configurations largely determine the ability of plants to acquire nutrients [23]. Photosynthesis is the material basis for crop yield formation and improving crop light energy utilization efficiency is one of the important ways to improve crop yield [24]. Therefore, incorporating tomato root growth, photosynthetic physiology, and other indicators into the comprehensive evaluation model may be reasonable and objective.

The objective of this paper is to study the effects of *W-Z* effect of alternate drip irrigation under mulch on tomato to construct a comprehensive evaluation model of tomato growth on the basis of growth, physiology, quality, yield, and *WUE*. The suitable W and Z amount for tomato growth were clarified to provide theoretical guidance for tomato planting under the condition of alternate drip irrigation under mulch.

2. Materials and Methods

2.1. Experimental Site

The test was carried out in the greenhouse of the High-efficiency Water-saving Demonstration Base of Shanxi Water Conservancy and Hydropower Research Institute. The average altitude of the area is 763–780 m, and the site has warm temperate continental climate. The annual average precipitation is about 468.4 mm, annual average evaporation is 1812.7 mm, and annual average temperature is 9.5 °C. The test soil was clayey loam with a saturated water content of 0.44 $cm^3 \cdot cm^{-3}$, and the field water holding rate was 0.28 $cm^3 \cdot cm^{-3}$. The experimental irrigation water source was the fresh water well in the base.

2.2. Experimental Design

This experiment studied the growth characteristics of tomato under different *W* and *Z* amounts. Here, *W* was set at three levels: W_{50}, W_{75}, and W_{100}, which were 50%, 75%, and 100% water surface evaporation (*E*), respectively. Meanwhile, *Z* was set at four levels: Z_0, Z_3, Z_6, and Z_9, which were 0, 3, 6, and 9 $t \cdot ha^{-1}$, respectively. The experiment adopted a comprehensive experimental design with 12 treatments in total, and each treatment was repeated three times. The experimental tomato (Ao-guan No. 8) was planted in a film-mulched ridge planting mode. The ridge length was 6 m, the ridge surface width was 0.8 m, the ridge height was 0.15 m, and the ridge width was 0.6 m. A total of 30 tomato plants were planted in each row with a row spacing of 0.4 m × 0.4 m. According to local planting habits, all plots were managed in the same field. The irrigation method was alternate irrigation under mulch. Two drip irrigation belts were laid in each ridge, the distance between drip heads was 40 cm, and the working flow was 1.2 L·h^{-1}. Irrigation was performed once every 4 days, and drip irrigation was conducted with one-side irrigation. The type of zeolite used in this study was the 4A zeolite purchased from Shanxi Taiheng Technology Co., Ltd. (Shanxi, China). Before tomato planting, zeolite was mixed into the soil to a depth of 30 cm. Tomato plants were transplanted on 4 June 2020, and the end time was 7 October 2020.

2.3. Measurement Parameters and Methods

(1) *Ph* and *St*: three tomatoes with similar growth rates at the seedling stage were selected and *Ph* was measured from the stem base of each tomato to the growth point with a tape measure. The accuracy was 1 mm. *St* below the first lateral branch of tomato stem base was measured with a digital vernier caliper with an accuracy of 0.01 mm.

(2) Root characteristic parameters: in the harvest period, three tomato plants with the same growth rates were randomly selected from each treatment. With the plants as the center, the whole root was excavated in the quadrats with an area of 40 cm × 40 cm and depth of 60 cm, and the whole root was removed and washed with water for the removal of debris. A V700 scanner (EPSON company, Nagano, Japan) was used for root scanning, and Win RHIZO 2003 software was used in analyzing and obtaining root characteristic parameters, such as total root length (*RL*), total root surface area (*RS*), and total root volume (*RV*).

(3) *LAI*: in each growth period, three representative plants were selected from each treatment, the leaf area was measured, and the average value was taken.

(4) Physiological indexes: during the expansion period of tomato fruit, clear and cloudless weather was selected, and photosynthesis was measured with an Li-6400 portable photosynthetic instrument. The measurement time was 9:00–11:00. The leaf selection principle was at the same position and the same leaf age.

(5) *Ff*: *Ff* was measured with GYJ-4 hardness tester.

(6) Nutritional quality: determination of *OA* by NaOH titration [25], determination of *VC* with molybdenum blue colorimetry [26], determination of *NC* by sulfuric acid salicylic acid method [27], and determination of *SS* with PAL-1 handheld refractometer [28]. The average of three measurements was obtained as the final measurement.

(7) Yield: yield was measured with an electronic scale and had an accuracy of 0.01 kg.

2.4. Data Processing and Statistical Analysis

Microsoft Office 2020 was used for data calculation, IBM SPSS statistics 25 was used for two-way ANOVA and principal component analysis, GSTA V7.0 for gray correlation analysis, and Origin 2018 was used for drawing.

3. Results

3.1. Effects of W-Z on the Soil Moisture Dynamics

Figure 1 shows the effects of *W-Z* on soil moisture dynamics subjected to alternate drip irrigation under mulch. As shown in Figure 1, soil moisture content with different treatments, ranging from 16.84% to 33.65%, showed sawtooth fluctuations with time growth. In the range of seedling stage and fruit expansion stage (19 July to 12 September 2020), which belongs to the vegetative and reproductive growth process, soil moisture content decreased with increased time, attributed to increased water transpiration consumption, caused by the rapid development of the rhizosphere system and functional leaves. In the range of the fruit harvest stage (12 to 30 September 2020), soil moisture content stabilized with the increased time, attributed to the stabilization of plant growth and transpirational water consumption. As shown in Figure 1, soil moisture content increased by an average of 0.69–10.38% with the increased W from W_{50} to W_{100}, and increased by an average of 0.98–5.79% with the increased Z from Z_0 to Z_9, respectively. This indicates that the increase in W and Z have different degrees of promoting effects on soil water content.

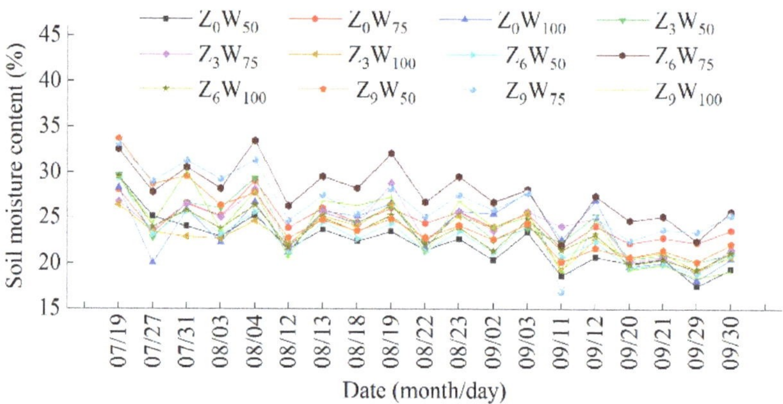

Figure 1. Effects of *W-Z* on variations of soil moisture during alternate drip irrigation under mulch.

3.2. Effects of W-Z on the Growth and Physiological Characteristics of Tomato

3.2.1. Effects of W-Z on Tomato Growth

Figure 2 shows the effect of *W-Z* on tomato growth indicators during alternate drip irrigation under mulch. Figure 2 shows that under the conditions of Z_0, Z_3, Z_6, and Z_9, when W increased from W_{50} to W_{100}, tomato *Ph*, *St*, *RL*, *RS*, *RV*, and *LAI* increased monotonically by 19.78–31.30%, 13.57–20.43%, 29.10–50.84%, 35.46–47.88%, 53.91–77.96%, and 14.55–35.67%, respectively. This result showed that an increase in W can significantly ($p < 0.01$) promote tomato growth between W and Z and have synergistic effects on tomato growth. Figure 2 also shows that under the conditions of W_{50}, W_{75}, and W_{100}, when Z increased from Z_0 to Z_6, *Ph*, *St*, *RL*, *RS*, and *RV*, these increased monotonically by 8.32–16.55%, 5.14–8.42%, 3.51–24.70%, 9.08–16.98%, and 4.67–20.13%, respectively. When Z increased from Z_6 to Z_9, *Ph*, *St*, *RL*, *RS*, and *RV* decreased by 2.08–4.91%, 1.13–3.85%, 0.07–11.64%, 2.49–11.04%, and 2.19–9.84%, respectively. However, under the conditions of W_{50}, W_{75} and W_{100}, when the amount of zeolite changed, the change in *LAI* was not regular. This result showed that the effects of increased Z on *Ph* ($p < 0.01$), *St*, and root system growth ($p < 0.01$) were first accelerated and then suppressed and the response trends and intensities of *Ph* and *St* to Z were relatively affected by W factors. In the Z_0–Z_6 range, the amounts of W and Z had a synergistic effect on tomato growth but exerted antagonistic effects in the Z_6–Z_9 range. Two-way ANOVA showed that the *W-Z* effect had a significant effect ($p < 0.01$) on *Ph*, *RL*, and *RS*, had a significant effect ($p < 0.05$) on *LAI*, but had no significant effect on *St* and *RV*. We found that the sum of the squares was for *Ph* (2778.95, 557.89, 170.54), *St* (20.38, 3.09, 0.54), *RL* (5,290,316.26, 348,892.52, 157,083.72), *RS* (170,236.82, 16,600.85, 3818.61), *RV* (28.87, 1.54, 0.55), and *LAI* (136.28, 16.10, 3.94) under the effects of W, Z, and *W-Z*, respectively. The effects of W, Z, and *W-Z* on *Ph*, *St*, *RL*, *RS*, *RV*, and *LAI* of tomato were as follows: $W > Z > W\text{-}Z$. The W factor played a leading role in tomato growth (Table S1).

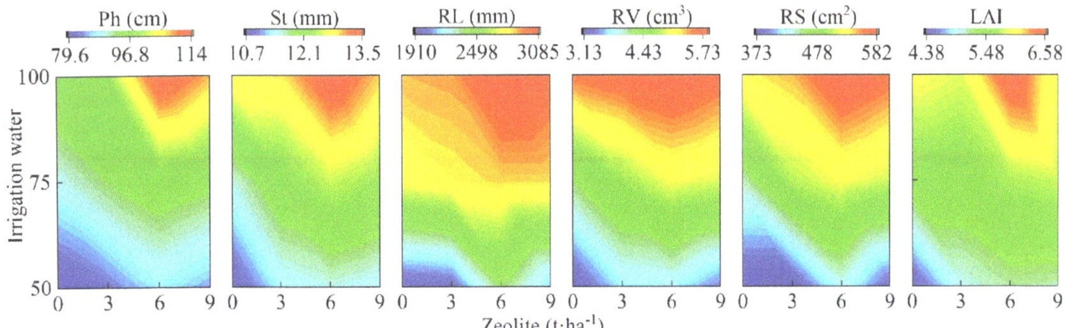

Figure 2. Effects of *W-Z* on tomato growth during alternate drip irrigation under mulch. *Ph*, *St*, *RL*, *RV*, *RS*, and *LAI* represent plant height, stem thickness, root length, root volume, root surface area, and leaf area index, respectively.

3.2.2. Effects of *W-Z* on Tomato Physiological Characteristics

Figure 3 shows the effect of *W-Z* on the physiological indexes of tomatoes subjected to alternate drip irrigation under mulch. Under the conditions of Z_0, Z_3, Z_6, and Z_9, when *W* increased from W_{50} to W_{100}, tomato *Ci* decreased monotonically by 8.96–12.57%, *Pn*, *Tr*, and *Gs* monotonically increased by 16.67–28.43%, 22.82–44.51%, and 20.37–26.16%, respectively. This result showed that an increase in *W* can significantly inhibit *Ci* ($p < 0.01$) and promote *Pn*, *Tr*, and *Gs*. *W* and *Z* exerted antagonistic effects on *Ci* and synergistic effects on *Pn*, *Tr*, and *Gs*. Under the conditions of W_{50}, W_{75}, and W_{100}, when *Z* increased from Z_0 to Z_6, *Ci* decreased monotonically by 2.11–9.02%, whereas *Pn*, *Tr*, and *Gs* monotonically increased by 6.49–15.20%, 8.95–27.34%, and 5.22–16.51%, respectively. When *Z* increased from Z_6 to Z_9, *Ci* increased by 0.60–2.01%, whereas *Pn*, *Tr*, and *Gs* decreased by 0.43–3.45%, 1.44–9.18%, and 0.16–5.77%, respectively. This result showed that in the Z_0–Z_6 range, *W* and *Z* had an antagonistic effect on *Ci* and a synergistic effect on *Pn*, *Tr*, and *Gs*, and opposite results were obtained in a range of Z_6–Z_9. Two-way ANOVA calculation showed that *W-Z* had a significant effect ($p < 0.05$) on *Ci*, *Tr*, and *Gs* but had no significant effect on *Pn*. As follows, the sum of the squares was for *Ci* (7994.00, 2138.08, 812.00), *Pn* (127.48, 28.14, 4.19), *Tr* (20.14, 4.58, 1.00), and *Gs* (0.10, 0.02, 0.01) under the effects of *W*, *Z* and *W-Z*, respectively. The effects of *W*, *Z*, and *W-Z* on tomato physiological indicators were as follows: *W* > *Z* > *W-Z*. The *W* factor played a leading role in tomato growth (Table S1).

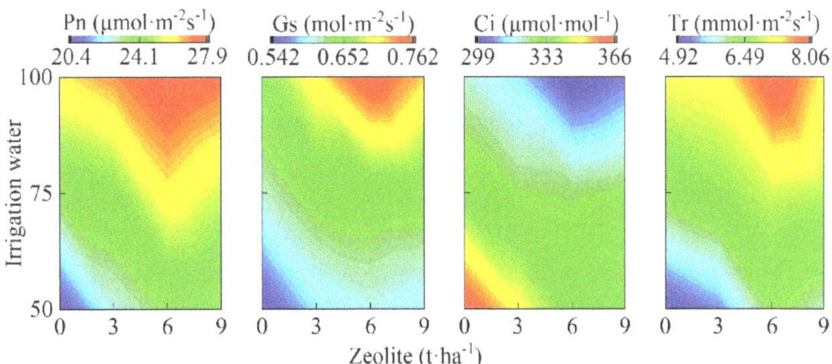

Figure 3. Effects of W-Z on tomato physiological indexes during alternate drip irrigation under mulch. *Pn*, *Gs*, *Ci*, and *Tr* represent photosynthetic rate, stomatal conductance, intercellular CO_2 concentration, and transpiration rate, respectively.

3.3. Effects of W-Z on Tomato Quality

Figure 4 shows the effect of *W-Z* on tomato quality during alternate drip irrigation under mulch. Under the conditions of Z_0, Z_3, Z_6, and Z_9, when *W* increased from W_{50} to W_{100}, tomato *OA* monotonically increased by 13.51–16.90%, and *NC*, *VC*, *SS*, and *Ff* monotonically decreased by 25.12–29.11%, 14.44–17.81%, 17.24–28.57%, and 23.16–33.69%, respectively. These results showed that an increase in *W* had a significant effect ($p < 0.01$) on tomato quality and *W* and *Z* had synergistic effects on *OA* and antagonistic effects on *NC*, *VC*, *SS*, and *Ff*. Under the conditions of W_{50}, W_{75}, and W_{100}, when *Z* increased from Z_0 to Z_6, *NC*, *VC*, *SS*, and *OA* monotonically increased by 6.42–16.23%, 3.65–10.18%, 3.57–20.00%, and 6.80–8.10%, respectively, and *Ff* decreased monotonically by 4.05–17.20%. When *Z* increased from Z_6 to Z_9, *NC*, *VC*, *SS*, and *OA* decreased by 4.00–7.23%, 2.69–3.40%, 1.72–12.50%, and 1.52–4.23%, *Ff* increased by 4.55% and 6.27% under W_{50} and W_{75} treatments, respectively, whereas it decreased by 2.82% under W_{100} treatment. These results showed that under the conditions of W_{50} and W_{75}, the significant effect ($p < 0.01$) of increasing *Z* on the quality of tomatoes was first suppressed and then enhanced, and under the W_{100} condition, the effects of *NC*, *VC*, *SS*, and *OA* were first accelerated and then suppressed, and only *Ff* was suppressed. In a range of Z_0–Z_6, *W* and *Z* exerted synergistic effects on tomato quality and antagonistic effects in a range of Z_6–Z_9 (except W_{100} *Ff*). The two-way ANOVA calculation showed that the *W-Z* effect had no significant effect on *OA* and *Ff* but had a significant effect ($p < 0.01$) on *NC*, *VC*, and *SS*. The sum of the squares was for *NC* (244.37, 22.07, 5.26), *VC* (44.32, 6.00, 2.04), *SS* (10.20, 1.15, 0.58), *OA* (0.01, 0.01, 0.00), and *Ff* (1.36, 0.13, 0.08) under the effects of *W*, *Z* and *W-Z*, respectively. The effects of *W*, *Z*, and *W-Z* on tomato physiological indicators were as follows: *W* > *Z* > *W-Z*. The *W* factor played a leading role in tomato quality (Table S1).

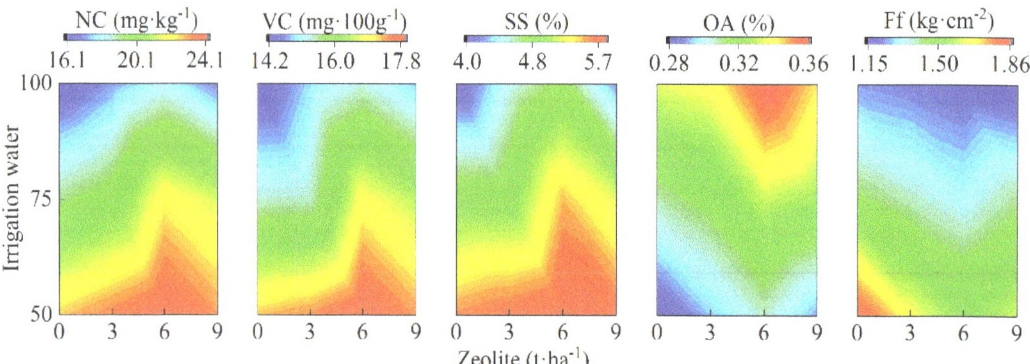

Figure 4. Effects of W-Z on tomato quality during alternate drip irrigation under mulch. *NC*, *VC*, *SS*, *OA* and *Ff* represent nitrate, vitamin C, soluble solid, organic acid, and fruit firmness, respectively.

3.4. Effects of W-Z on Tomato Yield and WUE

Figure 5 shows the effect of *W-Z* on tomato yield and *WUE* during alternate drip irrigation under mulch. Under the conditions of Z_0, Z_3, Z_6, and Z_9, when *W* increased from W_{50} to W_{100}, tomato yield monotonically increased by 16.56%, 16.31%, 16.25%, and 8.48%, respectively, whereas *WUE* monotonically decreased by 5.80%, 6.77%, 7.47%, and 16.17%, respectively. An increase in *W* had a significant promoting effect ($p < 0.01$) on the yield, and the strength of this promoting effect gradually decreased with increasing *Z*, and *WUE* showed opposite results. Under the conditions of W_{50}, W_{75}, and W_{100}, when *Z* increased from Z_0 to Z_6, the yield increased by 9.21%, 15.06%, and 8.91%, respectively, whereas *WUE* increased by 45.20%, 47.06%, and 42.63%, respectively. When *Z* increased from Z_6 to Z_9, the yield under W_{50} treatment increased by 3.68%, the yield under W_{75} and W_{100} treatments decreased by 2.25% and 3.24%, respectively, and *WUE* decreased by 16.60%, 20.68%, and

24.44%. These results showed that an increase in Z had a significant effect ($p < 0.01$) on tomato yield and *WUE*. In the Z_0–Z_6 range, W and Z had synergistic effects on yield and *WUE*. In the Z_6–Z_9 range, W and Z exerted antagonistic effects on yield and *WUE*. The two-way ANOVA calculation showed that the W-Z had no significant effect on yield and *WUE*. The effects of W, Z, and W-Z on the yield were 2574.23, 231.50, and 76.92, respectively. The effects of W, Z, and W-Z on *WUE* were 57.05, 830.44, and 11.44, respectively. The effects of W, Z, and W-Z on tomato yield were $W > Z > W$-Z, whereas the effect on WUE was as follows: $Z > W > W$-Z. The W factor played a dominant role in tomato yield, whereas Z factor played a dominant role in tomato *WUE* (Table S1).

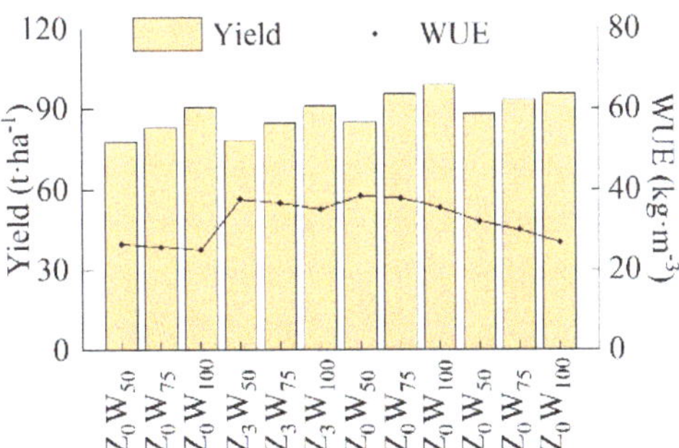

Figure 5. Effects of W-Z on tomato yield and *WUE* during alternate drip irrigation under mulch.

3.5. Comprehensive Evaluation and Analysis of Tomato

Our goal is to achieve high yield and water efficiency while ensuring normal growth of tomato plants and high fruit quality. Therefore, the factors of W and Z have different effects on tomato growth, physiology, quality, yield, and *WUE*. Determined based on a single index, the analysis results of the optimal W coupling strategy were inconsistent, and objectively and comprehensively meeting the goals of high-quality and high-efficiency planting and cultivation of tomatoes was difficult. Therefore, a comprehensive evaluation of various tomato indicators based on principal component analysis is necessary. A total of 17 indexes was used in this study, larger than the number of experimental treatments by 12, and forms a non-positive definite matrix in statistical analysis, making it difficult to carry out comprehensive evaluation. The gray correlation analysis method can be used to screen various tomato indicators [29]. The gray correlation analysis results (Table S2) showed that the gray correlation ranking was $RV > St > Tr > LAI > VC > NC > Pn > SS > WUE > Ff > $ yield $> Ph > Gs > Ci > OA > RS > RL$. However, because the *KMO* statistic of the first 11 indicators was 0.390, when the principal component analysis was performed, it did not meet the principal component analysis standard. Therefore, these 11 indicators were selected for the comprehensive evaluation of principal components: $RV(X_1)$, $St(X_2)$, $Tr(X_3)$, $VC(X_4)$, $NC(X_5)$, $Pn(X_6)$, $WUE(X_7)$, $SS(X_8)$, $Ff(X_9)$, yield(X_{10}), and $Ph(X_{11})$. After statistical calculation, the *KMO* statistic in this study was 0.575, and $p < 0.001$ for Bartlett test. Therefore, the data samples in this study were suitable for *PCA*. Based on the extraction criteria with eigenvalues of ≥ 1 [30], two principal components were obtained in this study (Tables S3 and S4). The contribution rate of the variance in the first principal component F_1 (named as growth factor) was 77.521%, which was a comprehensive reflection of tomato growth and quality. The contribution rate of the variance in the second principal component F_2 (named as water use factor) was 16.310%, which was a comprehensive reflection of tomato water use. The score function of each principal component was as follows:

$$F_1 = 0.335X_1 + 0.333X_2 + 0.326X_3 - 0.291X_4 - 0.301X_5 + 0.329X_6 - 0.047X_7 - 0.293X_8 - 0.335X_9 + 0.286X_{10-} + 0.324X_{11} \quad (1)$$

$$F_{2-} = 0.053X_1 + 0.153X_2 + 0.180X_3 + 0.352X_4 + 0.333X_{5-} + 0.184X_6 + 0.660X_7 + 0.364X_8 - 0.084X_9 + 0.282X_{10} + 0.135X_{11} \quad (2)$$

After the proportion of the variance contribution rate corresponding to each principal component was taken to the cumulative contribution rate of the principal component variance as the weight, the comprehensive model of the principal component score F was obtained using Formula (3).

$$F = (0.775F_1 + 0.163F_2)/0.938 \tag{3}$$

The comprehensive score F can be used as an objective evaluation index for evaluating the pros and cons of different *W-Z* strategies. Submitting each parameter under different treatments into Formula (3) can generate a comprehensive evaluation score of each treatment, as shown in Figure 6. The Z_0W_{50} had the lowest comprehensive evaluation score at -1.529, whereas Z_6W_{100} had the highest result of 1.295. The coupling planting strategy of Z of 6 t·ha^{-1} and W of 100% E was recommended.

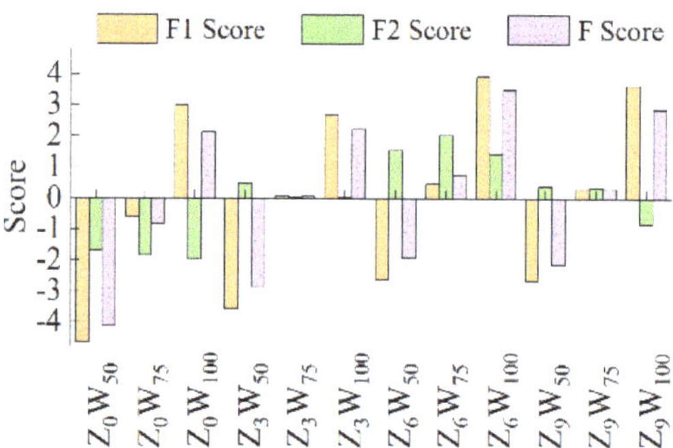

Figure 6. Comprehensive evaluation score of tomato planting under different *W-Z* conditions.

4. Discussion

4.1. Effects of Water Level on Tomato Growth

The present results showed that when W increased from 50% E to 100% E, tomato *Ph* and *St* increased by 19.78–31.30% and 13.57–20.43%, respectively. Wei et al. [31] found that tomato *Ph* under 4950, 4750, and 4500 m^3·ha^{-1} irrigation amount increased by 3.3%, 5.5%, and 5.7%, whereas the *St* increased by 1.9%, 11.4%, and 7.2%, respectively, compared with the 4050 m^3·ha^{-1} irrigation treatment. *Ph* and *St* increased first and then decreased with increasing irrigation amount, and the response intensity of *Ph* and *St* to *W* was lower than that in this study. In previous studies, soil texture was sandy soil and sandy clay, and the soil moisture content was maintained at 5.5–34.5% [31]. In this study, the soil texture was clay loam, and the soil moisture content was maintained at 16.84–33.65% (Figure 1), which was higher than that in previous reports. More sufficient soil moisture will help accelerate root water absorption and improve water absorption and utilization and then increase plant growth accumulation.

The present results showed that with an increase in *W*, tomato *RL*, *RV*, *RS*, and *OA* increased by 29.10–50.84%, 53.91–77.96%, 35.46–47.88%, and 13.51–16.90% respectively, whereas *Ff*, *NC*, *VC*, and *SS* decreased by 23.16–33.69%, 25.12–29.11%, 14.44–17.81%, and 17.24–28.57%, respectively. Cabello et al. [32] found that when the amount of irrigation increased from 60% to 140% *ETc* (crop evapotranspiration), the firmness of melon decreased by 7.27%, and the response intensity of hardness to *W* was significantly lower than that in this study. This result may be due to the different reduction ranges in cell wall pressure and cell wall swelling pressure when crops were subjected to different degrees of water stress [33], which affected the change in *Ff*.

Yang Hui et al. [15] showed that under different nitrogen application conditions, with an increase in irrigation amount, tomato *RL*, *RS*, *RV*, *VC*, *OA*, and *NC* first increased by 30.21%, 66.70%, 47.07%, 9.72%, 21.90%, and 22.04% on average, respectively, and then decreased by 10.48%, 7.53%, 10.13%, 12.73%, 6.38%, and 29.89% on average, respectively, whereas *SS* decreased by 10.81% on average. The response intensity and trend of *RL*, *RS*, *RV*, *NC*, *SS*, *VC*, and *OA* to *W* were different from the results of this study. This difference may be caused by different basic experimental conditions, such as experimental design and soil texture. Previous studies reported that the pot experiment was adopted, the soil texture was heavy loam, the soil total nitrogen content was 0.81 $g \cdot kg^{-1}$, and the nitrogen application rate was 0.18–0.42 $g \cdot kg^{-1}$ (equal to 565–1320 $kg \cdot ha^{-1}$) [15]. The greenhouse experiment was used in this study, the soil texture was clay loam, the soil total nitrogen content was 1.12 $g \cdot kg^{-1}$, and the nitrogen application rate was 350 $kg \cdot ha^{-1}$. In previous studies, the amount of nitrogen application was 1.61–3.77-times that in this study. Excessive nitrogen input significantly inhibits root growth while causing waste of resources [34]. In addition, the water-holding performance of soil in the present study was better than that in previous reports. Appropriate soil nutrients and sufficient water can stimulate root growth, improve the ability of roots to absorb water and nutrients, and improve tomato quality.

The present results showed that with increasing Irrigation, the *Pn*, *Tr*, and *Gs* of tomatoes increased by 16.67–28.43%, 22.82–44.51%, and 20.37–26.16%, respectively, and *Ci* decreased by 8.96–12.57%. Guo et al. [35] showed that with the weakening of drought degree, the *Pn*, *Tr*, *Gs*, and *Ci* of drought-resistant soybeans increased by 91.06%, 90.30%, 89.01%, and 50.71%, respectively. The response intensity of *Pn*, *Tr*, and *Gs* to *W* was significantly higher than that in this study, and the response trend of *Ci* to *W* was the opposite to that in this study. This finding may be attributed to different degrees of water stress and restrictions on photosynthesis. The previous irrigation levels were 40%, 70%, and 85% of the field water capacity, and the water stress intensity was higher than that in this study. The water potential in the cells increased with the water absorption ability of tomato guard cells because of high water stress intensity, enabling guard cells to absorb more water, increasing the degree of stomata opening and stomatal conductance [36], accelerating water loss in leaves, and increasing the values of *Tr* and *Pn*. The photosynthesis of tomatoes was limited by nonstomatal factors because of water stress [37], which increased the *Ci* of leaves.

Water significantly affected the yield and *WUE* of tomatoes throughout the growth period. The present results showed that after irrigation increased from 50% *E* to 100% *E*, the tomato yield increased by 8.48–16.56%, whereas the *WUE* decreased by 5.8–16.17%. Xia et al. [38] found that as the relative water content of the substrate increased from $50 \pm 5\%$ to $95 \pm 5\%$, the yield of tomato per square meter increased by 48.05% and *WUE* decreased by 28.23%. Abdel-Razzak et al. [11] found that with an increase in irrigation amount from 50% *eTc* to 100 *eTc*, tomato yield increased by 89.09% and *WUE* decreased by 14.80%. Xia et al. reported that the response intensity of yield and *WUE* to *W* were higher than that in this study possibly because of a difference in substrate. The substrate was organic matter mixed with mushroom residue and peanut shell mixed fermentation material and sheep manure, which contained 185.38 $mg \cdot kg^{-1}$ N, 122.46 $mg \cdot kg^{-1}$ P_2O_5, and 3.92 $g \cdot kg^{-1}$ K_2O (equal to 180 kg $N \cdot ha^{-1}$, 119 kg $P_2O_5 \cdot ha^{-1}$, and 3812 kg $K_2O \cdot ha^{-1}$) [38]. In this study, natural soil was used for cultivation, which had 350 kg $N \cdot ha^{-1}$, 200 kg of $P_2O_5 \cdot ha^{-1}$, and 400 kg $K_2O \cdot ha^{-1}$. Compared with organic matrix cultivation, the application amount of potassium fertilizer was significantly reduced, and, thus, the amounts of nutrients required for crop growth were reduced to a certain extent. However, the response intensity of yield to water in the Abdel-Razzak study [11] was significantly higher than that in this study, which may be because the soil texture in the previous study was sandy loam [11], and the soil texture in this study was clay loam, and the soil water-holding capacity was better than the predecessors. Therefore, the soil osmotic pressure increased significantly when the predecessors were irrigated in deficit, inhibiting crop growth and, thus, reducing yield [39].

4.2. Effects of Zeolite Amount on Tomato Growth

This paper showed that after *Z* increased from Z_0 to Z_6, tomato *Ph*, *St*, *NC*, *VC*, *SS*, and *OA* increased by 8.32–16.55%, 5.14–8.42%, 6.42–16.23%, 3.65–10.18%, 3.57–20%, and 6.8–8.1%, respectively, whereas *Ff* was reduced by 4.05–17.20%. After *Z* increased from

Z_6 to Z_9, *Ph*, *St*, *NC*, *VC*, *SS*, and *OA* were reduced by 2.08–4.91%, 1.13–3.85%, 4.0–7.23%, 2.69–3.4%, and 1.72–12.50%, respectively, whereas *Ff* increased by 4.55–6.27%, except for W_{100}. Obregon-Portocarrero et al. [40] found that after Z increased from Z_0 to Z_{15}, maize *Ph* and *St* increased by 2.96% and 6.06%, respectively. When Z increased from Z_{15} to Z_{35}, *Ph* decreased by 10.07%, whereas *St* showed no significant difference. Petropoulos et al. [41] found that compared with the treatment without Z, the addition of Z increased *Ff* and *OA* by 2.16% and 78.13%, respectively, and *SS* was reduced by 6.29%. This is due to the fact that adding Z can improve soil structure by improving the availability of fertilizer and the buffer capacity of soil [42], promoting crop root development, and improving crop growth and quality. However, the excessive application of Z caused a large amount of sodium ions from Z to invade the soil, which poisoned the roots of tomato and inhibited the normal growth of tomato [43]. The response intensities of *Ph* and *St* to Z in the study of Obregon-Portocarrero [40] and the response trends of *Ff*, *OA*, and *SS* to Z in the study of Petropoulos et al. [41] were different from those in this study, possibly because of differences in application amount and period of nitrogen fertilizer addition. In the study of Obregon-Portocarrero, nitrogen application was carried out at two time points: 30 kg·ha^{-1} at 15 days and 70 kg·ha^{-1} at 70 days after sowing. In the study of Petropoulos, conventional fertilizer (N:P:K = 21:0:0) was applied twice. In this study, nitrogen was applied three times: half of the nitrogen was applied at the base fertilizer stage, and one-fourth of nitrogen was applied with a drip irrigation system at the first and third ear fruit expanding stages. The amount of nitrogen was 350 kg·ha^{-1}. Sufficient nitrogen fertilizer and rational distribution of nitrogen fertilizer can supply nitrogen for crop growth effectively and continuously and then promote crop growth accumulation.

This paper showed that when Z increased from Z_0 to Z_6, tomato *RL*, *RV*, and *RS* increased by 3.51–24.70%, 4.67–20.13%, and 9.08–16.98%, respectively. When Z increased from Z_6 to Z_9, *RL*, *RV*, and *RS* decreased by 0.07–11.64%, 2.19–9.84%, and 2.49–11.04%, respectively. Wu et al. [44] found that compared with treatment without Z, adding Z increased *RL* and *RV* by 9.90–15.54% and 5.45–15.04%, respectively. The response trends of *RL* and *RV* to Z were different from those in this study, possibly because of differences in test conditions, such as application amount of Z and crop type. Previous studies set Z at two levels: 0 and 10 t·ha^{-1}, and the test crop was rice [44]. In the present study, Z was set at four levels: 0, 3, 6, and 9 t·ha^{-1}, and the test crop was tomato. Previous studies revealed that adding Z had a promoting effect on crop growth. The present study set multiple levels, which accurately described the effects of different Z levels on crops. Different crop types have different levels of response to Z levels. The current paper showed that after Z increased from Z_0 to Z_6, tomato *Ci* decreased by 2.11–9.02%, whereas *Pn*, *Tr*, and *Gs* increased by 6.49–15.20%, 8.95–27.24%, and 5.22–16.51%, respectively. When Z increased from Z_6 to Z_9, Ci increased by 0.60–2.01%, whereas *Pn*, *Tr*, and *Gs* decreased by 0.43–3.45%, 1.44–9.18%, and 0.16–5.77% respectively. Zheng et al. [45] showed that when Z increased from Z_0 to Z_{15}, the *Pn*, *Tr*, and *Gs* of rice increased by 0.52%, 17.13%, and 35% respectively, whereas *Ci* decreased by 17.02%. Chi et al. [46] showed that when Z increased from Z_0 to Z_{10}, the *Pn*, *Tr*, and *Gs* of rice increased by 0.95%, 3.93%, and 5.56% respectively, whereas *Ci* decreased by 4.58%. The response intensities of *Pn*, *Tr*, *Gs*, and *Ci* to Z were different from those in this study, possibly because of differences in irrigation methods and Z burial depths. Zheng et al.'s studies showed that Z was mixed into the soil to a depth of 5 cm under continuous flood irrigation [45]. Chi et al.'s studies showed that Z was applied to the soil along with the base fertilizer at one time and mixed with the soil evenly [46]. The present study adopted alternate drip irrigation under mulch, and Z was mixed into the soil to a depth of 30 cm. The apparent morphological characteristics of the crops were promoted as available nitrogen content in deep soil was reduced due to the increased buried depth of Z, increasing the chlorophyll content and leaf area index of crops and then promoting the net photosynthetic rate [47].

This paper showed that compared with no-Z treatment, the application of Z increased the yield and *WUE* of tomatoes by 0.52–15.06% and 7.77–47.06%, respectively. Previous studies showed that the application of Z significantly saved water by 4.8–11.4% and increased production by 9.7% [44]. The response intensities of yield and *WUE* to Z were less than those in this study, which may be caused by different crop types, irrigation methods, and soil texture. Previous studies adopted alternative irrigation, the test crop was rice, and the soil texture was sandy loam. The soil was affected by wind and sand in the entire year, poor fertilizer, and water-holding capacity [44]. This study adopted alternate drip irrigation under mulch, the test crop was tomato, and the soil texture was clay loam. The soil moisture content was 16.8–33.7%. The soil water-holding capacity in this study was better than that reported by previous studies, and sufficient soil water will help accelerate tomato growth. Compared with conventional alternate irrigation, alternate drip irrigation under mulch can stimulate and strengthen the root absorption compensation function [48] and thereby, improves root activity and *WUE*, and then promotes tomato growth and yield. These features were the reasons for the difference between the present study and previous reports.

4.3. Comprehensive Evaluation Analysis

Two traditional comprehensive evaluation models based on *Ph-St*-yield-*WUE* indicators and yield-*WUE*-quality indicators were used for validation in this study: PCA_{Bi} and PCA_{Wang} models [21,22]. Comprehensive evaluation results of tomato growth based on PCA_{Bi}, PCA_{Wang}, and the PCA_{Jv} model proposed by this paper are shown in Table 1. Z_6W_{100} was the best treatment, whereas Z_0W_{50} was the worst in the PCA_{Bi} and PCA_{Jv} models, and differences in the ranking of other treatments were observed. The PCA_{Wang} model showed that Z_6W_{50} and Z_0W_{100} were the best and worst treatment for tomatoes, respectively, in contrast to the PCA_{Jv} model. Considerable differences in ranking results were found between the PCA_{Wang} and PCA_{Jv} models. These differences may be due to differences in index factors among the evaluation methods and subsequent changes in principal component loads and contribution rates.

Table 1. Comprehensive evaluation results of tomato growth based on PCA_{Jv}, PCA_{Bi}, and PCA_{Wang} models.

Treament	F_{Jv}	Ranking	F_{Bi}	Ranking	F_{Wang}	Ranking
Z_0W_{50}	−1.53	12	−1.465	12	0.418	5
Z_0W_{75}	−0.41	8	−0.842	11	−0.505	9
Z_0W_{100}	0.59	4	0.007	7	−1.335	12
Z_3W_{50}	−0.95	11	−0.588	10	0.967	2
Z_3W_{75}	0.02	7	0.076	6	−0.008	7
Z_3W_{100}	0.76	3	0.587	3	−0.567	10
Z_6W_{50}	−0.54	9	−0.153	8	1.118	1
Z_6W_{75}	0.40	5	0.722	2	0.605	4
Z_6W_{100}	1.29	1	1.440	1	−0.371	8
Z_9W_{50}	−0.71	10	−0.476	9	0.698	3
Z_9W_{75}	0.14	6	0.116	5	0.052	6
Z_9W_{100}	0.93	2	0.576	4	−1.072	11

Note: F_{Jv}, F_{Bi}, and F_{Wang} were the comprehensive principal component scores of PCA_{Jv}, PCA_{Bi}, and PCA_{Wang} models, which can be calculated by Formulae (1)–(3), respectively.

In the PCA_{Bi} model, *Ph*, *St*, and yield in F_1 had large loads with values of 0.975, 0.979, and 0.949, respectively. In the PCA_{Wang} model, *NC*, *VC*, *SS*, and *OA* in F_1 had large loads with values of 0.954, 0.974, 0.963, and 0.848, respectively. The load values of *Ph*, *St*, yield, *NC*, *VC*, and *SS* in the PCA_{Jv} model F_1 were 0.945, 0.971, 0.836, 0.879, 0.851, and 0.857, respectively. In the F_2 of these three models, *WUE* dominated the largest load, with values of 0.998, 0.810, and 0.884. The variance contribution rates of F_1 and F_2 were 70.290% and 25.023% in the PCA_{Bi} model, 68.821% and 23.067% in the PCA_{Wang} model, and 77.521% and 16.310% in the PCA_{Jv} model. These results indicated that the variance contribution rate of F_1 increased and that of F_2 decreased compared with those in previous studies, which was due to the changing of the index load. The PCA_{Jv} model focuses much more on the effects of growth, physiological, yield, and quality indicators on the growth and development of tomatoes throughout the growth period. The PCA_{Wang} model showed that some treatments were ranked in the order of $Z_0W_{50} > Z_0W_{75} > Z_0W_{100}$ (Table 1), indicating under a condition without Z, tomato growth characteristics improve with decreasing moisture content. This finding contradicts the general law of tomato growth response to water content. It showed that the rationality and applicability of the model are closely related to the selection of evaluation indexes and crop types. Therefore, the typical indicators that can objectively reflect the growth status of crops should be reasonably selected in combination with specific crop types in the modeling process. These indicators can help build a reasonable model evaluation system and obtain reasonable evaluation results.

5. Conclusions

(1) Tomato Ph, St, root indexes, LAI, Pn, Tr, Gs, OA, and yield showed a positive response to W, whereas NC, VC, SS, Ci, Ff, and WUE showed opposite trends. The response of Ci and Ff to Z was first negative and then positive, whereas that of other indexes to Z showed an opposite trend (except W_{50} yield).

(2) The effects of W, Z, and W-Z on tomato growth physiological indexes, quality indexes, and yield were as follows: $W > Z > W$-Z; the effects on WUE were as follows: $Z > W > W$-Z.

(3) The two principal components of growth quality factor and water usage factor were extracted through analysis, and the cumulative variance contribution rate reached 93.831%. According to the comprehensive score evaluation of principal components, the optimum W of tomatoes was 100% E, and the amount of Z was 6 t·ha^{-1} during alternate drip irrigation under mulch.

Supplementary Materials: The following are available online at https://www.mdpi.com/article/10.3390/horticulturae8060536/s1, Table S1: Results of two-way ANOVA for tomato growth, physiology, quality, yield and WUE, Table S2: Grey relational analysis calculation, Table S3: Eigenvalues and Cumulative Variance Contribution Rates of Tomato Evaluation Factors, Table S4: Factor loading matrix of principal components on each index.

Author Contributions: Conceptualization, X.J.; methodology, X.S.; software, M.Z.; validation, X.J., T.L. and X.G.; formal analysis, R.L.; investigation, J.M.; resources, T.L.; data curation, X.G.; writing—original draft preparation, X.J.; writing—review and editing, T.L.; data analysis and visualization, X.G.; supervision, T.L.; project administration, T.L.; funding acquisition, T.L. All authors have read and agreed to the published version of the manuscript.

Funding: This research was funded by the National Natural Science Foundation of China (51809189; 51909184), the China Postdoctoral Science Foundation (2020M670693), and the Scientific and Technologial Innovation Programs of Higher Education Institutions in Shanxi (2019L0136).

Institutional Review Board Statement: Not applicable.

Informed Consent Statement: Not applicable.

Data Availability Statement: Not applicable.

Acknowledgments: We appreciate that postgraduates from the university of the first author, who investigated and collected data. We are also grateful to the editors and anonymous reviewers for their suggestions and comments.

Conflicts of Interest: The authors declare no conflict of interest.

References

1. Chand, J.B.; Hewa, G.; Hassanli, A.; Myers, B. Deficit Irrigation on Tomato Production in a Greenhouse Environment: A Review. *J. Irrig. Drain. Eng.* **2021**, *147*, 04020041. [CrossRef]
2. Fariasa, D.B.D.; da Silva, P.S.O.; Lucas, A.A.T.; de Freitae, M.I.; Santos, T.D.; Fontes, P.T.N.; de Oliveira, L.F.G. Physiological and productive parameters of the okra under irrigation levels. *Sci. Hortic.* **2019**, *252*, 1–6. [CrossRef]

3. Sepaskhah, A.R.; Barzegar, M. Yield, water and nitrogen-use response of rice to zeolite and nitrogen fertilization in a semi-arid environment. *Agric. Water Manag.* **2010**, *98*, 38–44. [CrossRef]
4. Cui, X.; Song, J.; Qu, M. Effect of soil water potential on hydraulic parameters of Fraxinus mandshurica seedlings. *J. Appl. Ecol.* **2004**, *15*, 2237–2244.
5. Wang, Y.S.; Liu, J.; Tang, S.; An, Z.X.; Guo, Z.L.; Ding, X.B.; Liu, F.J.; Cao, Z.L.; Zhang, T.; Zhang, Y. Modifications of chemically induced-enucleated nuclear transfer technique by reverse-order nuclear transfer in mouse. *Zygote* **2009**, *17*, 261–268. [CrossRef]
6. Wang, J.W.; Niu, W.Q.; Li, Y. Nitrogen and Phosphorus Absorption and Yield of Tomato Increased by Regulating the Bacterial Community under Greenhouse Conditions via the Alternate Drip Irrigation Method. *Agronomy* **2020**, *10*, 315. [CrossRef]
7. Sonmez, I.; Kaplan, M.; Demir, H.; Yilmaz, E. Effects of zeolite on seedling quality and nutrient contents of tomato plant (*Solanum lycopersicon* cv. Malike F1) grown in different mixtures of growing media. *Food Agric. Environ.* **2010**, *8*, 1162–1165.
8. Podkovyrov, I.; Kostin, M.; Dolgova, A.; Filipchuk, O.G.; Nesvat, A. Impact of zeolites on intensity of the vital processes of hybrid plants. *Vestn. Kazan State Agrar. Univ.* **2019**, *14*, 31–36. [CrossRef]
9. Urbina-Sanchez, E.; Baca-Castillo, G.A.; Nunez-Escobar, R.; Colinas-Leon, M.T.; Tijerina-Chavez, L.; Tirado-Torres, J.L. Tomato seedlings soilless culture on K^+, Ca^{2+} or Mg^{2+} loaded zeolite and different granule size. *Agrociencia* **2006**, *40*, 419–429.
10. Bernardi, A.C.D.; Monte, M.B.D.; Paiva, P.R.P.; Werneck, C.G.; Haim, P.G.; Barros, F.D. Dry matter production and nutrient accumulation after successive crops of lettuce, tomato, rice, and andropogongrass in a substrate with zeolite. *Rev. Bras. Cienc. Solo* **2010**, *34*, 435–442. [CrossRef]
11. Abdel-Razzak, H.; Wahb-Allah, M.; Ibrahim, A.; Alenazi, M.; Alsadon, A. Response of Cherry Tomato to Irrigation Levels and Fruit Pruning under Greenhouse Conditions. *J. Agric. Sci. Technol.* **2016**, *18*, 1091–1103.
12. Ajirloo, A.R.; Amiri, E. Responses of Tomato Cultivars to Water-Deficit Conditions (Case Study: Moghan Plain, Iran). *Commun. Soil Sci. Plant Anal.* **2018**, *49*, 2267–2283. [CrossRef]
13. Aydiner, E.; Tuzel, Y.; Tuzel, I.H.; Tunali, U.; Oztekin, G.B. Effects of Irrigation Based on Different Moisture Levels of growing Medium on Soilless Grown Greenhouse Tomatoes. *Int. Soc. Hort. Cult. Sci.* **2014**, *1142*, 93–98. [CrossRef]
14. Ullah, I.; Mao, H.P.; Rasool, G.; Gao, H.Y.; Javed, Q.; Sarwar, A.; Khan, M.I. Effect of Deficit Irrigation and Reduced N Fertilization on Plant Growth, Root Morphology and Water Use Efficiency of Tomato Grown in Soilless Culture. *Agronomy* **2021**, *11*, 228. [CrossRef]
15. Yang, H.; Cao, H.X.; Hao, X.M.; Guo, L.J.; Li, H.Z.; Wu, X.Y. Evaluation of tomato fruit quality response to water and nitrogen management under alternate partial root-zone irrigation. *Int. J. Agric. Biol. Eng.* **2017**, *10*, 85–94.
16. Liu, G.Y.; Du, Q.J.; Jiao, X.C.; Li, J.M. Irrigation at the level of evapotranspiration aids growth recovery and photosynthesis rate in tomato grown under chilling stress. *Acta Physiol. Plant.* **2018**, *40*, 2. [CrossRef]
17. Zhao, Z.L.; Li, B.; Feng, X.; Yao, M.Z.; Xie, Y.; Xing, J.W.; Li, C.X. Parameter estimation and verification of DSSAT-CROPGRO-Tomato model under different irrigation levels in greenhouse. *J. Appl. Ecol.* **2018**, *29*, 2017–2027.
18. Ozbahce, A.; Tari, A.F.; Gonulal, E.; Simsekli, N. Zeolite for Enhancing Yield and Quality of Potatoes Cultivated Under Water-Deficit Conditions. *Potato Res.* **2018**, *61*, 247–259. [CrossRef]
19. Nozari, R.; Tohidi-Moghadam, H.R.; Mashhadi-Akbar-Boojar, M. Effects of zeolite and cattle manure on growth, yield and yield components of soybean grown under water deficit stress. *Res. Crop.* **2012**, *13*, 920–927.
20. Ozbahce, A.; Tari, A.F.; Gonulal, E.; Simsekli, N.; Padem, H. The effect of zeolite applications on yield components and nutrient uptake of common bean under water stress. *Arch. Agron. Soil Sci.* **2015**, *61*, 615–626. [CrossRef]
21. Bi, Y.J.; Lv, P.P.; Su, R.D.; Wang, Y.M.; Wang, J.; Lei, M.J. Determination of the buried depth and pressure head under moistube irrigation based on principal component analysis. *Fresenius Environ. Bull.* **2020**, *29*, 5021–5028.
22. Wang, Y.; Zhang, F.-C.; Wang, H.-D.; Bi, L.-F.; Cheng, M.-H.; Yan, F.-L.; Fan, J.-L.; Xiang, Y.-Z. Effects of the frequency and amount of drip irrigation on yield, tuber quality and water use efficiency of potato in sandy soil of Yulin, northern Shaanxi, China. *J. Appl. Ecol.* **2019**, *30*, 4159–4168.
23. Zhang, G.W.; Yang, C.Q.; Liu, R.X.; Ni, W.C. Effects of p-hydroxybenzoic acid and phloroglucinol on mitochondria function and root growth in cotton (*Gossypium hirsutum* L.) seedling roots. *J. Appl. Ecol.* **2018**, *29*, 231–237.
24. Zhang, L.; Lu, C.; Peng, L.; Ma, W.; Qian, W. Progress in improving photosynthetic efficiency by synthetic biology. *Chin. J. Biotechnol.* **2017**, *33*, 486–493.
25. Colaric, M.; Stampar, F.; Hudina, M. Content levels of various fruit metabolites in the "Conferenc" pear response to branch bending. *Sci. Hortic.* **2007**, *113*, 261–266. [CrossRef]
26. Fan, B.H.; Ma, L.L.; Ren, R.D.; He, J.X.; Hamiti, A.; Li, J.M. Effects of irrigation frequency of organic nutrient solution and irrigation amount on yield, quality, fertilizer and water use efficiency of melon in facility. *J. Appl. Ecol.* **2019**, *30*, 1261–1268.
27. Tang, L.; Luo, W.J.; He, Z.L.; Gurajala, H.K.; Hamid, Y.; Khan, K.Y.; Yang, X.E. Variations in cadmium and nitrate co-accumulation among water spinach genotypes and implications for screening safe genotypes for human consumption. *Zhejiang Univ.-Sci. B* **2018**, *19*, 147–158. [CrossRef]
28. Tian, G.; Li, H.F.; Tian, M.; Liu, X.X.; Chen, Q.; Zhu, Z.L.; Jiang, Y.M.; Ge, S.F. Effects of different integration of water and fertilizer modes on the absorption and utilization of nitrogen fertilizer and fruit yield and quality of apple trees. *J. Appl. Ecol.* **2020**, *31*, 1867–1874.

29. Yin, S.X.; Wei, L.F.; Mei, Y.Q.; Liu, X.H.; Zou, L.S.; Cai, Z.C.; Yuan, J.H.; Ge, H.-T.; Wang, D.-G.; Wang, D.-D. Simultaneous determination of multiple bioactive constituents in Abelmoschi Corolla by UFLC-QTRAP-MS/MS. *China J. Chin. Mater. Med.* **2021**, *46*, 2527–2536.

30. Wang, X.; Cui, S.X.; Sun, Z.M.; Mu, G.J.; Cui, S.L.; Wang, P.C.; Liu, L.F. Ecological adaptability evaluation of peanut cultivars based on biomass and nutrient accumulation. *J. Appl. Ecol.* **2015**, *26*, 2023–2029.

31. Wei, C.L.; Zhu, Y.; Zhang, J.Z.; Wang, Z.H. Evaluation of Suitable Mixture of Water and Air for Processing Tomato in Drip Irrigation in Xinjiang Oasis. *Sustainability* **2021**, *13*, 7845. [CrossRef]

32. Cabello, M.J.; Castellanos, M.T.; Romojaro, F.; Martinez-Madrid, C.; Ribas, F. Yield and quality of melon grown under different irrigation and nitrogen rates. *Agric. Water Manag.* **2009**, *96*, 866–874. [CrossRef]

33. Guichard, S.; Gary, C.; Longuenesse, J.J.; Leonardi, C. *Water Fluxes and Growth of Greenhouse Tomato Fruits under Summer Conditions*; International Society for Horticultural Science (ISHS): Leuven, Belgium, 1999; pp. 223–230.

34. Song, Q.L.; Yue, S.C.; Cai, L.Q. Response of Maize Root Morphology to Nitrogen Application Under Film Mulch. *Res. Soil Water Conserv.* **2020**, *27*, 23–29. (In Chinese)

35. Guo, S.J.; Yang, K.M.; Huo, J.; Zhou, Y.H.; Wang, Y.P.; Li, G.Q. Influence of drought on leaf photosynthetic capacity and root growth of soybeans at grain filling stage. *J. Appl. Ecol.* **2015**, *26*, 1419–1425.

36. Luo, D.D.; Wang, C.K.; Jin, Y. Stomatal regulation of plants in response to drought stress. *J. Appl. Ecol.* **2019**, *30*, 4333–4343.

37. Winter, K.; Schramm, M.J. Analysis of Stomatal and Nonstomatal Components in the Environmental Control of CO_2 Exchange in Leaves of Welwitschia mirabilis. *Plant Physiol.* **1986**, *82*, 173–178. [CrossRef]

38. Xia, X.B.; Yu, X.C.; Gao, J.J. Effects of moisture content in organic substrate on the physiological characters, fruit quality and yield of tomato plant. *J. Appl. Ecol.* **2007**, *18*, 2710–2714.

39. Agbna, G.H.D.; She, D.L.; Liu, Z.P.; Elshaikh, N.A.; Shao, G.C.; Timm, L.C. Effects of deficit irrigation and biochar addition on the growth, yield, and quality of tomato. *Sci. Hortic.* **2017**, *222*, 90–101. [CrossRef]

40. Obregón-Portocarrero, N.; Díaz-Ortiz, J.E.; Daza-Torres, M.C.; Aristizabal-Rodríguez, H.F. Efecto de la aplicación de zeolita en la recuperación de nitrógeno y el rendimiento de maíz. *Acta Agron.* **2016**, *65*, 24–30. [CrossRef]

41. Petropoulos, S.A.; Fernandes, A.; Xyrafis, E.; Polyzos, N.; Antoniadis, V.; Barros, L.; Ferreira, I. The Optimization of Nitrogen Fertilization Regulates Crop Performance and Quality of Processing Tomato (*Solanum lycopersicum* L. cv. Heinz 3402). *Agronomy* **2020**, *10*, 715. [CrossRef]

42. Palanivell, P.; Ahmed, O.H.; Omar, L.; Majid, N.M.A. Nitrogen, Phosphorus, and Potassium Adsorption and Desorption Improvement and Soil Buffering Capacity Using Clinoptilolite Zeolite. *Agronomy* **2021**, *11*, 379. [CrossRef]

43. Ma, M.L. Effects of controlled root zoning and alternating sub-film drip irrigation on water consumption and yield of tomato. *Inn. Mong. Sci. Technol. Econ.* **2019**, *18*, 86–87. (In Chinese)

44. Wu, Q.; Chen, H.Y.; Wang, Y.Z.; Chi, D.C. Water-saving and Fertilizer-reducing Effect of Clinoptilolite in Water-saving Irrigated Paddy Fields in Semiarid Areas of Western Liaoning. *J. Agric. Mach.* **2021**, *52*, 305–313+406. (In Chinese)

45. Zheng, J.L.; Chen, T.T.; Wu, Q.; Yu, J.M.; Chen, W.; Chen, Y.L.; Siddique, K.H.M.; Meng, W.Z.; Chi, D.C.; Xia, G.M. Effect of zeolite application on phenology, grain yield and grain quality in rice under water stress. *Agric. Water Manag.* **2018**, *206*, 241–251. [CrossRef]

46. Chi, D.C.; Yu, J.M.; Chen, T.T.; Zheng, J.L.; Chen, W.; Yi, E.B. Effects of the Coupling of Nitrogen and Zeolite on Physiological Characteristics of Rice during Milky Ripe Stage. *J. Shenyang Agric. Univ.* **2017**, *48*, 745–750. (In Chinese)

47. Amiri, H.; Ghalavand, A.; Mokhtassi-Bidgoli, A. Growth, Seed Yield and Quality of Soybean as Affected by Integrated Fertilizer Managements and Zeolite Application. *Commun. Soil Sci. Plant Anal.* **2021**, *52*, 1834–1851. [CrossRef]

48. Kang, S.Z.; Pan, Y.H.; Shi, P.Z.; Zhang, J.H. Theory and Experiment on Alternate Irrigation of Root System of Controlled Crops. *J. Hydraul. Eng.* **2001**, 80–86. (In Chinese)

horticulturae

Article

The Impact of Insect-Proof Screen on Microclimate, Reference Evapotranspiration and Growth of Chinese Flowering Cabbage in Arid and Semi-Arid Region

Jiangli Wen [1,2], Songrui Ning [3], Xiaoming Wei [1], Wenzhong Guo [1], Weituo Sun [1], Tao Zhang [2,4] and Lichun Wang [1,*]

[1] Intelligent Equipment Research Center, Beijing Academy of Agriculture and Forestry Sciences, Beijing Nongke Mansion, No. 11 Shuguang Huayuan Middle Road, Haidian District, Beijing 100097, China; wenjiangli@bvca.edu.cn (J.W.); weixm@nercita.org.cn (X.W.); guowz@nercita.org.cn (W.G.); sunwt@nercita.org.cn (W.S.)

[2] Department of Water Conservancy and Construction Engineering, Beijing Vocational College of Agriculture, Beijing 102442, China; zhangtao8510@126.com

[3] State Key Laboratory of Eco-Hydraulics in Northwest Arid Region of China, Xi'an University of Technology, Xi'an 710048, China; ningsongrui@163.com

[4] Beijing Surround Line Management of Beijing South-to-North Water Diversion Project, Beijing Water Authority, Beijing 100176, China

* Correspondence: wanglc@nercita.org.cn

Citation: Wen, J.; Ning, S.; Wei, X.; Guo, W.; Sun, W.; Zhang, T.; Wang, L. The Impact of Insect-Proof Screen on Microclimate, Reference Evapotranspiration and Growth of Chinese Flowering Cabbage in Arid and Semi-Arid Region. *Horticulturae* **2022**, *8*, 704. https://doi.org/10.3390/horticulturae8080704

Academic Editors: Xiaohui Hu, Shiwei Song and Xun Li

Received: 6 July 2022
Accepted: 1 August 2022
Published: 3 August 2022

Publisher's Note: MDPI stays neutral with regard to jurisdictional claims in published maps and institutional affiliations.

Abstract: Despite the steadily increasing area under protected agriculture there is a current lack of knowledge about the effects of the insect-proof screen (IPS) on microclimate and crop water requirements in arid and semi-arid regions. Field experiments were conducted in two crop cycles in Ningxia of Northwest China to study the impact of IPS on microclimate, reference evapotranspiration (ET_0) and growth of Chinese Flowering Cabbage (CFC). The results showed that IPS could appreciably improve the microclimate of the CFC field in the two crop cycles. During the first crop cycle (C1), compared with no insect-proof screen (NIPS) treatment, the total solar radiation and daily wind speed under the IPS treatment were reduced by 5.73% and 88.73%. IPS increased the daily average air humidity, air, and soil temperature during C1 by 11.84%, 15.11% and 10.37%, respectively. Furthermore, the total solar radiation and daily wind speed under the IPS treatment during the second crop cycle (C2) were markedly decreased by 20.45% and 95.73%, respectively. During C2, the daily average air temperature and air humidity under the IPS treatment were increased slightly, whereas the daily average soil temperature was decreased by 4.84%. Compared with NIPS treatment, the ET_0 under the IPS treatment during C1 and C2 was decreased by 6.52% and 21.20%, respectively, suggesting it had great water-saving potential when using IPS. The plant height, leaf number and leaf circumference of CFC under the IPS treatment were higher than those under the NIPS treatment. The yield under the IPS treatment was significantly increased by 36.00% and 108.92% in C1 and C2, respectively. Moreover, irrigation water use efficiency (*IWUE*) was significantly improved under the IPS treatment in the two crop cycles. Therefore, it is concluded that IPS can improve microclimate, reduce ET_0, and increase crop yield and *IWUE* in arid and semi-arid areas of Northwest China.

Keywords: solar radiation; humidity; temperature; irrigation water use efficiency; yield

1. Introduction

Drought has long been the primary factor limiting crop production due to the shortage and uneven distribution of water resources in arid and semi-arid regions of China [1,2]. More than 90% of Ningxia lies in arid or semi-arid zones with annual precipitation of less than 400 mm. The temporal distribution of precipitation is uneven, with 70% of the annual precipitation mainly concentrated from July to September, and there is a high rate of evaporation [3]. With long radiation time, and a large temperature difference between day and night, it has become an important base for inland leafy vegetables in Hong Kong and Macau. The area of vegetables in the region is approximately 131,000 ha, and the area is increasing year by year [4]. There are a series of problems, such as low utilization efficiency of water resources, and unreasonable structure (i.e., agricultural water use accounts for about 90% of the total) [5]. The shortage of irrigation water restricts the improvement of agricultural productivity and economic development in this area [6,7]. Therefore, improving water use efficiency (*IWUE*) through reasonable measures is an important guarantee for the sustainable development of local agriculture.

Cultivation technology covered by an insect-proof screen (IPS) is one of the important measures for the production of pesticide-free vegetables, which is of great significance to exclude the penetration of insects, such as sweet potato whitefly (*Bemisia tabaci*) (Insecta: Hemiptera: Homoptera: Sternorrhyncha: Aleyrodoidea: Aleyrodidae: Aleyrodinae) or leaf miner [8–10]. The cultivation under the screens allows significant reductions in pesticide application (if high-mesh screens are used) and increases marketable yield by reducing the radiation and wind damage to the fruit [11,12]. However, it may have an important impact on microclimate and crop water requirements [13]. Screens impede the exchanges of radiation, mass, heat and momentum between the crop and the atmosphere, thus modifying crop microclimate and reducing water requirements [14]. Experimental evidence has been presented that suggests that significant reductions in water use can be achieved when crops are grown under protective screens and nets [15,16]. Low evaporation demand under shading can increase the stomatal conductance of plants in a way that may reduce irrigation demand, resulting in water saving, and promoting vegetable growth, edible yield, nutritional quality, and pest control [17,18]. Measurements have shown a reduction of about 50–70% in ventilation rate, and about 50% reduction in crop water use, as compared to estimated values for an open pepper field [9]. At present, studies on crop water requirements with IPS are mainly concentrated in the Mediterranean area, which is characterized by hot and dry summers and mild and rainy winters. The Ningxia area belongs to the arid and semi-arid temperate continental monsoon climate, which has four distinct seasons, characterized by a dry climate, strong evaporation, and concentrated precipitation. The climate of Ningxia is quite different from the Mediterranean climate, but there are few studies on the effect of IPS on crop water demand in a temperate continental monsoon climate. Therefore, it is of great significance to study the influence of IPS on water demand for improving irrigation management.

In addition, the studies of IPS were mainly focused on crops such as tomato [19,20], sweet pepper [15], cucumber [18], and banana [17]. Compared with other vegetables, leafy vegetables have fast growth, short growth periods, multiple cropping, are rich in minerals, and are vulnerable to pests [21]. Furthermore, leafy vegetables require high-frequency irrigation because of shallow roots and the large demand for water and fertilizer. Consequently, IPS technology confirmed to be beneficial to improving crop production [8,11], has been introduced and considered a potential agriculture practice for improving the *IWUE* of leafy vegetables in Northwest China. The Chinese Flowering Cabbage (CFC) is one of the most popular and frequently consumed leafy vegetables and has a large planting scale in this and many other areas [22]. Moreover, CFC has a large planting scale in many areas at home and abroad. At present, there was no study on the CFC under IPS in the arid and semi-arid regions.

This study was motivated by the demands for reducing ineffective transpiration evaporation and improving *IWUE* in the arid and semi-arid areas of Northwest China. The objectives of this research were to (i) determine the effects of IPS on microclimate parameters and ET_0 in the arid and semi-arid region, and (ii) quantify the effects of IPS on the growth, yield, and *IWUE* of CFC.

2. Materials and Methods

2.1. Experimental Field

The experiments were conducted at Wuzhong National Agricultural Science and Technology Park Management Committee (37°57′ N, 106°6′ E, 1130 m above sea level), Wuzhong city, Ningxia province in northwest China. The area belongs to the arid and semi-arid temperate continental monsoon climate with annual rainfall of 184.6–273.5 mm and annual evaporation of 1000–4000 mm. With sufficient sunshine, strong evaporation, sparse precipitation, and large temperature difference between day and night, the annual average sunshine duration of the site was 3000 h. The mean annual temperature was 6 °C, and the frost-free period was 176 days. In 2016, the maximum, minimum and average air temperatures were 34.7, 10.2 and 21.7 °C, respectively. The soil texture of the experiment site was sandy loam, which is well drained.

On the day before planting, undisturbed soil samples at three locations of the field were taken at 0–40 cm soil for determination of the basic physical and chemical properties of soil. The bulk density was measured by 100 cm^3 rings and field capacity was followed by the method by Veihmeyer and Hendrickson [23]. The masses of organic matter, total nitrogen, available potassium and available phosphorus were analyzed in accordance with the protocol described by Lu [24]. The bulk density and field capacity of soil were 1.65 g cm^{-3} and 23.7 cm^3 cm^{-3}, respectively. The masses of organic matter and total nitrogen and available potassium (K_2O) and available phosphorus (P_2O_5) of soil were 1.57, 0.20, 130.2, and 12.9 mg kg^{-1}, respectively.

2.2. Experimental Treatments

Field experiments were conducted from May to August (in two growth cycles) in 2016. It had two treatments (insect-proof screen and no insect-proof screen). The experiment of insect-proof screen (IPS) treatment was carried out in an insect-proof (60 meshes, white net) screenhouse that stretch out from east to west with a length of 100 m, a span of 10 m, a ridge height of 1.3 m, a side wall shoulder height of 2.2 m (Figure 1). The experimental field of no insect-proof screen (NIPS) treatment was arranged adjacent to the field of IPS treatment, and the size was consistent with the IPS treatment.

Figure 1. Photograph of the screenhouse (**a**) external view, (**b**) internal view.

In the experiment, Chinese Flowering Cabbage (*Brassica parachinensis* L. ssp. *Chinensis* var. *utilis* Tsen et Lee, Youlv 702) was seeded using a hand-push type planter on 13 May and 18 July in the first crop cycle (C1) and second crop cycle (C2), respectively. The row spacing was 15 cm. Thinning took place at growing stage of two leaves and one heart and the average plant spacing after thinning was 10 cm. According to the customary dosage of local vegetable farmers, organic fertilizer application rate of 110 t hm^{-2} was evenly scattered on the soil surface of study area before the first crop sowing, and then plowed into a 0–20 cm soil layer with a micro-tiller (LKNZ Farm Machinery, Jining, China). Finally, in order to make it fully mixed, turn up the soil three times. Furthermore, unified pest and weed control management for all treatments followed conventional practices in the region. The CFC was harvested on 23 June and 18 August in C1 and C2, respectively.

2.3. Irrigation Schedule

The irrigation method was drip irrigation and the emitter spacing was 10 cm, and the nominal flow rate for each emitter (IrriGreen, Beijing, China) was 2.8 L h^{-1} at 0.1 MPa pressure. In order to ensure that water was irrigated evenly, a probe pipe (AZ-Trime, Beijng, China) with a depth of 1 m was installed between the two treatments. The water content in the 0–60 cm soil layer was measured every 2–3 days with a TDR meter (PICO-BT, ANDRES Industries AG, Etlingen, Baden-Württemberg, Germany). The measurement frequency was increased after irrigation or rainfall. CFC plants were irrigated to 90% of field capacity (θ_f) when the soil water content of the experiment was depleted to 75% of the θ_f. During the experimental season, there was no precipitation recorded at the experimental site. The irrigation amount of each treatment is consistent, and the water meter for each irrigation amount is accurately measured. Irrigation water amounts (*Ir*, mm) were determined as follows [25]:

$$Ir = 100 \times (\theta_f - \theta_i) \times \gamma \times H \times p \tag{1}$$

where θ_f is the soil field capacity (cm^3 cm^{-3}), θ_i is the average value of the actual mass water content measured by TDR (cm^3 cm^{-3}); γ is the soil bulk density (g cm^{-3}); H is the planned wet depth of soil (m), and H is 0.3 m in this experiment; p is drip irrigation water use efficiency 0.95 [26].

2.4. Measurements

The microclimate parameters such as solar radiation (R_s), air humidity (RH), wind speed (u), air temperature (T_a), and soil temperature (T_s) at the 10 cm layer were monitored by the weather station (HOBO U30 station, Bourne, MA, USA) which had been installed at 2 m above ground in the middle of the IPS and NIPS treatments and had been well calibrated before installation. The microclimate parameters were measured every second and average values were recorded every 30 min on the HOBO data logger (HOBOware Pro, Bourne, MA, USA).

To examine the potential effect of the IPS on crop water use, reference crop evapotranspiration (ET_0) was estimated using the FAO Penman–Monteith method [27], using the IPS and NIPS measured meteorological data such as T_a, R_s, and u measured. Assuming that the daily soil heat flux density is zero, the ET_0 is calculated from:

$$ET_0 = \frac{0.408\Delta R_n + \gamma \frac{900}{T_a+273} u\left(e_s - e_a\right)}{\Delta + \gamma(1 + 0.34u)} \tag{2}$$

where Δ is the slope pressure curve (kPa C^{-1}), γ is the psychrometric constant (kPa C^{-1}), R_n is the net radiation at the crop surface, ($e_s - e_a$) is the vapour pressure deficit of the air (kPa). The use of Equation (2) implicitly implies that bulk surface resistance was 70 s m^{-1}, aerodynamic resistance (s m^{-1}) was equal to $208/u$, and crop albedo was fixed at 0.23.

During each growing season, three representative CFC plants per treatment were selected for plant height, leaf number, and leaf circumference area determination. The growth parameters were measured every seven days. The plant height was measured from the vegetable sprout base to the tip using a ruler with accuracy of 0.01 mm. Leaf number was determined by the number of actual green leaves per plant. In the measurement, the longest canopy length and width of the marked plants were recorded and leaf circumference area was calculated by multiplying the longest canopy length and width for each plant.

The yield of CFC for each treatment was determined from two rows of three equally distributed locations in each field. At each location, 1.0 m^2 of CFC plants were manually harvested and the samples were cut from the third green leaf at the base of the plant with a flat cut. The weight of plants was measured using an electronic balance (Leqi battery balance, Suzhou, China) with precision of 0.01 g. The average yield of the three locations samples for each treatment was used to represent the value of treatment.

Irrigation water use efficiency (*IWUE*, kg m^{-3}) was calculated based on the following equation:

$$IWUE = \frac{100 \times GY}{I} \tag{3}$$

where *GY* is the yield (t ha^{-1}) and *I* is the irrigation amount (mm).

2.5. Statistical Analysis

One-way analysis of variance (ANOVA) with three replications was used to test whether the IPS had a significant effect on plant growth, yield, and *IWUE* at the probability levels of 0.05 or 0.01. These statistical tests were performed by using the SPSS version 18.0 software (version 18.0, SPSS, Chicago, IL, USA).

3. Results

3.1. Daily Average Solar Radiation, Air Humidity and Wind Speed

IPS had a great influence on daily solar radiation (R_s), air humidity (*RH*) and wind speed during C1 and C2 (Figure 2). The R_s under the IPS and NIPS treatments during C1 varied from 41.31 to 275.25 W m^{-2} and 53.23 to 287.75 W m^{-2}, respectively. The average daily R_s under the IPS and NIPS treatments were 198.27 W m^{-2} and 210.33 W m^{-2} during C1, respectively. The total R_s under the IPS treatment was 8327.2 W m^{-2}, which was 5.73% lower than that under the NIPS treatment during C1 (Figure 2a). Moreover, the range of R_s under the IPS and NIPS treatments during C2 varied from 45.38 to 275.30 W m^{-2} and 59.73 to 327.93 W m^{-2}, respectively. The average R_s under the IPS and NIPS treatments were 198.47 W m^{-2} and 249.49 W m^{-2} during C2, respectively. The total R_s under the IPS treatment during C2 was 7740.33 W m^{-2}, which was 20.45% lower than that under the NIPS treatment (Figure 2d). It can be concluded that IPS can markedly reduce R_s, especially during C2.

During C1, the daily average *RH* under the IPS and NIPS treatments were 26.38–89.27% (mean = 47.22%) and 21.38–84.27% (mean = 42.22%), respectively (Figure 2b). The daily average *RH* under the IPS and NIPS treatments during C1 showed a gradual downward trend, which may be due to the gradual increase in T_a. The daily average *RH* under the IPS treatment was increased by 11.84% during C1 (Figure 2b). During C2, the daily average *RH* under the IPS and NIPS treatments were 35.48–90.83% (mean = 60.58%) and 32.33–90.51% (mean = 59.35%), respectively (Figure 2e). The daily average *RH* during C2 showed a trend of first decreasing and then gradually increasing. This may be due to the gradual increase in T_a and then the gradual decrease, resulting in changes in daily average *RH*. The daily average *RH* under the IPS treatment was 2.08% higher than that under the NIPS treatment during C2 (Figure 2e). It can be seen that IPS can markedly increase the daily average *RH*, especially in C1.

The daily average wind speed during C1 under the IPS and NIPS treatments were 0.14 (0–1.16 m s^{-1}) and 1.28 (0–2.83 m s^{-1}) m s^{-1}, respectively (Figure 2c). During C2, the average wind speed under the IPS and NIPS treatments were 0.06 (0–0.56 m s^{-1}) and 1.41 (0.42–3.65 m s^{-1}) m s^{-1}, respectively (Figure 2f). Compared with NIPS treatment, the daily average wind speed under the IPS treatment was decreased by 88.73% and 95.73% during C1 and C2, respectively. It can be found that the wind speed under IPS treatment was greatly reduced and IPS gave wind efficient protection.

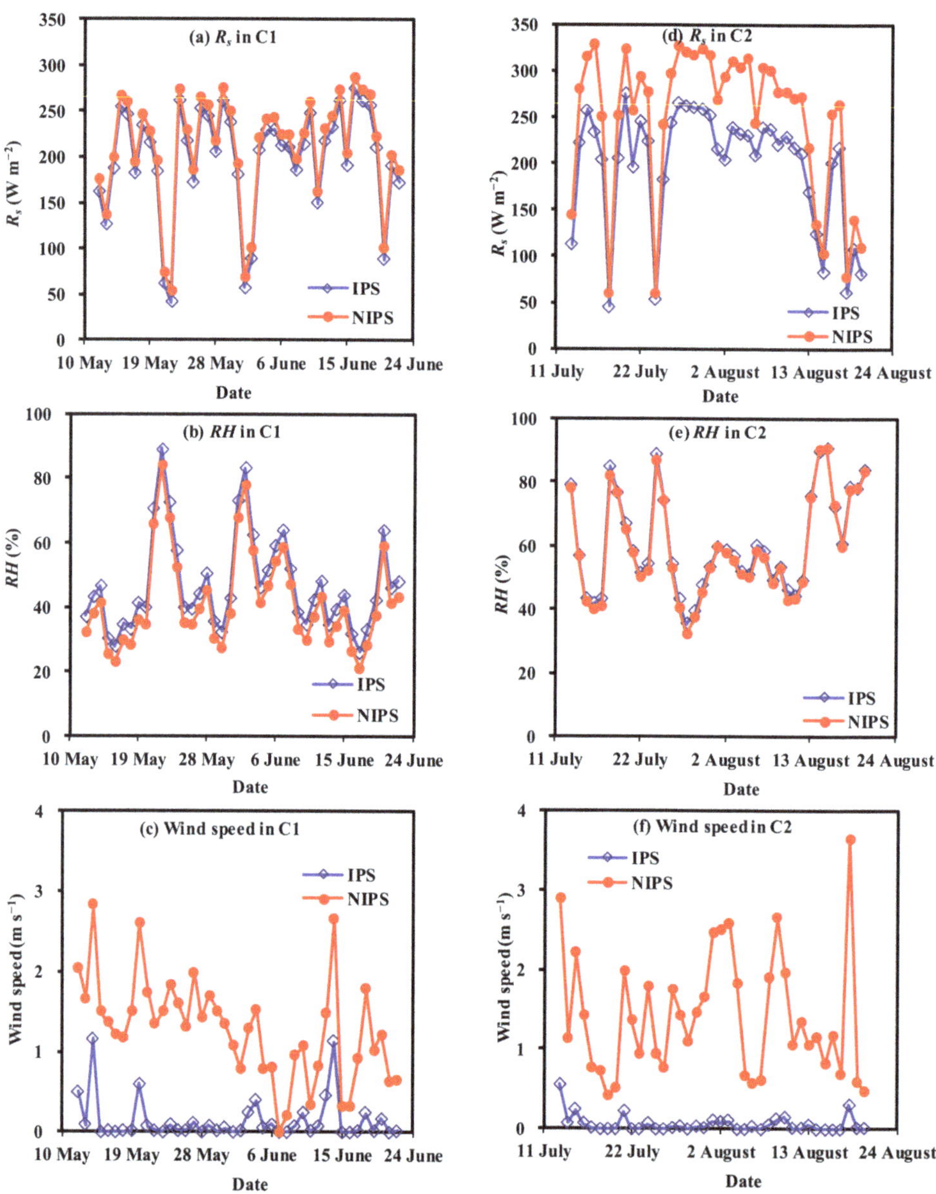

Figure 2. Effects of IPS on (**a**,**d**) daily solar radiation (R_s), (**b**,**e**) air humidity (RH), and (**c**,**f**) wind speed.

3.2. Air and Soil Temperature

The Earth's surface is the interface between the soil and the atmosphere for heat exchange, and its temperature is directly affected by changes in air temperature (T_a). There is a positive correlation between T_a and R_s. During C1 and C2, the IPS had large effects on the T_a and soil temperature (T_s) (Figures 3 and 4). The trends of change in daily T_a and T_s were similar during C1 and C2. The daily average T_a ($T_{a\text{-}avg}$) and T_s ($T_{s\text{-}avg}$) were markedly correlated ($R^2 = 0.992$ under the IPS treatment and $R^2 = 0.989$ under the NIPS treatment during C1, $R^2 = 0.996$ under the IPS treatment, $R^2 = 0.998$ under the NIPS treatment during C2). The $T_{a\text{-}avg}$ under the IPS and NIPS treatments were well correlated with corresponding maximum T_a ($T_{a\text{-}max}$) during C1 and C2, with $R^2 \geq 0.85$.

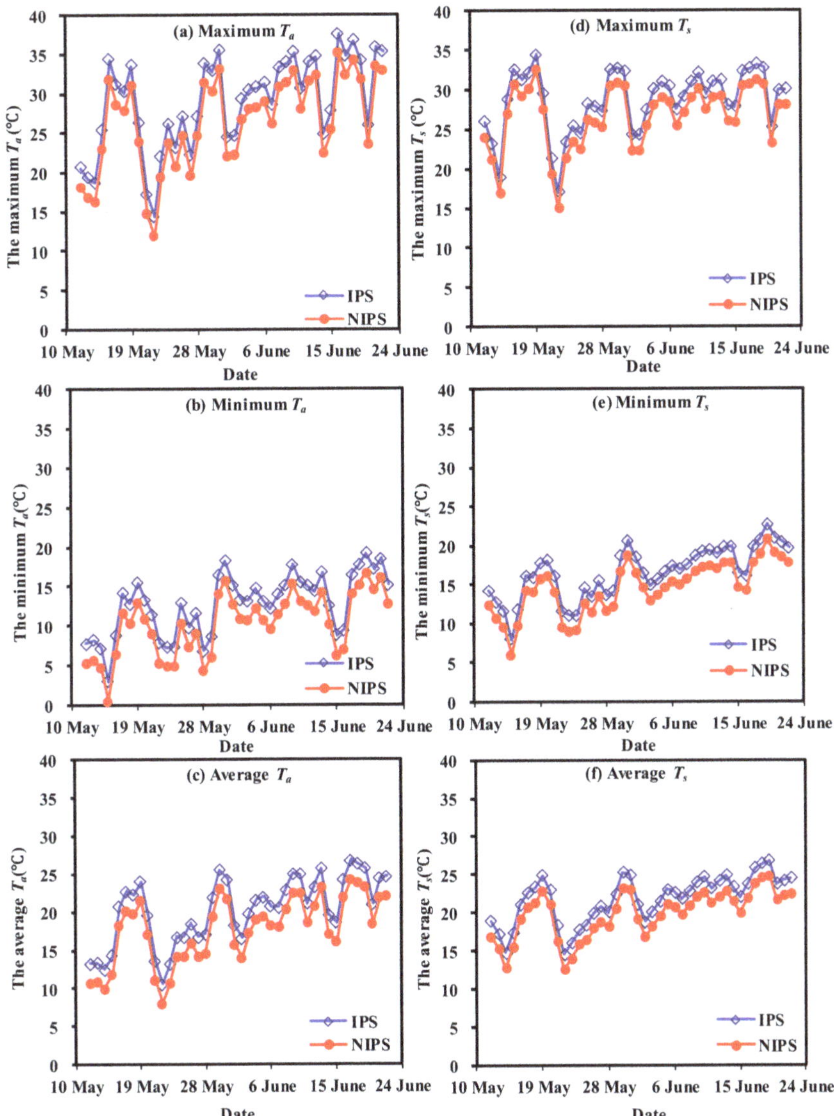

Figure 3. Effects of IPS on air temperature T_a (**a**) maximum T_a, (**b**) minimum T_a, (**c**) average T_a and soil temperature T_s (**d**) maximum T_s, (**e**) minimum T_s, (**f**) average T_s during C1 of Chinese Flowering Cabbage (CFC).

During C1, daily T_a and T_s showed a gradually increasing trend. The variation range of daily $T_{a\text{-}max}$ and minimum T_a ($T_{a\text{-}min}$) were 14.34–37.56 °C and 3.01–19.25 °C under the IPS treatment, and 11.84–35.06 °C and 0.51–16.75 °C under the NIPS treatment, respectively (Figure 3a,b). The daily maximum T_s ($T_{s\text{-}max}$) and minimum T_s ($T_{s\text{-}min}$) were 17.01–34.47 °C and 7.95–22.71 °C under the IPS treatment, and 15.01–32.47 °C and 5.95–20.71 °C under the NIPS treatment, respectively (Figure 3d,e). The daily $T_{a\text{-}avg}$ and $T_{s\text{-}avg}$ were 10.63–26.83 °C (mean = 20.32 °C) and 14.59–26.83 °C (mean = 21.83 °C) under the IPS treatment, and 8.13–24.33 °C (mean = 17.82 °C) and 12.59–24.83 °C (mean = 19.83 °C) under the NIPS treatment, respectively (Figure 3c,f). It can be found that the variation range of daily $T_{a\text{-}max}$, $T_{a\text{-}min}$, and $T_{a\text{-}avg}$ under the NIPS and IPS treatments were larger than the daily $T_{s\text{-}max}$, $T_{s\text{-}min}$, and $T_{s\text{-}avg}$, while the daily $T_{s\text{-}max}$, $T_{s\text{-}min}$, and $T_{s\text{-}avg}$ at 10 cm layer were relatively stable. Compared with NIPS treatment, the daily $T_{a\text{-}max}$, $T_{a\text{-}min}$, and $T_{a\text{-}avg}$ under the IPS treatment were increased by 7.13–21.12% (mean = 10.09%), 14.93–57.29% (mean = 27.81%), and 10.28–30.75% (mean = 15.11%), respectively (Figure 3a–c). The daily $T_{s\text{-}max}$, $T_{s\text{-}min}$, and $T_{s\text{-}avg}$ under the IPS treatment were increased by 6.16–13.32% (mean = 7.73%), 9.66–33.64% (mean = 14.72%), and 8.06–15.89% (mean = 10.37%), respectively (Figure 3d–f). Therefore, IPS greatly increased the daily T_a and T_s during C1, especially the daily $T_{a\text{-}min}$ and $T_{s\text{-}min}$.

During C2, daily T_a and T_s showed a trend of gradual increased first and then decreased with the development of the Chinese Flowering Cabbage (CFC). The daily $T_{a\text{-}max}$ and $T_{a\text{-}min}$ were 21.82–41.24 °C and 14.05–22.42 °C under the IPS treatment, and 21.22–37.18 °C and 14.89–23.33 °C under the NIPS treatment, respectively (Figure 4a,b). The daily $T_{s\text{-}max}$ and $T_{s\text{-}min}$ were 23.50–33.63 °C and 20.32–26.97 °C under the IPS treatment, and 23.50–40.40 °C and 17.87–26.13 °C under the NIPS treatment, respectively (Figure 4d,e). The daily $T_{a\text{-}avg}$ and $T_{s\text{-}avg}$ were 19.10–29.56 °C (mean = 25.60 °C) and 22.94–29.69 °C (mean = 26.91 °C) under the IPS treatment, and 19.28–29.02 °C (mean = 25.11 °C) and 22.52–31.94 °C (mean = 28.38 °C) under the NIPS treatment, respectively (Figure 4c,f). There was no obvious difference in the daily $T_{a\text{-}max}$, $T_{a\text{-}min}$, and $T_{a\text{-}avg}$ between IPS and NIPS treatments. However, compared with the NIPS treatment, the daily $T_{s\text{-}max}$ and $T_{s\text{-}avg}$ under the IPS treatment were decreased by 12.30% and 4.84%, respectively (Figure 4d,f). While the daily $T_{s\text{-}max}$ and $T_{s\text{-}avg}$ under the NIPS treatment were generally higher than that under the IPS treatment, this decline was very similar to what happened to the IPS treatment during C2. The high similarity hinted at the strong impact that external T_a had on conditions inside the screenhouse. However, the daily $T_{s\text{-}min}$ under the IPS treatment was increased by 4.92% (Figure 4e).

3.3. Diurnal Variation of Microclimate

Figures 5 and 6 show the diurnal courses of the average values of T_a, T_s, RH and variation amplitude of these three parameters in C1 and C2, respectively. Each set of internal conditions represents the average of data from 8 days. The data presented in Figures 5 and 6 were collected during 8 d (1–8 June in C1, 6–13 August in C2). It can be seen that the diurnal variation of T_a, T_s and RH under the IPS treatment was consistent with those under the NIPS treatment. The diurnal changes of T_a and T_s under the NIPS and IPS treatments in C1 and C2 were approximately sine functions.

In C1, from AM 0:00 to PM 24:00, the T_a, T_s, and RH under the IPS treatment were markedly higher than those under the NIPS treatment (Figure 5). The highest T_a and T_s under the IPS and NIPS treatments approximately appeared at PM 15:30 and PM 16:00, respectively. The lowest T_a and T_s appeared at AM 5:00 and AM 6:30, respectively. The results showed that the change of T_s lagged behind T_a, and the average lag time was one hour. The highest T_a (27.23 °C) and lowest T_a (14.56 °C) during all hours of the day under the IPS treatment were increased by 10.11% and 20.73%, respectively, compared with NIPS treatment (Figure 5a). The highest T_s (27.33 °C) and lowest T_s (17.01 °C) during all hours of the day under the IPS treatment were increased by 7.89% and 13.32%, respectively, compared with NIPS treatment (Figure 5b). Obviously, the effect of increasing the lowest T_a and T_s during all hours of the day in C1 was greater than the effect on the highest T_a and T_s, which would benefit crop growth especially in early spring. The diurnal variation of RH under the IPS and NIPS treatments was opposite to the T_a. The RH during all hours of the day under the IPS treatment was always higher than that under the NIPS treatment. The average RH under the IPS treatment over a day was 61.61%, which was 8.83% higher than that under the NIPS treatment (Figure 5c). The amplitude variation of T_a change, T_s change and *RH* change (measured by the ratios of the difference value of IPS and NIPS treatments to the value of NIPS treatment) over a day was just opposite to the variation trend of T_a, T_s, and *RH* over the day in C1 (Figure 5d). These values of T_a change (10.11–20.73%), T_s change (7.89–13.33%), and *RH* change (6.64–13.75%) during all hours of the day in C1 were positive. It can be seen that IPS could increase T_a, T_s, and *RH* during all hours of the day in C1.

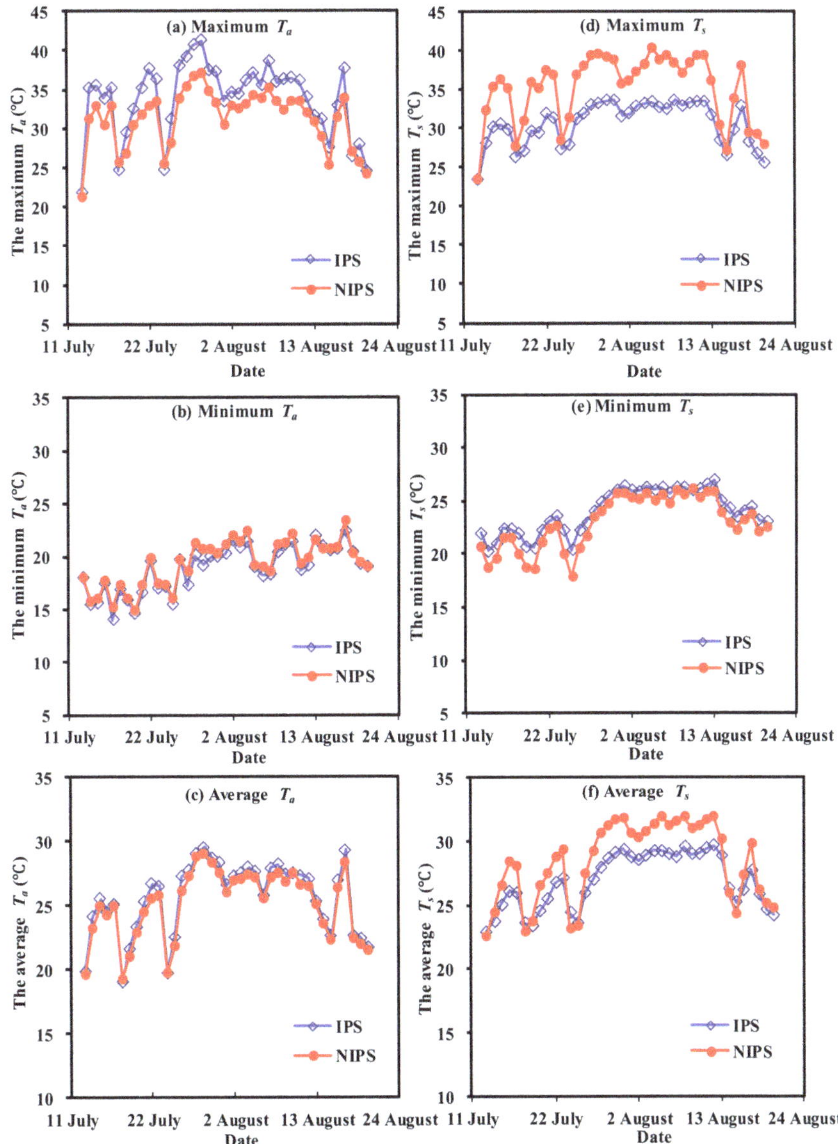

Figure 4. Effects of IPS on air temperature T_a (**a**) maximum T_a, (**b**) minimum T_a, (**c**) average T_a and soil temperature T_s (**d**) maximum T_s, (**e**) minimum T_s, (**f**) average T_s during C2 of CFC.

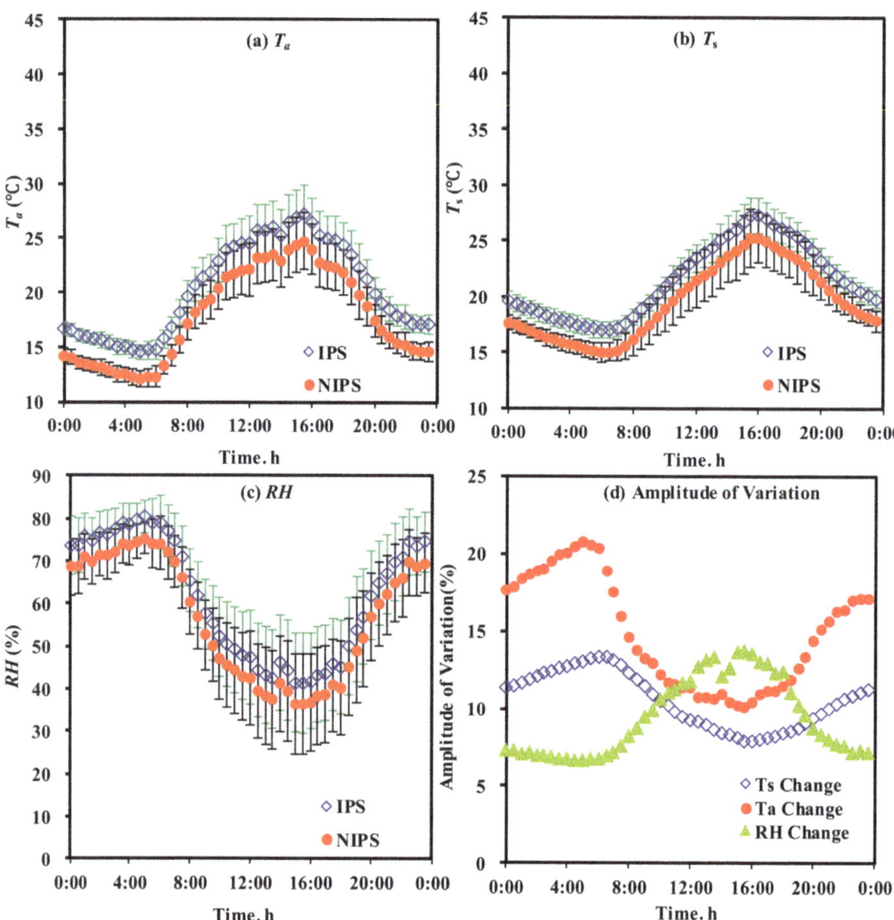

Figure 5. Diurnal courses of (**a**) T_a, (**b**) T_s, (**c**) RH, and (**d**) amplitude of variation, under the IPS and NIPS treatments during C1. Average over 8 days (1–8 June 2016). The vertical bar represents one standard deviation.

In C2, from AM 8:30 to PM 18:30, the T_a under the IPS treatment was higher than that under the NIPS treatment and they were slightly lower than that under the NIPS treatment at other times (Figure 6a). From AM 2:30 to AM 8:00, the T_s under the IPS treatment was higher than that under the NIPS treatment and they were lower than that under the NIPS treatment at other times (Figure 6b). Compared with NIPS treatment, the highest T_s (32.78 °C) under the IPS treatment over a day was decreased by 13.70%, whereas the lowest T_s (26.34 °C) over a day was increased by 1.86%. Obviously, the IPS treatment had a greater effect on reducing the $T_{s\text{-}max}$ in C2 than the $T_{s\text{-}min}$. The results showed that the IPS treatment had a certain buffer effect on high temperatures. From AM 8:00 to PM 17:00, the RH under the IPS treatment was slightly lower than that under the NIPS treatment and they were obviously higher than that under the NIPS treatment at other times (Figure 6c). The average RH during all hours of the day under the IPS treatment (54.63%) was 2.57% higher than that under the NIPS treatment (53.26%). From AM 8:00 to PM 18:00, the values of amplitude variation of T_a change were positive; nevertheless, the values were negative at other times (Figure 6d). From AM 8:30 to midnight (AM 0:00), the values of T_s change were negative. The trend of amplitude variation of *RH* change was opposite to that of T_a change. Hence, IPS could reduce T_s when the T_a was high during the daytime, and increase the T_s and *RH* when the T_a was low at night in C2.

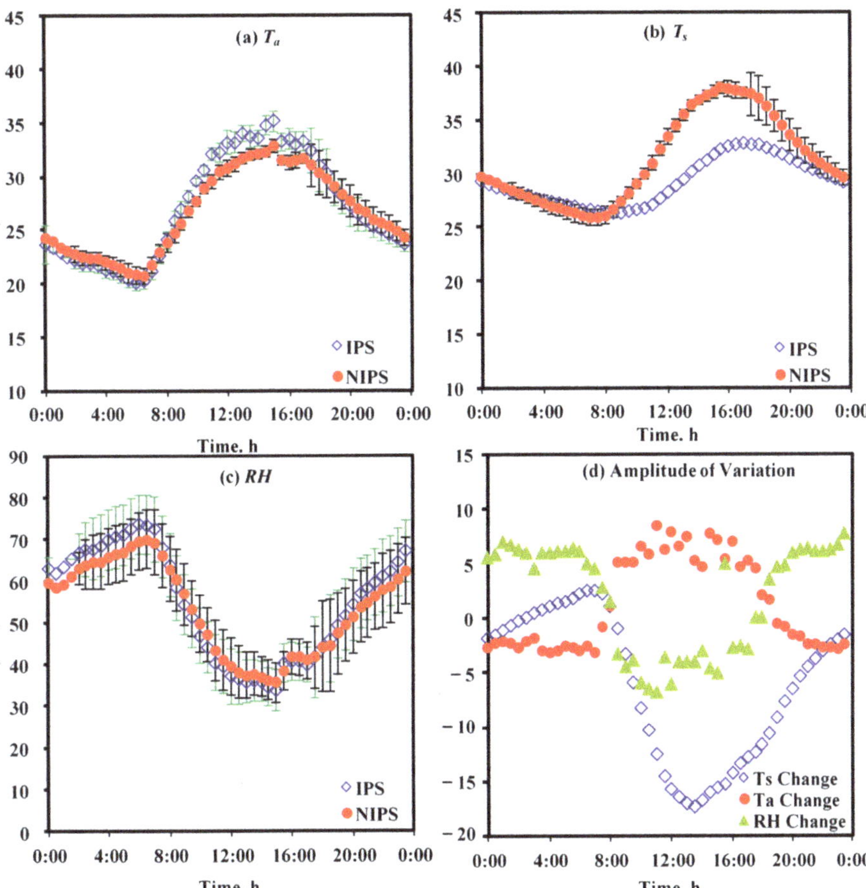

Figure 6. Diurnal courses of (**a**) T_a, (**b**) T_s, (**c**) *RH* and (**d**) amplitude of variation, under the IPS and NIPS treatments during C2. Average over 8 days (6–13 August 2016). The vertical bar represents one standard deviation.

3.4. Crop Evapotranspiration

Figure 7 shows the change process of the reference crop evapotranspiration (ET_0) under the IPS and NIPS treatments during the two growth periods. It can be seen that the daily ET_0 under the NIPS treatment was higher than that under the IPS treatment during C1. The maximum daily ET_0 under the IPS treatment in C1 was 7.32 mm, which was 11.02% lower than the maximum daily ET_0 under the NIPS treatment (8.23 mm). In C1, the accumulation of ET_0 under the IPS treatment (214.51 mm) was decreased by 6.52%, compared with that under the NIPS treatment (229.46 mm) (Figure 7a).

The trend of change in daily ET_0 in C2 under the IPS treatment was more consistent with that under the NIPS treatment, but the difference in daily ET_0 between the two treatments was more obvious compared with the C1. The maximum daily ET_0 under the IPS treatment was 7.40 mm, which was 24.94% lower than the maximum daily ET_0 under the NIPS treatment (9.86 mm) (Figure 7b). The accumulation of ET_0 under the IPS treatment during the growth period in C2 was 205.88 mm, which decreased by 21.20% compared with that under the NIPS treatment (261.26 mm). It can be seen that the IPS treatment can appreciably reduce the ET_0.

Figure 8 shows the relationship between ET_0 under the NIPS and IPS treatments. It can be seen that the ET_0 under the NIPS and IPS treatments was linearly correlated (shown with a solid line in Figure 8). The ET_0 under the IPS treatment was 91.22% of that under the NIPS treatment in C1, and the ET_0 under the IPS treatment was 78.09% of that under the NIPS treatment in C2. It can be seen that under the condition of selecting the same crop coefficient, the crop water requirement under the IPS treatment in C1 and C2 will be reduced by about 8.78% and 21.91%, respectively, compared with NIPS treatment. This may be due to IPS reduced incoming R_s and wind speed. Therefore, the IPS treatment had an important effect on ET_0 and crop water requirements.

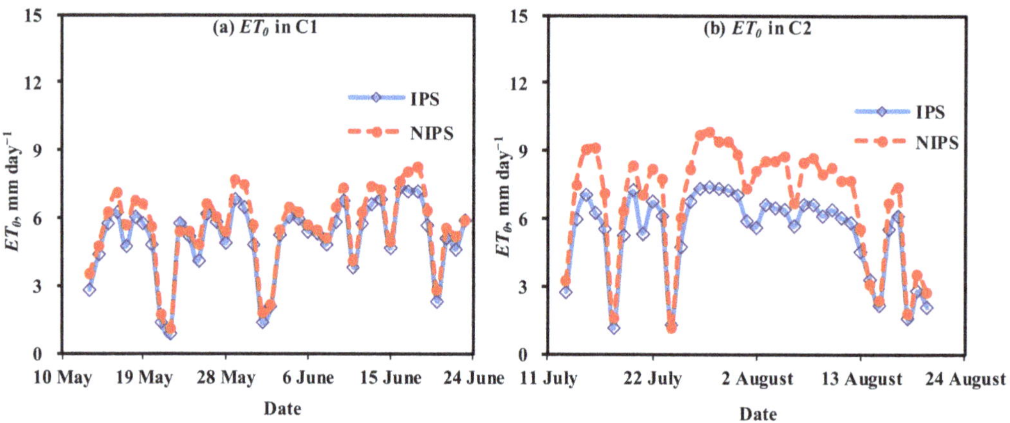

Figure 7. Changes of ET_0 under the IPS and NIPS treatments during C1 (**a**) and C2 (**b**) of CFC.

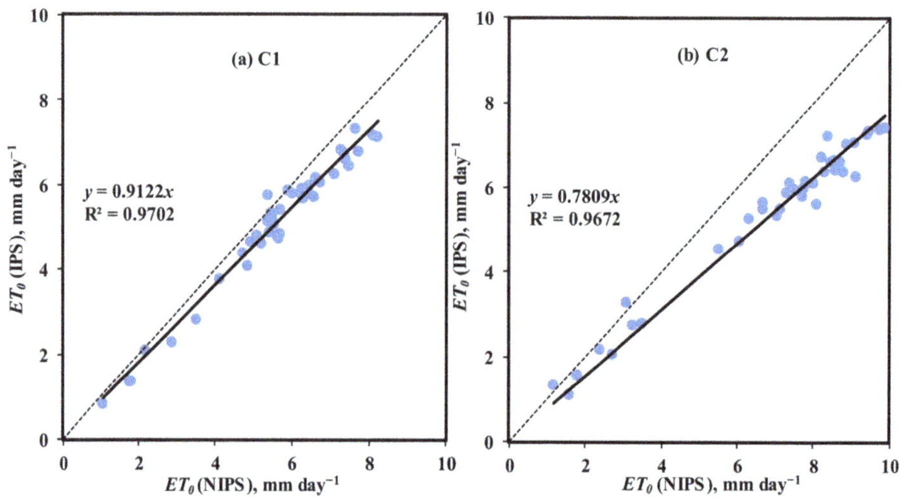

Figure 8. Relationship of ET_0 between IPS and NIPS treatments during C1 (**a**) and during C2 (**b**) of CFC.

3.5. The Growth of CFC

The changes in plant height, leaf number, and leaf circumference area of CFC under the NIPS and IPS treatments are shown in Figure 9. It can be seen that the plant height, leaf number, and leaf circumference area of CFC under the IPS treatment were higher than that under the NIPS treatment. As the growth period progressed, the difference between the two treatments was gradually significant. In C1, the plant heights of CFC under the IPS treatment on 31 May, 8 June, 15 June, and 21 June were 3.29%, 15.01%, 31.11%, and 8.33% higher than that under the NIPS treatment, respectively (Figure 9a). In C2, the plant heights of CFC under the IPS treatment on 27 July, 3 August, 10 August, and 18 August were 5.66, 7.38, 10.60, and 21.85 cm, respectively, which were increased by 23.6%, 8.49%, 40.71%, and 44.22% compared with that under the NIPS treatment (Figure 9d). The results of variance analysis showed that the IPS had a significant impact on the plant heights on June 15 and 21 June in C1, and the plant height on 18 August in C2. The result showed that IPS could significantly increase the plant height of CFC (Table 1).

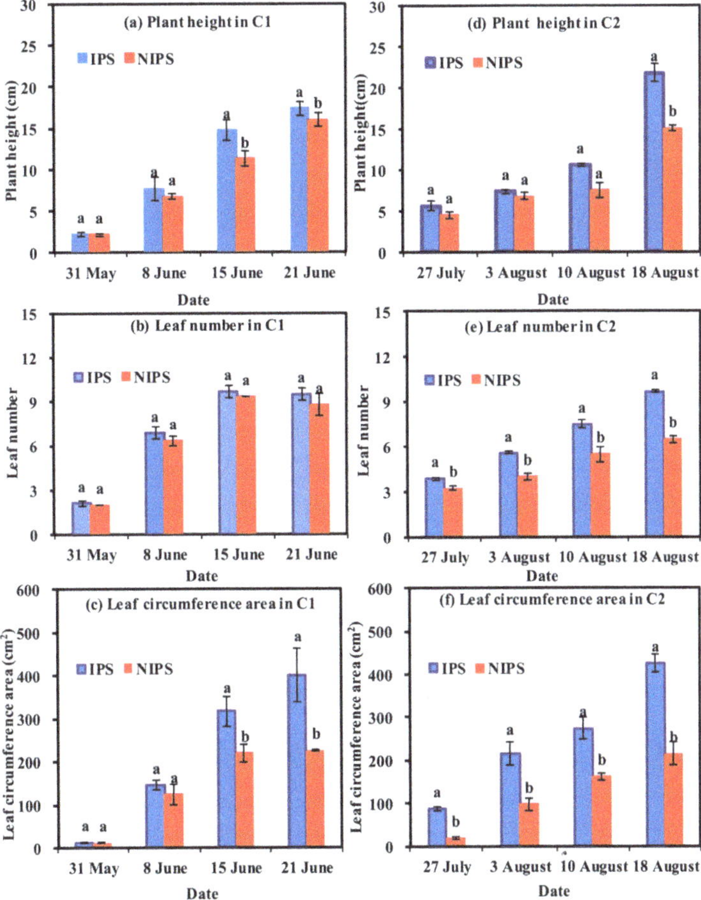

Figure 9. The variation of (**a**,**d**) the plant height, (**b**,**e**) leaf number, and (**c**,**f**) leaf circumference area during C1 and C2. The vertical bar represents one standard deviation. Different letters above bars indicate significant differences among treatments at the level of 0.05.

Table 1. Effect of IPS on growth and development of CFC.

Stubble Number	Date		Plant Height	Leaf Number	Leaf Circumference Area
C1	DAS18	31 May	NS ($p = 0.740$)	NS ($p = 0.332$)	NS ($p = 0.184$)
	DAS26	8 June	NS ($p = 0.230$)	NS ($p = 0.161$)	NS ($p = 0.423$)
	DAS33	15 June	** ($p = 0.003$)	NS ($p = 0.085$)	* ($p = 0.027$)
	DAS39	21 June	** ($p = 0.035$)	NS ($p = 0.335$)	* ($p = 0.044$)
C2	DAS12	27 July	NS ($p = 0.249$)	** ($p = 0.007$)	** ($p = 0.001$)
	DAS19	3 August	NS ($p = 0.339$)	* ($p = 0.036$)	* ($p = 0.027$)
	DAS26	10 August	NS ($p = 0.056$)	* ($p = 0.036$)	* ($p = 0.021$)
	DAS34	18 August	** ($p = 0.004$)	** ($p = 0.000$)	** ($p = 0.006$)

The values followed by the same letter in the column are not significantly different at the level of 0.05. NS = not significant at the 0.05 level. * = significant at the 0.05 level; ** = very significant at the 0.01 level.

It can be seen that the leaf number of CFCs under the IPS treatment was higher than that under the NIPS treatment. For example, the leaf number of CFC on 21 June under the IPS treatment (9.5) was 8.57% higher than that under the NIPS treatment (8.75) in C1, and the leaf number of CFC under the IPS treatment (9.7) was 49.23% higher than that under the NIPS treatment (6.5) on 18 August in C2 (Figure 9b,e). The variance results showed that the difference in leaf number between the two treatments in C1 at different stages was not significant. There were significant differences in leaf number between the two treatments at different growth stages in C2. These results showed that IPS could promote the increase in the number of leaves in the later growth period of CFC, especially during C2 (Table 1).

The changes in the leaf circumference area of the CFC under the NIPS and IPS treatments are shown in Figure 9c,f, respectively. It can be seen that the differences in leaf circumference area of CFC between the two treatments gradually increased with the growth period in C1. On 21 June, the leaf circumference area of CFC under the IPS treatment was significantly higher than that under the NIPS treatment. On 31 May, 8 June, 15 June and 21 June, the leaf circumference area of the CFC under the IPS treatment were 13.93, 145.25, 316.33, and 401.00 cm^2, which were 12.68%, 17.04%, 44.03%, and 77.96% higher than that under the NIPS treatment, respectively. The leaf circumference area of the CFC under the IPS treatment was significantly higher than that under the NIPS treatment in C2, and it was more obvious in the early growth stage of CFC. For example, on 27 July, 3 August, 10 August, and 18 August, the leaf circumference area of the CFC under the IPS treatment were 87.78, 215.27, 274.16, and 425.14 cm^2, which were 339%, 120%, 69.15%, and 98.36% higher than that under the NIPS treatment, respectively. The results of variance analysis showed that IPS had a significant impact on the leaf circumference area during C1 and C2. Therefore, IPS can significantly promote the increase in leaf circumference area, and it may be related to a certain degree of insect damage to the CFC under the NIPS treatment, especially in C2.

3.6. Yield and Irrigation Water Use Efficiency

Table 2 shows the effect of IPS on yield and irrigation water use efficiency (*IWUE*) of the CFC. The result showed that IPS had significant effects on the yield and *IWUE* in C1 and C2. The yield of CFC under the IPS treatment (10,031.68 kg ha^{-1}) was 36.00% higher than that under the NIPS treatment (7376.13 kg ha^{-1}) in C1. The *IWUE* of IPS and NIPS treatments were 2.33 and 1.71 kg m^{-1} in C1, respectively. The yield of CFC under the IPS treatment (13,707.59 kg ha^{-1}) was 108.92% higher than that under the NIPS treatment (6561.06 kg ha^{-1}) in C2. The *IWUE* of IPS and NIPS treatments were 3.01 and 1.44 kg m^{-1} in C2, respectively. The yield and *IWUE* of NIPS treatment in C2 both were the lowest, which may be due to the higher T_a and R_s. The result showed that the IPS treatment could save water and increase vegetable yield under the same irrigation conditions. The yield and *IWUE* of CFC were greatly increased under the IPS treatment, especially in C2.

Table 2. Effect of IPS on yield and IWUE of CFC.

Treatment	C1		C2	
	Yield (kg ha^{-1})	*IWUE* (kg m^{-3})	Yield (kg ha^{-1})	*IWUE* (kg m^{-3})
IPS	10031.68 + 310.15	2.33 + 0.07	13707.59 + 1536.59	3.01 + 0.34
NIPS	7376.13 + 40.25	1.71 + 0.01	6561.06 + 290.32	1.44 + 0.06
		ANOVA		
n	** ($p = 0.002$)	** ($p = 0.002$)	* ($p = 0.017$)	* ($p = 0.017$)

* = significant at the 0.05 level; ** = very significant at the 0.01 level.

4. Discussion

Although screenhouses have been used for many years, their popularity is mainly due to the desire to reduce the use of pesticides and requires an increased understanding of their climate. Previous studies were mostly under the conditions of the Mediterranean climate, with hot, dry summers and mild, wet winters. There have been few previous studies on the climate and water use of screenhouses in the arid and semi-arid temperate continental monsoon climate, and pressure on water resources makes this information timely. Early research on screen climate reported general characteristics, but did not combine crop growth, yield, and water consumption. The current study was conducted in the arid and semi-arid region of Northwest China to study the effects of IPS on the main microclimate, ET_0, crop growth, and *IWUE* of CFC.

4.1. Microclimate

In recent years, many researchers have found that R_s in the screen house decreased [28,29], the wind speed decreased [30,31], and *RH* increased, compared with those outside. The results of this study showed that IPS can appreciably reduce the daily average R_s. The daily average R_s under the IPS treatment during C1 and C2 were reduced by 4.34–22.39% and 8.42–30.57% compared with NIPS treatment, respectively (Figure 2). Ombódi et al. [32] observed that IPS decreased incoming radiation by 23–39%, while cultivated pepper in walk-in plastic tunnels (2.3 height, 5 m width, and 40 m length) was covered by external photo-selective shade nets in Hungary. The IPS obstructed the air flow and reduced the evaporation of moisture in the screen, which caused the *RH* in the screen higher than the outside [14,28]. During C1 and C2, the daily average *RH* under the IPS treatment was 11.84% and 2.08% higher than that under the NIPS treatment, respectively. Xing et al. [33] found that IPS can increase daily *RH* by 1.9–7.6% compared with that in the open field. Xu et al. [34] carried out a field experiment in Nanjing characterized by a subtropical monsoon climate and the area of each plot was 150 m^2. The daily *RH* in the screenhouse was 1.81% and 1.96% higher than that in the open field in 2014 and 2015, respectively. In our study, the average wind speed under the IPS treatment during C1 and C2 was decreased by 57.75–100% and 80.72–100%, respectively, compared with NIPS treatment. Previous studies had shown that IPS increased air resistance and reduced air flow, thereby reducing the wind speed [14,18]. A reduction in wind speed observed in a study conducted in the central of Israel was lower than 78–86% reported for a screenhouse made of a 50-mesh insect-proof screen [35].

Although the effect of the screen is always to reduce radiation and air velocity, its effect on temperature is much more complex [36]. The T_a is an integrated outcome of several simultaneous energy transfer processes, which include radiation exchange, convection (ventilation), and evapotranspiration. In this study, the daily average T_a and T_s under the IPS treatment during C1 increased by 10.28–30.75% and 8.06–15.89% compared with NIPS treatment, respectively (Figure 3). During C1, IPS can effectively increase the T_a and T_s, especially the T_{s-min}. However, the daily T_{a-avg} and T_{s-avg} under the IPS treatment during C2 were increased by −0.94–4.31% and −8.95–5.16% compared with NIPS treatment, respectively. Furthermore, IPS can distinctly reduce the T_{s-max} by 0.00–17.55% and increase the $T_{s\text{-}min}$ by 1.22–14.11% during C2, respectively (Figure 4). The IPS treatment had different effects on the T_a and T_s of the two trials because the study area was relatively cold during C1 (in spring) and hot during C2 (in summer and autumn).

By analyzing the diurnal changes of the microclimate, it was found that the T_a, T_s, and RH under the IPS treatment were higher than those outside from AM 0:00 to PM 24:00 within a day during C1 (Figure 5). However, the T_s under the IPS treatment was 0.09–2.52% higher than that under the NIPS treatment from AM 2:30 to AM 8:00, and it was 0.25–17.23% lower than that under the NIPS treatment at other times during C2 (Figure 6). From AM 8:00 to PM 18:30, the T_a under the IPS treatment was 1.11–8.53% higher than that under the NIPS treatment, and it was 0.59–3.15% lower than that under the NIPS treatment at other times during C2 (Figure 6). The RH was opposite to the trend of the T_a. Obviously, the IPS can improve the T_a, T_s, and RH during all hours of the day in C1, especially for T_a and T_s at night. The IPS can reduce T_s during the daytime in C2, and increase T_s and RH at night. These results indicated that the IPS had a certain buffering effect on high temperatures.

4.2. Crop Evapotranspiration

Atmospheric water requirement combines the effects of R_s, wind speed, T_a and RH on crop microclimate, and has the greatest influence on crop water requirement. Therefore, the corrected microclimate, especially the reduced R_s and wind speed, may reduce crop evapotranspiration (ET) [37,38]. The cumulative amounts of ET_0 in C1 and C2 were reduced by 6.52% and 21.20% compared with that under the NIPS treatment (Figure 7). It can be seen that IPS can effectively reduce ET_0. It was of greater significance to reduce the amount of irrigation, especially in summer and autumn. In the past two decades, there have been studies on the effects of IPS on crop microclimate and water use. In many regions, such as the Mediterranean, screens are normally fitted to all vents in intensive greenhouse production [15,19,20,39]. It reduced by about 30% the airflow through the insect-proof house and consequently the air exchange rate in southern Israel [20]. ET_0 under a shading screen was 38% lower than the estimated value for the open atmosphere in Northern Negev, Israel. The main reason for the observed decrease in ET_0 was the distinctly reduced R_s, while lower wind speeds inside the screenhouse have contributed to a smaller extent [13,40]. Similarly, Pirkner et al. [41] reported a 34% reduction in evapotranspiration for a table-grape vineyard cultivated in a 10%-shading screenhouse, as compared with estimations in open field vineyard in Israel. Möller et al. [13,15,16] showed that the ET of sweet pepper grown in a 50-mesh insect-proof screen was reduced by 30–50% compared to that in an open pepper field. In another study, ET_0 of sweet pepper was investigated by using three different screens, they found that the mean relative reduction in ET_0 with respect to outside ET_0 in pearl insect-proof screen, white insect-proof screen and green shade screen were 17.4%, 41.3%, and 42.6%, respectively [42].

4.3. Crop Growth, Yield and IWUE

The IPS can improve the microclimate of the farmland and reduce crop evapotranspiration, thereby promoting crop growth and increasing yield. The plant height of CFC under the IPS treatment at harvest of C1 and C2 were 8.33% and 44.22% higher than those under the NIPS treatment (Figure 9a,d). In our study, the IPS can significantly increase the growth and development of CFC, especially in C2. Ilić et al. [43] showed that shade application of nets to tomato plants was effective in substantially improving vegetative growth parameters, (i.e., leaf area index and leaf pigments) under excessive R_s during the summer period.

The yields of CFC under the IPS treatment in C1 and C2 were 36.00% and 108.92% higher than those under the NIPS treatment, respectively, and IPS can significantly improve *IWUE* in the two crop cycles (Table 2). In general, a properly higher temperature can positively affect crop growth. Sufficient radiation is needed for photosynthesis; however, under supra-optimal radiation plants may be under stress, close their stomata and reduce production. Vegetables exposed to high levels of direct radiation may suffer sunburn which significantly reduces the marketable yield [14]. In this study, there was no obvious difference in the average R_s under the IPS treatment in C1 (198.27 W m^{-2}) and C2 (198.47 W m^{-2}) (Figure 2). The daily $T_{a\text{-}avg}$ (25.60 °C) under the IPS treatment in C2 was 25.98% higher than that in C1 (Figures 3 and 4). It can be seen that the growth and yield of CFC under the IPS treatment in C2 were higher than that in C1 because of the increase in temperature. The IPS could appreciably reduce R_s, and reducing R_s may be favorable for improving the thermal climate and avoiding sunburn in hot climates. The total R_s under the IPS treatment in C1 and C2 were 20.45% and 5.73% lower than that under the NIPS treatment, respectively. For the NIPS treatment, the average R_s (249.49 W m^{-2}) in C2 was 18.62% higher than that in C1 (Figure 2). The CFC is a short-season crop and 70% of the root system distribute in the 10–30 cm soil layer [44]. To ensure that irrigation water can be absorbed by crops, high-frequency irrigation is required to keep the water in the main distribution area of the crop root system. The accumulation of ET_0 under the NIPS treatment during C2 was 261.26 mm, which was 13.86% higher than that in C1 (Figure 7). The yield of CFC under the NIPS treatment in C2 was lowest in all treatments for the two trials which could be due to the extremely high R_s and ET_0 under the NIPS treatment. IPS had the potential for water saving and retention, and the yield of CFC had also been greatly improved, which had a more obvious impact on summer and autumn crops. Earlier studies showed that IPS cultivation in semi-arid areas with strong radiation can reduce the physiological stress caused by strong radiation on plants by reducing R_s [45], prevent the decline of crop photosynthesis [46,47], and improve crop yield and quality [18,48]. Möller et al. [13] found that the *IWUE* of sweet pepper planted in the screenhouse was significantly higher than that in the open field in Northern Negev, Israel.

5. Conclusions

Field experiments with two growing crops of CFC were conducted in the arid and semi-arid areas of Ningxia to evaluate the effect of IPS on field microclimate, ET_0, vegetable growth, yield, and *IWUE*. The following conclusions were supported by our study:

Compared with NIPS treatment, IPS could appreciably decrease solar radiation and wind speed during C1 and C2, and protect CFC from sunburn and wind damage. Air temperature, soil temperature, and air humidity were obviously increased during C1, whereas soil temperature was distinctly decreased during C2. These effects have the potential of reducing crop evapotranspiration under IPN treatment, as illustrated by the estimates of ET_0. IPS could create a better air, soil water, and thermal environment, significantly promote the growth and development of CFC, and thereby significantly improve *IWUE* compared with NIPS treatment in the two crop cycles. The application of IPN was very important for improving the microclimate, ET_0, vegetable growth, yield, and *IWUE* of CFC.

IPS was recommended for leafy vegetable planting to both increase crop yield and improve water use efficiency in the arid and semi-arid regions of China.

Author Contributions: Conceptualization, S.N. and L.W.; Investigation, X.W.; Methodology, W.G. and T.Z.; Resources, W.S.; Writing—original draft, J.W.; Writing—review and editing, L.W. All authors have read and agreed to the published version of the manuscript.

Funding: This research was financially supported by the National Key Research and Development Program of China (grant no 2019YFE0125100), Ningxia Hui Autonomous Region Key R&D Program (grant no 2019BBF02010), National Key R&D Program of China (grant no 2020YFD1000300) and Science and Technology Innovation Project of Beijing Vocational College of Agriculture (XY-YF-22-07).

Institutional Review Board Statement: Not applicable.

Informed Consent Statement: Not applicable.

Data Availability Statement: The data are available upon request from the corresponding author.

Acknowledgments: We would like to express our thanks to the anonymous reviewers for their useful comments.

Conflicts of Interest: The authors declare no conflict of interest.

References

1. Wei, Y.; Zhong, F.L.; Luo, X.J.; Wang, P.L.; Song, X.Y. Ways to improve the productivity of oasis agriculture: Increasing the scale of household production and human capital? A case study on seed maize production in northwest China. *Agriculture* **2021**, *11*, 1218. [CrossRef]
2. Bai, J.; Chen, X.; Li, L.; Luo, G.; Yu, Q. Quantifying the contributions of agricultural oasis expansion, management practices and climate change to net primary production and evapotranspiration in croplands in arid northwest China. *J. Arid Environ.* **2014**, *100*, 31–41. [CrossRef]
3. Wu, X. Analysis on present situation and existing problems of water resources utilization in Ningxia Yellow River irrigation district. *Technol. Inf.* **2009**, *18*, 655–656. (In Chinese)
4. National Bureau of Statistics of the People's Republic of China (NBS). 2020. Available online: https://data.stats.gov.cn/ (accessed on 30 May 2022).
5. Du, J.; Yang, Z.H.; Wang, H.; Yang, G.Y.; Li, S.Y. Spatial–temporal matching characteristics between agricultural water and land resources in Ningxia, northwest China. *Water* **2019**, *11*, 1460. [CrossRef]
6. Al-Harbi, A.R.; Al-Omran, A.M.; Alharbi, K. Grafting improves cucumber water stress tolerance in Saudi Arabia. *Saudi J. Biol. Sci.* **2018**, *25*, 298–304. [CrossRef]
7. Wang, X.J.; Jia, Z.K.; Liang, L.Y.; Han, Q.F.; Yang, B.P.; Ding, R.X.; Cui, R.M.; Wei, T. Effects of organic fertilizer application on soil moisture and economic returns of maize in dryland farming. *Trans. CSAE* **2012**, *6*, 144–149. (In Chinese)
8. Teitel, M.; Barak, M.; Berlinger, M.J.; Lebiush-Mordechai, S. Insect proof screens: Their effect on roof ventilation and insect penetration. In Proceedings of the Third International Workshop on Models for Plant Growth and Control of the Shoot and Root Environments in Greenhouses, Bet Dagan, Israel, 21–25 February1999; Volume 507, pp. 29–37.
9. Tanny, J.; Cohen, S.; Israeli, Y. Screen constructions: Microclimate and water use in Israel. In Proceedings of the XXVIII International Horticultural Congress on Science and Horticulture for People (IHC2010): International Symposium on Greenhouse 2010 and Soilless Cultivation, Lisbon, Portugal, 22–27 August 2010; Volume 927, pp. 515–528.
10. Shamshiri, R. Measuring optimality degrees of microclimate parameters in protected cultivation of tomato under tropical climate condition. *Measurement* **2017**, *106*, 236–244. [CrossRef]
11. Ilić, S.Z.; Milenković, L.; Šunić, L.; Barać, S.; Mastilović, J.; Kevrešan, Ž.; Fallik, E. Effect of shading by coloured nets on yield and fruit quality of sweet pepper. *Zemdirbyste* **2017**, *104*, 53–62. [CrossRef]
12. Ilić, S.Z.; Milenković, L.; Dimitrijević, A.; Stanojević, L.; Cvetković, D.; Kevrešan, Ž.; Fallik, E.; Mastilović, J. Light modification by color nets improve quality of lettuce from summer production. *Sci. Hortic.* **2017**, *226*, 389–397. [CrossRef]
13. Möller, M.; Assouline, S. Effects of a shading screen on microclimate and crop water requirements. *Irrig. Sci.* **2007**, *25*, 171–181. [CrossRef]
14. Tanny, J. Microclimate and evapotranspiration of crops covered by agricultural screens: A review. *Biosyst. Eng.* **2013**, *114*, 26–43. [CrossRef]
15. Möller, M.; Tanny, J.; Li, Y.; Cohen, S. Measuring and predicting evapotranspiration in an insect-proof screenhouse. *Agric. For. Meteorol.* **2004**, *127*, 35–51. [CrossRef]
16. Möller, M.; Tanny, J.; Cohen, S.; Li, Y.; Grava, A. Water consumption of pepper grown in an insect proof screenhouse. In Proceedings of the VII International Symposium on Protected Cultivation in Mild Winter Climates: Production, Pest Management and Global Competition, Kissimmee, FL, USA, 23–27 March 2004; Volume 659, pp. 569–575.
17. Israeli, Y.; Zohar, C.; Arzi, A.; Nameri, N.; Shapira, O.; Levi, Y. Growing banana under shade screens as a mean of saving irrigation water: Preliminary results. In Proceedings of the fifteenth ACORBAT Meeting, Cartagena, Colombia, 27 October–2 November 2002; pp. 384–389.
18. Kitta, E.; Katsoulas, N.; Savvas, D. Shading effects on greenhouse microclimate and crop transpiration in a cucumber crop grown under Mediterranean conditions. *Appl. Eng. Agric.* **2012**, *28*, 129–140. [CrossRef]
19. Teitel, M. Diurnal energy-partitioning and transpiration modelling in an insect proof screen house with a tomato crop. *Biosyst. Eng.* **2017**, *160*, 170–178. [CrossRef]
20. Teitel, M.; Liang, H.; Tanny, J.; Garcia-Teruel, M.; Levi, A.; Ibanez, P.F.; Alon, H. Effect of roof height on microclimate and plant characteristics in an insect-proof screenhouse with impermeable sidewalls. *Biosyst. Eng.* **2017**, *162*, 11–19. [CrossRef]
21. Trdan, S.; Valič, N.; Znidarcic, D.; Vidrih, M.; Bergant, K.; Zlatić, E.; Milevoj, L. The role of Chinese cabbage as a trap crop for flea beetles (Coleoptera: Chrysomelidae) in production of white cabbage. *Sci. Hortic.* **2005**, *106*, 12–24. [CrossRef]
22. Wu, X.; Bai, M.; Li, Y.; Du, T.; Zhang, S.H.; Shi, Y.; Liu, Y. The effect of fertigation on cabbage (*Brassica oleracea* L. var. *capitata*) grown in a greenhouse. *Water* **2020**, *12*, 1076.
23. Veihmeyer, F.J.; Hendrickson, A.H. Methods of measuring field capacity and permanent wilting percentage of soils. *Soil Sci.* **1949**, *68*, 75–94. [CrossRef]
24. Lu, R. *Methods of Soil and Agrochemical Analysis*; China Agricultural Science and Technology Press: Beijing, China, 2000. (In Chinese)

25. Qiu, R.j.; Du, T.S.; Kang, S.Z.; Chen, R.Q.; Wu, L.S. Assessing the SIMDualKc model for estimating evapotranspiration of hot pepper grown in a solar greenhouse in Northwest China. *Agric. Syst.* **2015**, *138*, 1–9. [CrossRef]
26. Li, Y.K.; Guo, W.Z.; Xue, X.Z.; Qiao, X.J.; Wang, L.C.; Chen, H.; Zhao, Q.; Chen, F. Effects of different fertigation modes on tomato yield, fruit quality, and water and fertilizer utilization in greenhouse. *Sci. Agric. Sin.* **2017**, *50*, 3757–3765. (In Chinese)
27. Allen, R.G.; Pereira, L.S.; Raes, D.M.; Smith, M. Crop evapotranspiration, guidelines for computing crop water requirements. In *FAO Irrigation and Drainage Paper No. 56*; FAO: Rome, Italy, 1998.
28. Harmanto; Tantau, H.J.; Salokhe, V.M. Microclimate and air exchange rates in greenhouses covered with different nets in the humid tropics. *Biosyst. Eng.* **2006**, *94*, 239–253. [CrossRef]
29. Cohen, S.; Fuchs, M. Measuring and predicting radiometric properties of reflective shade nets and thermal screens. *J. Agric. Eng. Res.* **1999**, *73*, 245–255. [CrossRef]
30. Tanny, J.; Cohen, S.; Teitel, M. Screenhouse microclimate and ventilation: An experimental study. *Biosyst. Eng.* **2003**, *84*, 331–341. [CrossRef]
31. Tanny, J.; Cohen, S. The effect of a small shade net on the properties of wind and selected boundary layer parameters above and within a citrus orchard. *Biosyst. Eng.* **2003**, *84*, 57–67. [CrossRef]
32. Ombódi, A.; Pék, Z.; Szuvandzsiev, P.; Taskovics, Z.T.; Kőházi-Kis, A.; Kovács, A.; Darázsi, H.L.; Helyes, L. Effects of external coloured shade nets on sweet peppers cultivated in walk-in plastic tunnels. *Not. Bot. Horti Agrobot. Cluj-Napoca* **2015**, *43*, 398–403. [CrossRef]
33. Xing, H.Y.; Zhou, L.L.; Bo, G.L.; Zhang, Y.Y.; Yin, D.X. Change of meteorological factors and correlative analysis under the condition of net for protection against insects. *J. Chang. Veg.* **2007**, *2*, 41–42.
34. Xu, G.; Liu, X.; Wang, Q.; Xiong, R.; Hang, Y. Effects of screenhouse cultivation and organic materials incorporation on global warming potential in rice fields. *Environ. Sci. Pollut. Res.* **2017**, *24*, 6581–6591. [CrossRef]
35. Möller, M. The Effect of Insect-Proof Nets on Exchange of Mass and Momentum in a Screenhouse for Pepper Cultivation in Central Israel. Diploma Thesis, Department of Meteorology TU Dresden, Dresden, Germany, 2002.
36. Ashraf, M.; Hu, Y.; Josef, T.; Amoah, A.E. Effects of shading and insect-proof screens on crop microclimate and production: A review of recent advances. *Sci. Hortic.* **2018**, *241*, 241–251.
37. Rana, G.; Katerji, N.; Introna, M.; Hammami, A. Microclimate and plant water relationship of the "overhead" table grape vineyard managed with three different covering techniques. *Sci. Hortic.* **2004**, *102*, 105–120. [CrossRef]
38. Moratiel, R.; Martinez-Cob, A. Evapotranspiration of grapevine trained to a gable trellis system under netting and black plastic mulching. *Irrig. Sci.* **2012**, *30*, 167–178. [CrossRef]
39. Shilo, E.; Teitel, M.; Mahrer, Y.; Boulard, T. Air-flow patterns and heat fluxes in roof-ventilated multi-span greenhouse with insect-proof screens. *Agric. For. Meteorol.* **2003**, *122*, 3–20. [CrossRef]
40. Liu, H.; Cohen, S.; Lemcoff, J.H.; Israeli, Y.; Tanny, J. Sap flow, canopy conductance and microclimate in a banana screenhouse. *Agric. For. Meteorol.* **2015**, *201*, 165–175.
41. Pirkner, M.; Dicken, U.; Tanny, J. Penman-monteith approaches for estimating crop evapotranspiration in screenhouses–A case study with table-grape. *Int. J. Biometeorol.* **2014**, *58*, 725–737. [CrossRef]
42. Kitta, E.; Baille, A.; Katsoulas, N.; Rigakis, N. Predicting reference evapotranspiration for screenhouse-grown crops. *Agric. Water Manag.* **2014**, *143*, 122–130. [CrossRef]
43. Ilić, S.Z.; Milenković, L.; Šunić, L.; Fallik, E. Effect of colored shade-nets on plant leaf parameters and tomato fruit quality. *J. Sci. Food Agric.* **2015**, *95*, 2660–2667. [CrossRef]
44. Murakami, T.; Yamada, K.; Yoshida, S. Root distribution of field-grown Chinese cabbage (*Brassica campestris* L.) under different fertilizer treatment. *Soil Sci. Plant Nutr.* **2002**, *48*, 393–400. [CrossRef]
45. Stanhill, G.; Cohen, S. Global dimming: A review of the evidence for a widespread and reduction in global radiation with discussion of its probable causes and possible agricultural consequences. *Agric. For. Meteorol.* **2001**, *107*, 255–278. [CrossRef]
46. Medina, C.L.; Souza, R.P.; Machado, E.C.; Ribeiro, R.V.; Silva, J.A.B. Photosynthetic response of citrus grown under reflective aluminized polypropylene shading nets. *Sci. Hortic.* **2002**, *96*, 115–125. [CrossRef]
47. Kouchi, M.; Roche, A.F. The excess light energy that is neither utilized in photosynthesis nor dissipated by photoprotective mechanisms determines the rate of photoinactivation in photosystem II. *Plant Cell Physiol.* **2003**, *44*, 318–325.
48. Leonardi, C.; Baille, A.; Guichard, S. Predicting transpiration of shaded and non-shaded tomato fruits under greenhouse environments. *Sci. Hortic.* **2000**, *84*, 297–307. [CrossRef]

MDPI

St. Alban-Anlage 66

4052 Basel

Switzerland

Tel. +41 61 683 77 34

Fax +41 61 302 89 18

www.mdpi.com

Horticulturae Editorial Office

E-mail: horticulturae@mdpi.com

www.mdpi.com/journal/horticulturae